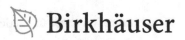

Birkhäuser Advanced Texts Basler Lehrbücher

Series Editors

Steven G. Krantz, Washington University, St. Louis, USA

Shrawan Kumar, University of North Carolina at Chapel Hill, Chapel Hill, USA

Jan Nekovář, Sorbonne Université, Paris, France

More information about this series at https://link.springer.com/bookseries/4842

Olivier Ramaré

Excursions in Multiplicative Number Theory

With contributions by Pieter Moree and Alisa Sedunova

 Birkhäuser

Olivier Ramaré
Institut de Mathématiques de Marseille
CNRS / Aix-Marseille University
Marseille, Bouches du Rhône, France

ISSN 1019-6242   ISSN 2296-4894   (electronic)
Birkhäuser Advanced Texts Basler Lehrbücher
ISBN 978-3-030-73171-7   ISBN 978-3-030-73169-4   (eBook)
https://doi.org/10.1007/978-3-030-73169-4

Mathematics Subject Classification: 11Nxx, 11Mxx, 11Lxx, 11Pxx, 11Yxx, 11Axx

This book is published under the imprint Birkhäuser, www.birkhauser-science.com by the registered
company Springer Nature Switzerland AG
The registered company address is: Gewerbestrasse 11, 6330 Cham, Switzerland

Preface

This book is concerned with the arithmetic of integers, i.e. with the behaviour of (usually positive) integers with respect to addition and multiplication. One major way to describe this is through the study of *arithmetical functions*. Such functions often behave in a locally erratic but globally predictable way. For example, their mean value up to n is generally close to some simple function of n. It is this feature of global regularity that allows one to study their behaviour using methods from analytic number theory.

The goal of this book is to teach the readers how *to establish* such results. Since we encourage an active attitude with a focus on methods rather than performance, we have included nearly 300 exercises that guide the readers towards proving results. These in themselves may not be the best available, but some exercises in the later chapters ask the readers to prove results that are close to the forefront of current research.

As mentioned above, the typical behaviour of an arithmetical function is irregular, especially when its definition depends on the factorization structure of n. Consider for instance the function

$$f_0(n) = \prod_{p|n}(p - 2). \tag{0.1}$$

The introduction of some notation is called for. Here, and in the whole book, the letter p (or p_1, p_2, etc.) stands for a prime number, while the notation "$p|n$" means that we are considering the set of primes p that divide n. For $n = 1, 2, \ldots, 54$, we find the following values of $f_0(n)$:

$$1, 0, 1, 0, 3, 0, 5, 0, 1, 0, 9, 0, 11, 0, 3, 0, 15, 0, 17, 0, 5, 0, 21, 0, 3, 0, 1, 0, 27,$$
$$0, 29, 0, 9, 0, 15, 0, 35, 0, 11, 0, 39, 0, 41, 0, 3, 0, 45, 0, 5, 0, 15, 0, 51, 0$$

This list does not give us much information (other than $f_0(n)$ vanishes when n is even), but we shall show that its *average* (*mean value*) behaves in a regular way.

The book in itself is split in five parts that we now describe briefly.

Approach.

Before embarking on computing average orders, we spend some time exploring arithmetic functions and notions of *arithmetical* interest. Multiplicativity is a fundamental notion there, and we develop a calculus on multiplicative functions that renders their handling very easy. The basics on Dirichlet series is then introduced and rough estimates concerning the growth of multiplicative functions are proved. Additional chapters on the Legendre symbol and its Dirichlet series; formulas akin to the Möbius inversion formula (some of which are new!) conclude the introductory part.

The Convolution Walk.

The first guided walk to higher ground exposes the readers to "elementary" methods, the focus of which is to prove the next theorem.

Theorem $\mathscr{A}$

For all real positive x we have

$$\frac{1}{x} \sum_{n \leq x} f_0(n) = \tfrac{1}{2}\mathscr{C}_0 x + O^*(3.5\, x^{3/5}),$$

where $\mathscr{C}_0$ is a constant given by

$$\mathscr{C}_0 = \prod_{p \geq 2}\left(1 - \frac{3}{p(p+1)}\right) = 0.29261\ 98570\ 45154\ 91401\cdots \qquad (0.2)$$

Here and in the sequel, we write $f = O(g)$ as a shorthand for there being a positive constant C such that $|f| \leq Cg$. This is the Landau big-O notation that is standard in mathematics. We supplement it with the notation O^*; we say $f = O^*(g)$ if $|f| \leq g$. This notation is very practical when one wants to compute explicit bounds for the intermediary error terms.

The above theorem shows that, by taking into account many values of f_0, the influence of the aberrant ones is swept under the rug and a simple regularity is brought to the fore. This theorem is proved by comparing the function f_0 to a simpler one (here $n \mapsto n$) that is easier to analyse.

This first walk is based on the relatively simple nature of f_0. The convolution method is folklore (though only a few systematic expositions are available, e.g. [2]). It is very flexible and allows one to obtain excellent error terms. We present several examples on how to use it. A similar philosophy is at work when one wants to compute *Euler products* (such as the product in (0.2)) and *Euler sums* and we dedicate a full chapter to this issue. We conclude this part with the Dirichlet hyperbola formula, as it is the second tool at our disposal to compute average orders elementarily.

The Levin–Faǐnleǐb Walk.

The general idea of this second walk is to *deduce* the evaluation of mean values of multiplicative functions from the behaviour of our function on prime powers. We give in particular a completely explicit version of the Levin–Faǐnleǐb Theorem. Here is the consequence that we have in store.

Theorem $\mathscr{B}$

When $x \geq \exp(20\,000)$ and $d(n)$ is the number of divisors of n, we have

$$\sum_{n \leq x} \frac{\sqrt{d(n)}}{n} = \mathscr{C}_1 (\log x)^{\sqrt{2}} \left(1 + O^*(20000/\log x)\right)$$

where
$$\mathscr{C}_1 = \frac{1}{\Gamma(1 + \sqrt{2})} \prod_{p \geq 2} \left\{ \left(1 + \sum_{v \geq 1} \frac{\sqrt{v+1}}{p^v}\right)\left(1 - \frac{1}{p}\right)^{\sqrt{2}} \right\}. \qquad (0.3)$$

Getting an accurate value of $\mathscr{C}_1$ is a challenge, we leave to the readers. In effect, this method transfers regularity of the function on the primes to regularity on the integers. But to be able to apply it, we need to detect the regularity on the primes! We devote two chapters to this question and prove several classical estimates of the sort concerning logarithmic averages. For instance, $\sum_{p \leq x}(\log p)/p$ is shown to be asymptotic to $\log x$. We also prove a similar estimate when p is restricted to a congruence class modulo 3 or 4, but we are not able at this level to dispense of the *logarithmic* average and to prove, for instance, the Prime Number Theorem, i.e. that $\sum_{p \leq x} \log p$ is asymptotic to x. We are thus unable to provide an asymptotic for $\sum_{n \leq x}\sqrt{d(n)}$. We however have enough material to compute some Euler products and sums where the prime is restricted to a congruence class modulo 3 or 4; a chapter is dedicated to this task. We conclude this part with an application of our accumulated expertise and prove an asymptotic for $\sum_{n \leq x} d(n^2 + 1)$.

The Mellin Walk.

The third walk goes through analytical landscape where the information on our function is taken from its *Dirichlet series*. One feature is that we introduce a regular weight function $F(t)$ and consider expressions of the form $\sum_{n \geq 1} f_0(n)F(n/x)$. As the sum now involves every natural integers, it captures some aspect of the global behaviour of f_0, provided, of course, that $F(t)$ is sufficiently smooth and tends quickly enough to zero as t tends to infinity in order for the sum to represent a well-defined function of x. Theorem $\mathscr{C}$ is a typical example of a result involving a weight function.

Theorem $\mathscr{C}$

Let x be real and positive. We have

$$\sum_{n\geq 1} f_0(n)e^{-n/x} = \mathscr{C}_0 x^2 + O^*(133 \cdot x^{7/4}),$$

where $\mathscr{C}_0$ is the constant defined in (0.2).

As a matter of fact, one can infer the asymptotic given in Theorem $\mathcal{A}$ from this result, and this is the theme developed in Chap. 22.

Mellin transforms are also a powerful tool to prove the Prime Number Theorem, and we finally achieve this in Chap. 23. We do so in a completely explicit manner in the form of an inequality satisfied by the summatory function of the Möbius function.

Higher Ground.

In the final part, we introduce extensions and applications based on the classical techniques that were gently introduced and commented on in the earlier parts of the book. Non-negative multiplicative functions were our main focus in the first four parts. In the final part, we consider mean values of some oscillating multiplicative functions, and in particular, we give an explicit bound for $\sum_{n\leq x}\mu(n)/\varphi(n)$. We also present P_k-numbers (numbers having at most k, not necessarily distinct, prime factors) and delve deeply enough into the theory of the Brun sieve to prove a *fundamental lemma*, i.e. a wide-ranging but rather sharp upper bound for the quantities considered in sieve, as the number of primes in an interval for instance. This will enable us to evaluate the two exponential sums $\sum_{p\leq x}\exp(2i\pi\rho p)$ and $\sum_{n\leq x}\mu(n)\exp(2i\pi\rho n)$ with modern techniques and to prove, for instance, that the two sequences $(\cos 2\pi\rho p)$, when p ranges, over the primes and $(\cos 2\pi\rho n)$, when n ranges over the integers having an even number of prime factors, are dense in $[-1, 1]$. We will end this journey with a short presentation of the large sieve and of a practical arithmetical form of it due to H.L. Montgomery.

An Active Teaching.

Since we aim at keeping the readers on their toes, and since we shall also explicitly bound most of the intermediary error terms, we have decided to give this monograph an algorithmical streak. This will help the readers in carrying out experiments. Computation is a way to get better mastery and insights into the mathematical topic one studies. B. Riemann, one of the greatest mathematicians of all time, had extremely large sheets of paper, on which he performed computations. For example, he computed by hand the location of the first few zeros of the Riemann zeta-function. Study of these sheets has shown that Riemann kept many findings up his sleeve, the formula now known under the name "Riemann-Siegel formula" being an example. Riemann developed it in order to have a faster and more accurate way of computing the Riemann zeta-function in the critical strip. Another brilliant mathematician known for carrying out many numerical calculations was C.F. Gauss. He became famous in 1802 for predicting the position of the planet-like object Ceres. The prediction required Gauss to do an enormous amount of calculation and its purpose was to help astronomers find Ceres again in the sky after it had been too close to the sun's glare for months to confirm the first observations by Piazzi in 1801.

Pari/GP [6] and Sage [8] are the two computer algebra packages that we shall use, in version 2.11 at the time of writing for Pari/GP and in version 9 (with Python 3) for Sage. They are freewares and developed by a dedicated community that guarantees the accuracy of the results. Numerical precision is always an issue and we do not dwell on it here, but the readers are expected to document themselves on this issue if they want to obtain trustworthy results. Furthermore, in Sage, we have *interval arithmetic* at our disposal: results are expressed as a couple (E^-, E^+) representing an (unknown) real number E in the interval $[E^-, E^+]$. Let us elaborate a bit on this issue here. The script

```
R = RealIntervalField(64)
sqrt(R(2))
sqrt(R(2)).upper()
sqrt(R(2)).lower()
```

answers successively 1.4142135623730950488? with the last digit 8 being maybe wrong: the ?-sign means that it may be 7 or 9, then 1.41421356237309505 followed by 1.41421356237309504 where this time, all the digits are correct: the inequalities

$$1.41421356237309504 \leq \sqrt{2} \leq 1.41421356237309505$$

are *certified*. A word is surely required on the Sage syntax: `sqrt(R(2))` is an *object* that contain several fields: here we have an upper bound and a lower bound; to have access to these fields, we have *accessory functions* at our disposal, here `upper()` and `lower()` that come as further specification, which explains the syntax `sqrt(R(2)).upper()`. Pari/GP sometimes also uses this syntax.

Neither Pari/GP nor Sage are extremely fast and for specific purposes it is better to use some C-code directly; Pari/GP has the advantage that one may automatically derive some more efficient C-code from most Pari/GP scripts by using the free software `gp2c`. Furthermore, the ease of usage of both packages compensates for the relative loss of speed.

Notation.

We end this introduction by commenting on our notation.

- It is common in multiplicative number theory to write $\sum_{p \geq 2} f(p)$ in the case that the variable p is restricted to *prime* values. The same applies to more intricate expressions and to products as well (cf. our definition (0.1) of f_0). As a rule of thumb, a variable named p is assumed to be prime unless specified otherwise.
- We shall often need to refer to the gcd ("greatest common divisor") of two integers, say a and b. The notation $\gcd(a, b)$ is explicit, but is often shortened to (a, b). The readers may thus find expressions like $f((a, b))$ referring to the value of the function f on the gcd of a and b, or $\sum_{n \leq q, (n,q)=1} 1$ to denote the number of integers below q that are coprime with q (and this number is exactly the value of the Euler φ-function at q).
- Although we try to be very precise, we sometimes prefer to loosen our methodological constraints and use $\mathcal{O}$ rather than $\mathcal{O}^*$. We shall also use the expression $f \ll g$ to mean that $f = \mathcal{O}(g)$. When we add subscripts, like in $f \ll_r g$, it means that there exists a constant C that may depend on r such that $|f| \leq C \cdot g$.

Acknowledgment.

THE BOOK HAS ITS ROOTS in three courses in french given by the author, who with the help of his contributors, Pieter and Alisa, translated the original course material into a more structured write-up. The author did much of the writing and decided the layout of the book, while the contributors added exercises, edited language, put references, wrote some passages and proofread earlier versions. Their work supported the author tremendously; he felt like a frontiersman in uncharted territory, with two companions keeping his back free.

THANKS are due to Kam Hung Yau for his careful checking of this book and for his precious advices.

Further Reading

Different and complementary approaches to arithmetic can be found in several textbooks that can be read without too much prerequisite. Let us mention the classical [1] of T. Apostol, the book [9] of W. Schwarz and J. Spilker dedicated to arithmetical functions, and the more advanced book [5] by H.L. Montgomery and R.C. Vaughan. Other rich sources of information are the books [7] by P. Pollack and [3] by O. Bordelles. A major reference on mathematical computations with Sage is the open access book [4], written and maintained up-to-date by a community of researchers. Finally, we direct the reader to the collection of exercices [10] by W. Sierpinski.

Marseille, France   Olivier Ramaré

References

[1] T.M. Apostol, *Introduction to Analytic Number Theory. Undergraduate Texts in Mathematics*. (Springer-Verlag, New York, 1976), pp. xii+338 (cit. on p. x)

[2] P. Berment, O. Ramaré, Ordre moyen d'une fonction arithmétique par la méthode de convolution. in *Revue de la filière mathématiques (RMS)* 122.1 (2012), pp. 1–15 (cit. on p. vi)

[3] O. Bordellès, *Arithmetic Tales. Universitext*. (Springer London Heidelberg New York Dordrecht, 2012) (cit. on p. x)

[4] P. Zimmermann et al, *Computational Mathematics with SageMath*. 2018. url: `sagebook.gforge.inria.fr/english.html` (cit. on p. x).

[5] H.L. Montgomery, R.C. Vaughan, *Multiplicative Number Theory: I. Classical Theory. Vol. 97. Cambridge Studies in Advanced Mathematics*. (Cambridge University Press, 2006), pp. xviii+552 (cit. on p. x)

[6] PARI/GP, version 2.11. `pari.math.u-bordeaux.fr/`. The PARI Group. Bordeaux, 2018 (cit. on p. viii)

[7] P. Pollack, Not always buried deep. A second course in elementary number theory. Am. Math. Soc., Providence, RI, pp. xvi+303. (2009), doi: `10.1090/mbk/068` (cit. on p. x)

[8] The Sage Developers, *SageMath, the Sage Mathematics Software System (Version 9.0)*. (2019). url: `www.sagemath.org` (cit. on p. viii)

[9] W.K. Schwarz, J. Spilker. *Arithmetical functions, an introduction to elementary and analytic properties of arithmetic functions and to some of their almost-periodic properties. Vol. 184. Lectures Notes Series*. (Cambridge, London, 1994) Math. Soc., (cit. on p. x)

[10] W. Sierpinski, *250 problems in elementary number theory. Modern Analytic and Computational Methods in Science and Mathematics, No. 26*. (American Elsevier Publishing Co., Inc., New York; PWN Polish Scientific Publishers, Warsaw, 1970), pp. vii+125 (cit. on p. x)

Using this Book for a Course

This chapter is meant for teachers who wish to use
(all or part of) this book for a course.

This book may be used from Chaps. 1 to 29 linearly; this gives a long course requiring about 100–120 hours, though it naturally depends on the familiarity of the students with arithmetical notions.

It may also be used for 36-hour courses, which is how these lessons came to be. To build such a course, the *Multiplicativity part* is unavoidable and, in my experience, should always be at least recalled. The *unitary convolution* may be skipped for beginners but is useful to let the more advanced students acquire a new notion.

Once this basic core is over, one may divert to the two light chapters on *Möbius Inversions* and on *Dirichlet Hyperbola Principle*, depending on how much time is available; these chapters may also be kept for term papers.

From there onwards, there are three possibilities: using the *Convolution Method*, applying the *Levin–Faĭnleĭb method*, or invoking the *Mellin Transform*. Some hours should be reserved at the end to delve into one of the last chapters that are grouped in four blocks: the *Selberg Formula* is one, using the *Convolution Method with non-positive multiplicative functions* is the second one and the two heavier ones concerns respectively *Rankin's trick, Brun's sieve and some arithmetical exponential sums*, and the *Montgomery's arithmetical version of the large sieve*. These last lectures will be challenging, all the more so since the students will have less time to acquire the material, but the results therein may be presented at the beginning as the goal of the course. This is how the author has gone about it several times, with his choice of core blocks being

- *Multiplicativity*, followed by the main course of the *Convolution Walk* and the chapter on *Convolution and Non-negative functions*. This gives a light course that can be spiced with some addenda.

- *Multiplicativity*, followed by the main course of the *Levin–Faĭnleĭb Walk* and closing with *Rankin and exponential sums*. This gives a rather strong series on combinatorial methods, showing a path to good results without using the Prime Number Theorem.
- *Multiplicativity*, followed by the main course of the *Mellin Walk* and ending with the *Large sieve/Montgomery's sieve*. This gives a more traditional course.

We have tried to represent this architecture in the diagram displayed next page. The core block of *Multiplicativity* is in yellow, the main chapters for each of the three ways are displayed below in green, while the more advanced topics at the bottom of the diagram are being displayed in blue.

Once such a path is chosen, it is possible to dress them with additional material, all displayed in pink. We have ordered them in a way suitable for a full course but there is a large margin of freedom there. It is interesting in the *Levin–Faĭnleĭb Walk* to add information on the distribution of primes in arithmetic progressions, as this extends sizeably the power of this theorem.

Of special interest are the units concerning scientific computations; these are Chaps. 9, 16 and 17. They use the arithmetical background and fit there, but have a different flavour. The arithmetical prerequisite are rather modest, only the following basic definitions are needed: the Euler product representation of the Riemann zeta-function is essential, and later the one concerning Dirichlet series, and finally, the property saying that the Möbius function is the convolution inverse of the constant function $\mathbb{1}$.

The level of difficulty increases from the top to the bottom of the diagram with the chapters not directly depending on their predecessors in the diagram.

To maintain a proper level of independence, some material and often the definitions are retold. We have also tried to be as extensive as we could with cross-references, so that a reader skipping some chapters may still be able to follow.

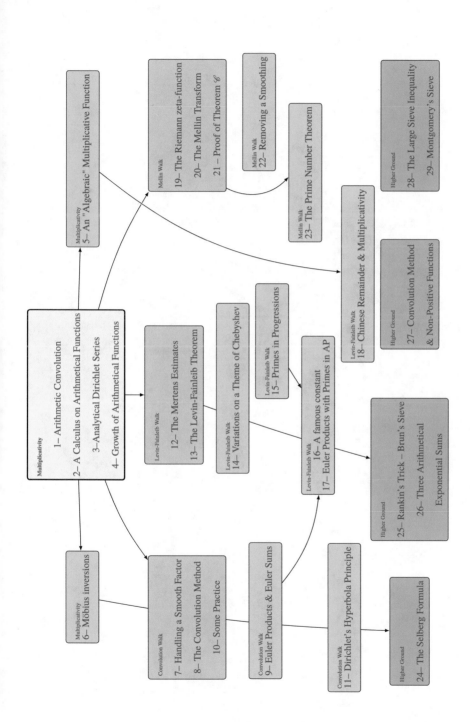

Multiplicativity

5– An "Algebraic" Multiplicative Function

Mellin Walk
19– The Riemann zeta-function
20– The Mellin Transform
21 – Proof of Theorem $\mathscr{C}$

Mellin Walk
22– Removing a Smoothing

Mellin Walk
23– The Prime Number Theorem

Higher Ground
28– The Large Sieve Inequality
29– Montgomery's Sieve

Multiplicativity
1– Arithmetic Convolution
2– A Calculus on Arithmetical Functions
3–Analytical Dirichlet Series
4– Growth of Arithmetical Functions

Levin-Fainleib Walk
18– Chinese Remainder & Multiplicativity

Higher Ground
27– Convolution Method
& Non-Positive Functions

Levin-Fainleib Walk
12– The Mertens Estimates
13– The Levin-Fainleib Theorem

Levin-Fainleib Walk
14– Variations on a Theme of Chebyshev

Levin-Fainleib Walk
15– Primes in Progressions

Levin-Fainleib Walk
16– A famous constant
17– Euler Products with Primes in AP

Higher Ground
25– Rankin's Trick – Brun's Sieve
26– Three Arithmetical Exponential Sums

Multiplicativity
6– Möbius inversions

Convolution Walk
7– Handling a Smooth Factor
8– The Convolution Method
10– Some Practice

Convolution Walk
9– Euler Products & Euler Sums

Convolution Walk
11– Dirichlet's Hyperbola Principle

Higher Ground
24– The Selberg Formula

Contents

Part I
Approach: Multiplicativity

Chapter 1
Arithmetic Convolution

1.1 Multiplicative Functions

A function $f : \mathbb{N} \setminus \{0\} \to \mathbb{C}$ is called *multiplicative* if it satisfies

$$\begin{cases} f(1) = 1, \\ f(mn) = f(m)f(n), \qquad \text{when } n \text{ and } m \text{ are coprime.} \end{cases} \tag{1.1}$$

Equivalently, f is multiplicative if it satisfies

$$f\left(\prod_p p^{\alpha_p}\right) = \prod_p f\left(p^{\alpha_p}\right), \tag{1.2}$$

where the product is taken over all primes and the exponents α_p are non-negative integers. The latter identity shows that a multiplicative function f is completely determined by its values on prime powers. Conversely, the values on prime powers determine a multiplicative function f by the identity above.

Here is a result that is essential in this setting and indeed in all of number theory.

Lemma 1.1 (Euclid's Lemma)

Let p be a prime number. If p divides the product of two integers x and y, then p divides either x or y (and may divide both of them).

(A bit of *ring* theory: this is a consequence of the fact that $\mathbb{Z}$ is a *factorial* ring).

Exercise 1-1. Prove that the function f_1 given by $f_1(n) = \prod_{p|n}(p + 1)/(p + 2)$ is multiplicative.

Here is an important immediate consequence.

O. Ramaré, *Excursions in Multiplicative Number Theory*, Birkhäuser Advanced Texts Basler Lehrbücher, https://doi.org/10.1007/978-3-030-73169-4_1

Lemma 1.2

If two multiplicative functions g and h take the same values on prime powers, then they are equal.

Exercise 1-2. Prove that if x, y, z are integers and x is coprime to z, then the greatest common divisor of xy and z is equal to the greatest common divisor of y and z, namely, $\gcd(xy, z) = \gcd(y, z)$.

Exercise 1-3. Let a and b be two coprime integers.

1 ⋄ Prove that $\gcd(a + b, a - b)$ is either 1 or 2. Give an example for both cases.

2 ⋄ Prove that $a + b$ and ab are coprime as well.

The notion of multiplicativity is fundamental. In particular, we remark that many a priori mysterious looking arithmetic functions can be understood much better by looking at their values on prime powers.

Exercise 1-4. Let f be a multiplicative function and m, n be two positive integers Show that $f([m, n])f((m, n)) = f(m)f(n)$, where $[m, n]$ and (m, n) denote the least common multiple and greatest common divisor of m and n, respectively.

Exercise 1-5. Show that when n_1, n_2 and n_3 are positive integers, we have
$$[n_1, n_2, n_3] = \frac{n_1 n_2 n_3 (n_1, n_2, n_3)}{(n_1, n_2)(n_1, n_3)(n_2, n_3)}.$$

A general formula expressing the lcm of r integers in terms of partial gcd's has been obtained by V.-A. Le Besgue in 1862 in [4]. It may be found for instance in [7, Lemma 2] by N. Sedrakyan and J. Steinig.

1.2 Some Members of the Zoo

We present in this section a list of functions we are going to deal with in this book.

1. Given two integers a and b, we denote the *greatest common divisor* of a and b by $\gcd(a, b)$ or by (a, b). For example, $\gcd(6, 14) = 2$ and $\gcd(2, 5) = 1$.
2. Similarly, the *least common multiple* of two integers a and b is denoted by $\mathrm{lcm}(a, b)$ or by $[a, b]$. For instance, $\mathrm{lcm}(6, 14) = 42$ and $\mathrm{lcm}(2, 5) = 10$.
3. $\varphi(n)$ is *Euler's totient function*, namely, the number of integers between 1 and n that are coprime to n. For instance, $\varphi(5) = 4$ and $\varphi(6) = 2$.
4. $d(n)$ is the number of (positive) divisors of n, i.e. $d(3) = 2$ and $d(6) = 3$.
5. $d(n^2)$ is the number of (positive) divisors of n^2. We will show that it is also a *multiplicative function* of n.

6. $\sigma(n)$ is the sum of (positive) divisors of n, i.e. $\sigma(3) = 1 + 3 = 4$ and $\sigma(6) = 1 + 2 + 3 + 6 = 12$.

7. The constant function $\mathbb{1}$ is uniformly equal to 1, and, more generally, functions $X^\alpha : n \mapsto n^\alpha$. We follow some authors and use the notation Id for the somewhat less explicit X.

8. $\mathbb{1}_{X^2}$ denotes the characteristic function of squares.

9. $\delta_{n=1}$ or $\delta_1(n)$ is the function that is 1 if $n = 1$ and 0 otherwise.

10. The function $\omega(n)$ is the number of distinct prime divisors of n and, for example, $\omega(12) = 2$ since 2 and 3 are the only prime numbers dividing 12. One can also say "without multiplicities", as, for example, 2^2 also divides 12. The number of prime divisors with multiplicities is denoted by $\Omega(n)$. In our example, we have $\Omega(12) = \Omega(2 \cdot 2 \cdot 3) = 3$. Note that $\omega(nm) = \omega(n) + \omega(m)$ if $(n, m) = 1$ and that the same is true for Ω. An arithmetic function f satisfying $f(nm) = f(n) + f(m)$ when $(n, m) = 1$ is called *additive*.

11. The Möbius function $\mu(n)$ is equal to $(-1)^r$ if n is a product of exactly r distinct prime factors $n = p_1 \cdots p_r$ and $\mu(n) = 0$ otherwise. We use the convention $\mu(1) = 1$. Although this function was introduced by Euler, it is named in the honor of A.F. Möbius, who studied it later.

12. The Liouville function λ defined by $\lambda(n) = (-1)^{\Omega(n)}$ where $\Omega(n)$ given above is the number of prime divisors counted with multiplicity. It is a *multiplicative function*, meaning that $\lambda(mn) = \lambda(m)\lambda(n)$ when $(m, n) = 1$.

13. The characteristic function of *square-free integers* is 0 if n is divisible by a square (strictly) bigger than 1 and is equal to 1 otherwise. We readily discover that this function equals the square of the Möbius function $\mu(n)$.

14. The *von Mangoldt function* $\Lambda(n)$ is the function that equals $\log p$ if $n = p^a$ and is 0 otherwise (here p is a prime and a is a positive integer). This function is obviously neither multiplicative nor additive.

15. The prime counting function $\pi(x)$ counts the number of primes less or equal than x. For example, $\pi(3) = 2$, because the only primes less or equal than 3 are 2 and 3.

In addition, we will also encounter more exotic functions such as follows:

16. The function φ_2 that associates to an integer n the number of integers modulo n that are coprime to $n(n + 2)$.

17. The function that associates to an integer n the number of squares modulo n.

Exercise 1-6. Let σ be the sum-of-divisors function. Let p be a prime number and $a \geq 1$ an integer. Show that $\sigma(p^a) = (p^{a+1} - 1)/(p - 1)$, where $\sigma(n)$ is the sum of the (positive) divisors of n.

Exercise 1-7.

1 ⋄ Show that if two distinct integers n and m are square-free (as introduced in the previous page), then $\dfrac{\varphi(m)}{m} \neq \dfrac{\varphi(n)}{n}$, thus the map $n \mapsto \varphi(n)/n$ is *injective* on square-free integers.

2 ⋄ Show that the function $n \mapsto \sigma(n)/n$ is also injective on the square-free integers.

3 ⋄ What do you think about the property above for $n \mapsto \sigma(n)/\varphi(n)$?

4 ⋄ What do you think about the property above for $n \mapsto \displaystyle\prod_{p|n} \dfrac{p+2}{p+1}$?

1.3 The Divisor Function

We start by showing that the function $d(n)$ assigning to an integer n its number of divisors is indeed multiplicative. This property has its origin in the structure of the set $\mathcal{D}(n) = \{\ell \in \mathbb{N} : \ell | n\}$ of divisors of n. Clearly, for any n, we have $1 \in \mathcal{D}(n)$. One verifies that

$$\mathcal{D}(p^\alpha) = \{1, p, p^2, \ldots, p^{\alpha-1}\} . \tag{1.3}$$

Next, if p_1 and p_2 are two distinct primes, then all the divisors of $p_1^{\alpha_1} p_2^{\alpha_2}$ are of the form $p_1^{\beta_1} p_2^{\beta_2}$ with $0 \leq \beta_1 \leq \alpha_1$ and $0 \leq \beta_2 \leq \alpha_2$. Conversely, every integer of that form is a divisor of $p_1^{\alpha_1} p_2^{\alpha_2}$. This brings us to

$$\mathcal{D}\left(p_1^{\alpha_1} p_2^{\alpha_2}\right) = \mathcal{D}\left(p_1^{\alpha_1}\right) \cdot \mathcal{D}\left(p_2^{\alpha_2}\right) , \tag{1.4}$$

where *the product of two sets* is defined as

$$\mathcal{D}(m) \cdot \mathcal{D}(n) = \{\ell \in \mathbb{N} : \ell = d_1 d_2, \ d_1 \in \mathcal{D}(m), d_2 \in \mathcal{D}(n)\}.$$

In the same manner, we prove that $\mathcal{D}(mn) = \mathcal{D}(m) \cdot \mathcal{D}(n)$ if m and n are coprime. Hence the following function is one to one:

$$\begin{aligned}
\mathcal{D}(mn) &\to \mathcal{D}(m) \cdot \mathcal{D}(n) \\
d &\mapsto \left((d, m), (d, n)\right) .
\end{aligned} \tag{1.5}$$

This is a form of multiplicativity at the level of sets, which we exploit in the following manner: for any arithmetic function F and for m and n coprime integers, we have

$$\sum_{d|mn} F(d) = \sum_{d_1|m} \sum_{d_2|n} F(d_1 d_2) . \tag{1.6}$$

Proof Recall that *the Cartesian product of two sets* is defined by $A \times B = \{(a, b) : a \in A, b \in B\}$. Let $\mathcal{D}(\ell)$ be the set of positive divisors of ℓ. Suppose we are given two coprime integers m and n. Consider

$$f : \mathcal{D}(m) \times \mathcal{D}(n) \to \mathcal{D}(mn) \ , \quad g : \mathcal{D}(mn) \to \mathcal{D}(m) \times \mathcal{D}(n)$$
$$(u, v) \mapsto uv \qquad\qquad\qquad w \mapsto (\gcd(w, m), \gcd(w, n))$$

We show that $g \circ f = \mathrm{Id}$. In fact, $(g \circ f)(u, v) = (\gcd(uv, m), \gcd(uv, n))$. When v divides n and n is coprime to m, the integers v and m are coprime. Then Exer. 1-2 gives us that $\gcd(uv, m) = \gcd(u, m) = u$ and $\gcd(uv, n) = \gcd(v, n) = v$. That is what we wanted to prove. $\qquad\square$

Exercise 1-8. Show that we have $d(n) \le 2\sqrt{n}$ for all $n \ge 1$.

Exercise 1-9. Show that we have $d(n) \le 4 n^{1/3}$ for all $n \ge 1$.

Exercise 1-10.

$1 \diamond$ Let f be a multiplicative function such that $f(p^a)$ tends to 0 when p^a tends to infinity (here, p is a prime and a is a positive integer). Show that $f(n)$ tends to 0 as n goes to infinity.

$2 \diamond$ Deduce that for all $\varepsilon > 0$, there is a constant $C_\varepsilon > 0$ such that, for all integers $n \ge 1$, we have $d(n) \le C_\varepsilon n^\varepsilon$.

See Theorem 4.1 for more on this last point. Figure 1.1 displays the graph of the values of the divisor function compared to $4n^{1/3}$ in blue and to $6n^{1/4}$ in dotted green, the integer n ranging between 1 and 50000.

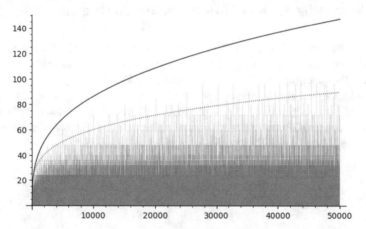

Fig. 1.1: The number of divisors at the bottom in red versus $4n^{1/3}$ in blue and $6n^{1/4}$ in dotted green

Here is the Sage script that we have used to produce this graph.

```
def getList(end):
    mylist = [(0,0)]
    for n in range(1, end+1):
        aux  = number_of_divisors(n)
        mylist += [(n, 0), (n, aux), (n+1, aux), (n+1, 0)]
    return(mylist)

mygraph = polygon(getList(50000),color='red',thickness=0.05)
y = var('y')
mygraph += plot(4*y^(1/3), (y, 0, 50000), color='blue')
mygraph += plot(6*y^(1/4), (y, 0, 50000), color='green',
                linestyle=':')

mygraph.show(aspect_ratio=200)
```

Exercise 1-11. Show that we have $\varphi(n) \geq \sqrt{n/2}$ for all $n \geq 1$.

Exercise 1-12. Show that we have $\varphi(n) \geq (9/2)^{1/3} n^{2/3}$ for all $n \geq 1$.

Exercise 1-13. Show that we have $\sigma(n)\varphi(n) \leq n^2$ for all $n \geq 1$.

1.4 Computing Values of Multiplicative Functions

How can we bring Pari/GP or Sage to produce for us values of an arithmetical function? When they are already implemented this is easy, and we immediately have access to the functions given in the Chart of Chap. 30.

To be sure that the computations are absolutely correct, it is required to use default(factor_proven, 1) in Pari/GP and proof.arithmetic(True) in Sage, see the Pari/GP manual for further explanations. Let us now create new functions. Here is how to compute $g(n) = \mu^2(n) f_0(n)$ with Pari/GP:

```
{g(n) =
    my(res = 1.0, dec = factor(n), P = dec[,1], E = dec[,2]);
    for(i = 1, #P,
        my(p = P[i]);
        if(E[i] != 1, return(0));
        res *= p-2);
    return(res);}
```

The result is returned as a real number with a precision that is defined by the global parameters. This code can be shortened as follows

```
g(n) = issquarefree(n)*vecprod([p-2 | p<-factor(n)[,1]])
```

This last code is slightly less efficient in this precise example but is so much faster to write! The same function is equally easily programmed in Sage:

```
R = RealIntervalField(100)

def g(n):
    res = 1
    l = factor(n)
    for p in l:
        if p[1] > 1:
            return(R(0))
        else:
            res *= (p[0]-1)/p[0]^2
    return(R(res))
```

In the above script, the result is returned as an interval real number on 100 bits. It is also possible to write a shorter form.

```
def g(n):
    return(is_squarefree(n)*prod([p-2 for p in prime_factors(n)]))
```

The readers will find Chap. 30 a chart of the common functions and tools to compute them in Pari/GP as well as in Sage.

Exercise 1-14. Find all the n satisfying $\sigma(n) = 24$. Do the same for the equation $\sigma(n) = 57$.

1.5 Convolution of Multiplicative Functions

We define *the Dirichlet convolution $f \star g$* of two arithmetic functions f and g by

$$(f \star g)(n) = \sum_{d|n} f(n/d)g(d), \tag{1.7}$$

where the sum is taken over the divisors d of n. We denote by $\mathbb{1}$ the function that is 1 for all integers. The readers can check that this product is commutative and associative (in case of difficulties with these notions, they may consult Sect. 5.2). The function $\delta_{n=1} = \delta_1(n)$ is the neutral element for the Dirichlet convolution, since for any arithmetic function g, we have

$$(\delta_1 \star g)(n) = \sum_{\ell m=n} \delta_1(\ell)g(m) = g(n) .$$

The convolution operation is distributive with respect to the addition of two arithmetic functions. These two rules endow the set of arithmetic functions with the structure of commutative algebra on $\mathbb{C}$, which is formalized in the following theorem.

Theorem 1.1

Let $\mathscr{F}$ be the $\mathbb{C}$-vector space of functions from $\mathbb{N} \setminus \{0\}$ to $\mathbb{C}$. The triple $(\mathscr{F}, +, \star)$ is a commutative unitary algebra on $\mathbb{C}$ with the unit element $\delta_{n=1}$. The function $\mathbb{1}$ and the Möbius function μ are inverses of each other.

We can also extend this setting by considering the pointwise product

$$f \cdot g : n \mapsto f(n)g(n)$$

and the derivation rule

$$\partial : (f(n))_{n \geq 1} \mapsto (f(n) \log n)_{n \geq 1} ,$$

which is linear and satisfies the identity $\partial(f \star g) = (\partial f) \star g + f \star (\partial g)$. The readers can find a detailed study of this enriched structure in the book [1] of P.T. Bateman & H. Diamond.

The following general theorem gives us the multiplicativity of a large class of functions.

Theorem 1.2

If f and g are two multiplicative functions, then $f \cdot g$ and $f \star g$ are also multiplicative.

Proof The case of $f \cdot g$ is trivial and we leave it to the interested readers. Let us work with the convolution product. Clearly, $(f \star g)(1) = f(1)g(1) = 1$. Let m and n be two coprime integers. By definition, we have

$$(f \star g)(mn) = \sum_{d \mid mn} f(mn/d)g(d) .$$

On invoking (1.6), we obtain

$$(f \star g)(mn) = \sum_{d_1 \mid m} \sum_{d_2 \mid n} f\left(\frac{mn}{d_1 d_2}\right) g(d_1 d_2)$$

$$= \sum_{d_1 \mid m} \sum_{d_2 \mid n} f\left(\frac{m}{d_1}\right) f\left(\frac{n}{d_2}\right) g(d_1)g(d_2) = (f \star g)(m)(f \star g)(n) ,$$

as required.   □

This theorem provides us with a tool to detect the multiplicativity of many functions by starting from $\mathbb{1}$ or, more generally, from $X^a : n \mapsto n^a$. In particular, the readers will check that

$$d(n) = (\mathbb{1} \star \mathbb{1})(n), \quad \sigma(n) = (\mathbb{1} \star X)(n).$$

This convolution also allows us to express certain relations in a simple manner. For example, $\mu^2(n) = (\mathbb{1} \star \mathbb{1}_{X^2})(n)$ with $\mathbb{1}_{X^2}$ being the characteristic function of the squares. Pari/GP proposes the function `direuler` to compute the first coefficients of a series whose local factors are rational.

We say that an arithmetic function f is *invertible with respect to the convolution product* if there exists an arithmetic function g such that $(f \star g)(n) = \delta_{n=1}$ for every integer n. We call g the *convolution inverse* of f.

Exercise 1-15.

1 ◇ Show that a function f is invertible with respect to the convolution product if and only if $f(1) \neq 0$.

2 ◇ Let f be an invertible multiplicative function. Prove that its inverse is also a multiplicative function.

Exercise 1-16. Find four values of n for which $\sum_{m|n} d(m) = n$ and show that they are the only ones.

Exercise 1-17. Given a function f, we define $\tilde{f}(n) = \dfrac{1}{d(n)} \sum_{m|n} f(m)$ as *the average value of f on the divisors of n.*

1 ◇ Prove that $\tilde{f}$ is multiplicative, when f is.

2 ◇ Let $f(n) = 2^{\omega(n)}$. Show that $\tilde{f}(n) = d(n^2)/d(n)$.

3 ◇ Let $f(n) = 2^{\Omega(n)}$. Show that $\tilde{f}(n) = \prod_{p^a \| n} \dfrac{2^{a+1} - 1}{a + 1}$, where the notation $p^a \| n$ means that $p^a \geq p$ is the *exact power* of the prime p that divides n.

4 ◇ Show that, if we have $\tilde{f} = f$, then f remains constant.

Exercise 1-18. Show that the function that associates to $n > 1$ two times the sum of the integers between 1 and n that are coprime to n, and is 1 for $n = 1$, is multiplicative.

1.6 Unitary Convolution

The definition of the convolution product from the previous section is the most usual one, though several other closely related products can be defined. We only mention the one introduced by R. Vaidyanathaswamy in [10] and studied in [2] by E. Cohen, as it appears to be useful in practice, see Theorem 3.5. Here is its definition:

$$(f \square g)(n) = \sum_{\substack{d|n \\ (d,n/d)=1}} f(n/d)g(d) \,. \tag{1.8}$$

This is similar to the definition of the Dirichlet convolution of f and g with the extra condition that "d and n/d are coprime", and hence usually results in a sum having a smaller number of summands. One finds that this *unitary product* is associative and commutative. When d and n/d are coprime, we say that d is a *unitary divisor* of n. Equivalently, we say that d *divides n exactly* which we denote by $d\|n$. The unitary divisors of 12 are 1, 3 and 4. As a matter of notation, we add a star to objects that correspond to unitary divisors: $d(n)$ is the number of divisors of n and $d^*(n)$ is the *number of unitary divisors of n*. Similarly, $\sigma(n)$ and $\sigma^*(n)$ stand, respectively, for the sum-of-divisors function and *the sum of unitary divisors function*. The following theorem is the unitary analog of Theorem 1.2.

Theorem 1.3

If f and g are two multiplicative functions, then $f \square g$ is also multiplicative.

Proof Notice that $(f \square g)(1) = 1$. Further, by induction on the number of prime factors of $n = \prod_i p_i^{a_i}$, we now show that

$$(f \square g)(n) = \prod_i (f \square g)(p^{a_i}) \,.$$

Write $n = p_r^{a_r} m$, where m is coprime to p_r and $a_r \geq 1$. By definition, we have

$$(f \square g)(n) = \sum_{\substack{d|p^a m \\ (d,n/d)=1}} f(d)g(n/d) \,.$$

On prime powers p^a, we clearly have

$$(f \square g)(p^a) = f(p^a) + g(p^a) \,.$$

A unitary divisor d of $p_r^{a_r} m$ is either a divisor of m, (in which case it is enough to show that $(d, m/d) = 1$ to get $(d, n/d) = 1$), or it can be written as $p_r^b d'$, where d' is a divisor of m. Clearly, we have $(d', m/d') = 1$. At the same time, the condition $(d, n/d) = 1$ ensures that $b = a_r$ or $b = 0$, allowing us to write

$$(f \square g)(n) = \sum_{\substack{d|m \\ (d,m/d)=1}} f(d)g(p_r^{a_r} m/d) + \sum_{\substack{d'|m \\ (d',m/d')=1}} f(p_r^{a_r} d')g(m/d') \,.$$

The multiplicativity of f and g gives us

$$(f \square g)(n) = g(p_r^{a_r})(f \square g)(m) + f(p_r^{a_r})(f \square g)(m)$$
$$= \left(f(p_r^{a_r}) + g(p_r^{a_r}) \right) \prod_{1 \le i \le r_1} (f \square g)(p_i^{a_i})$$
$$= (f \square g)(p_r^{a_r}) \prod_{1 \le i \le r-1} (f \square g)(p_i^{a_i}),$$

where $m = \prod_{1 \le i \le r-1} p_i^{a_i}$ thus completing the proof. $\square$

Exercise 1-19. Show that $d^*(n) = 2^{\omega(n)}$ and give a formula for $\sigma^*(n)$.

Exercise 1-20. Prove that $d^*(n)$ is also the number of square-free divisors of n.

Exercise 1-21. Let $\mathscr{F}$ be the $\mathbb{C}$-vector space of functions from $\mathbb{N} \setminus \{0\}$ to $\mathbb{C}$.

1 ◇ Show that $f \square (g \square h) = (f \square g) \square h$.

2 ◇ Show that $(\mathscr{F}, +, \square)$ is a commutative algebra on $\mathbb{C}$.

3 ◇ Find the unit element for $\square$.

4 ◇ Recall that $\omega(n)$ stands for the number of prime factors of n counted without multiplicities. Show that $\mathbb{1}$ (the constant function that is equal to 1) and the function $n \mapsto (-1)^{\omega(n)}$ are inverse of each other for $\square$.

Exercise 1-22.

1 ◇ Show that for all square-free integers d the number of solutions in d_1 and d_2 of the equation $[d_1, d_2] = d$ is $3^{\omega(d)}$.

2 ◇ Let q be a positive integer. We define $f(q)$ to be the number of solutions in q_1 and q_2 of an equation $[q_1, q_2] = q$. Show that f is multiplicative.

3 ◇ Show that $(\mathbb{1} \star f)(n)$ is the number of pairs (q_1, q_2) such that $[q_1, q_2] | n$.

Exercise 1-23. Define $\sigma_k(n) = \sum_{d|n} d^k$. For $k > 1$ define *the Riemann zeta-function* as $\zeta(k) = \sum_{m=1}^{\infty} m^{-k}$. Show that $1 \le \sigma_k(n)/n^k \le \zeta(k)$ for $k > 1$.

1.7 Taxonomy

A rapid typology allows us to distinguish three ways of appearances of multiplicative functions:

1. *Multiplicative by definition.*

a. The constant function $\mathbb{1}$, which is equal to 1, or, more generally, the function $X^{\alpha} : n \mapsto n^{\alpha}$.
b. The Möbius function and the Liouville function are multiplicative because they are defined in terms of the prime factor decomposition.
c. The functions connected to the Dirichlet characters that we introduce later.

2. *Multiplicative by construction.* Three constructions are the pointwise product (for example, the function $d \mapsto \mu(d)d$ is multiplicative), the convolution product (that gives us the multiplicativity of the divisor function $d(n)$ and of the sum-of-divisors function $\sigma(n)$) and the unitary convolution product. There is another class of constructions: given an integer $k \geq 1$ and a multiplicative function f, the function $n \mapsto f(n^k)$ is also multiplicative. We often encounter such constructions with $k = 2$. In Exer. 10.7, a similar mechanism is used.

3. *Multiplicativity of an algebraic type.* Functions that count numbers of solutions, like Euler's totient function $\varphi(n)$, fall into this class. When P is a polynomial with coefficients in $\mathbb{Z}$, the number of roots $\rho_P(d)$ of P modulo d is a multiplicative function of d, as proved in Cor. 18.1. Chaps. 5 and 18 offer a slow introduction to this subject. We may also define more exotic functions, for example, the *Ramanujan sums* introduced by the S. Ramanujan in 1918[*]. They are defined by

$$c_d(n) = \sum_{\substack{a \bmod d \\ (a,d)=1}} e(na/d), \qquad (1.9)$$

where $e(\alpha) = \exp(2i\pi\alpha)$. The Kluyver identity states that

$$c_d(n) = \sum_{\substack{\delta \mid n, \\ \delta \mid d}} \delta\mu(d/\delta). \qquad (1.10)$$

Finally, we discuss one of the most celebrated and intriguing multiplicative functions: the *Ramanujan τ function* (the notation is historical and it should not be confused with the number of divisors counting function). We can define $\tau(n)$ as the nth coefficient in the following Taylor series:

$$x\prod_{n=1}^{\infty}(1 - x^n)^{24} = \sum_{n=1}^{\infty} \tau(n)x^n . \qquad (|x| < 1) \qquad (1.11)$$

It is not at all clear from this definition that τ is multiplicative! This was conjectured by S. Ramanujan and proved by L. Mordell in 1917. The coefficients $\tau(n)$ are Fourier coefficients of the most basic example of a so-called cusp form. The theory of modular forms (a cusp form is a special modular form) allows one to produce many identities, a typical one (due to B. van der Pol in [6]), being

---

[*] As a historical aside, we note that they had already been studied by Kluyver in 1906. They also appear in a different form in the work of Daublebsky von Sterneck.

$$\tau(n) = n\sigma_7(n) - 540 \sum_{1 \leq m \leq n-1} m(n-m)\sigma_3(m)\sigma_3(n-m),\qquad(1.12)$$

where $\sigma_k(n) = \sum_{d|n} d^k$. However, τ itself remains quite mysterious. We know from the epoch making work of P. Deligne in 1974, that it satisfies the inequality $|\tau(p)| \leq 2p^{11/2}$.

Exercise 1-24. Prove that if d is square-free, then for any integer $n \geq 1$ we have $|c_d(n)| \leq \varphi(\gcd(n,d))$.

In the next exercise, we define $\mathcal{P}(n) = \{p : p|n\}$ to be the set of the prime factors of the integer n. P. Erdős and A.R. Woods [3] conjectured that, given any two integers m and n, there exists an integer k_0 such that, when $\mathcal{P}(n+k) = \mathcal{P}(m+k)$ for every $k \leq k_0$, then $m = n$. This conjecture is still open and a disproof of it would have important consequences. The next exercise proposes a first result.

Exercise 1-25. Let n and m be two positive integers such that $\mathcal{P}(n+k) = \mathcal{P}(m+k)$ for all $k \leq 3\min(n,m)$. Show that $m = n$.

Exercise 1-26. We denote by $k(n)$ the product of distinct prime factors of n (i.e its so-called *radical* or *square-free kernel*). Write a Pari/GP or Sage script to compare $\log(a+b)$ with $\log k(ab(a+b))$ when a and b are coprime integers that range $\{1, \ldots, 3000\}^2$ and deduce that in this range we have $a+b \leq k(ab(a+b))^2$

No one has ever been able to disprove this bound. The *abc-conjecture* of J. Oesterlé and D. Masser predicts that even more is true. The a priori surprising fact that $a+b$ can be bounded above solely in terms of $k(ab(a+b))$ has been proven to be true by C. Stewart and R. Tijdeman in [9]. The *abc*-conjecture has been shown to imply the Erdős-Woods conjecture by T. N. Shorey and R. Tijdeman in [8].

Further Reading

The book [5] of P.J. McCarthy contains more than *four hundreds* exercises on the topic of arithmetical functions, and includes further developments as well.

References

[1] P.T. Bateman and H.G. Diamond. *Analytic number theory*. An introductory course. World Scientific Publishing Co. Pte. Ltd., Hackensack, NJ, 2004, pp. xiv+360 (cit. on p. 10).

[2] E. Cohen, "Arithmetical functions associated with the unitary divisors of an integer". In: *Math. Z.* 74 (1960), pp. 66–80 (cit. on p. 11).

[3] P. Erdős. "Research Problems: How Many Pairs of Products of Consecutive Integers Have the Same Prime Factors?" In: *Amer. Math. Monthly* 87.5 (1980), pp. 391–392. https://doi.org/10.2307/2321216 (cit. on p. 14).

[4] V.-A. Le Besgue. *Introduction à la Théorie des nombres*. Mallet-Bachelier, 1862, p. 104 (cit. on p. 4).

[5]   P. J. McCarthy. *Introduction to arithmetical functions*. Universitext. Springer-Verlag, New York, 1986, pp. vii+365. https://doi.org/10.1007/978-1-4613-8620-9 (cit. on p. 15).

[6]   B. van der Pol. "On a non-linear partial differential equation satisfied by the logarithm of the Jacobian theta-functions, with arithmetical applications. I, II". In: *Nederl. Akad. Wetensch. Proc. Ser. A.* **54** = *Indagationes Math.* 13 (1951), pp. 261–271, 272–284 (cit. on p. 14).

[7]   N. Sedrakian and J. Steinig. "A particular case of Dirichlet's theorem on arithmetic progressions". In: *Enseign. Math. (2)* 44.1-2 (1998), pp. 3–7 (cit. on p. 4).

[8]   T. N. Shorey and R. Tijdeman. "Arithmetic properties of blocks of consecutive integers". In: *From arithmetic to zeta-functions*. Springer, [Cham], 2016, pp. 455–471 (cit. on p. 15).

[9]   C.L. Stewart, R. Tijdeman, "On the Oesterlé-Masser conjecture". In: *Monatsh. Math.* 102.3 (1986), pp. 251–257. https://doi.org/10.1007/BF01294603 (cit. on p. 15).

[10]   R.S. Vaidyanathaswamy. "The theory of multiplicative arithmetic functions". English. English. In: *Trans. Am. Math. Soc.* 33 (1931), pp. 579–662. https://doi.org/10.2307/1989424 (cit. on p. 11).

Chapter 2
A Calculus on Arithmetical Functions

The previous chapter introduced the concept of arithmetical convolution, unitary or otherwise, and the basics of a new type of *calculus* appeared. We take this project to its next stage and develop it into a powerful working tool. We do not dwell on the algebraic aspects, but focus on its practical usage. The readers who worry about convergence issues may consult the next chapter and restrict themselves to absolutely convergent Dirichlet series.

2.1 The Ring of Formal Dirichlet Series

To each prime p, we associate an *indeterminate* p^{-s}, just as in the case of polynomials. However, we now have a denumerable number of variables, though we consider only finite products of them. The monomials are $p_1^{-k_1 s} p_2^{-k_2 s} \cdots p_r^{-k_r s}$, which we shorten to n^{-s}, where $n = \prod_{1 \le i \le r} p_i^{k_i}$. We define a *formal Dirichlet series* to be a series of the form

$$\mathcal{D}(f, s) = \sum_{n \ge 1} f(n) n^{-s} ,$$

where the coefficients $f(n)$ are arbitrary complex numbers. This is only a different manner to denote the sequence $(f(n))$, or equivalently the function f. The choice of the symbol p^{-s} is to prepare for the next chapter, where it will indeed be replaced by p to some power $-s$, with s a complex variable. So far, so good, but our formal series is merely a way of writing the same thing differently... Here is the crux: the usual product of polynomial series (equally named *formal power series*) corresponds to the arithmetical convolution. This is so important that we declare this result, although not deep, to be a theorem.

> ### Theorem 2.1
>
> For any two arithmetical functions f and g, we have
>
> $$\mathcal{D}(f, s)\, \mathcal{D}(g, s) = \mathcal{D}(f \star g, s) \,.$$

Pari/GP proposes the function `dirmul` to compute the first coefficients of a convolution product as well as the reverse function `dirdiv`.

Proof The product of $\mathcal{D}(f, s)$ by $\mathcal{D}(g, s)$ is the formal series of the sum of all the products $f(k)k^{-s}g(\ell)\ell^{-s}$. Since $k^{-s}\ell^{-s} = (k\ell)^{-s}$, the coefficient of n^{-s} in $\mathcal{D}(f \star g, s)$ reads

$$\sum_{k\ell=n} f(k)g(\ell)\,,$$

which is exactly $(f \star g)(n)$.   □

The neat effect of going from the function to its formal Dirichlet series is that now the arithmetical convolution can be treated like the usual product! The readers will readily check that

$$\mathcal{D}(\delta_{n=1}, s) = 1 \,.$$

Well, this is to be expected: the unit of the arithmetical convolution gets carried by our transformation to the unit of the usual product. Note also that the polynomials with several variables form an *algebra*: we can meaningfully consider inverses. For instance (as we shall often use),

$$(1 - X)(1 + X + X^2 + X^3 + \ldots) = 1 \,,$$

which means that the inverse of $1 + X + X^2 + X^3 + \ldots$ is $1 - X$, i.e. that we have

$$\frac{1}{1 - X} = 1 + X + X^2 + X^3 + \ldots$$

We may replace X by any p^{-s} to get a useful tool. We have detailed this computation to convince the readers that familiar algebraic manipulations are also valid. More will be said at the beginning of Chap. 17.

Exercise 2-1.

1 ⋄ Show that $\mathcal{D}(\mathbb{1}, s) = \prod_{p \geq 2}(1 - p^{-s})^{-1}$.

2 ⋄ Deduce from $\mu \star \mathbb{1} = \delta_{n=1}$ that $\mathcal{D}(\mu, s) = \prod_{p \geq 2}(1 - p^{-s})$.

The power of Theorem 2.1 is largely enhanced by the next two theorems.

Theorem 2.2

For any multiplicative function f, we have

$$\mathcal{D}(f, s) = \prod_{p \geq 2}\left(1 + \sum_{k \geq 1} f(p^k)p^{-ks}\right),$$

where the product is over all primes p and the sum is over all positive integers k.

We call the factors on the right-hand side *the local factors of the Dirichlet series of f* at the prime p. Since $f(1) = 1 = p^{-0s}$, we could also write

$$\mathcal{D}(f, s) = \prod_{p \geq 2}\sum_{k \geq 0} f(p^k)p^{-ks},$$

but practice teaches that this last expression leads to computational errors. It is thus better to use the slightly more cumbersome form that we stated.

Proof The proof is immediate: expanding the product on the right-hand side, we get finite products of the form

$$f(p_1^{k_1})f(p_2^{k_2}) \cdots f(p_r^{k_r})p_1^{-k_1 s}p_2^{-k_2 s} \cdots p_r^{-k_r s} \tag{2.1}$$

and this is exactly $f(n)n^{-s}$ by the multiplicativity of f. By the uniqueness of prime factorization, each integer $n \geq 1$ appears exactly once. □

We have now a powerful tool for computing the values of multiplicative functions at our disposal. Indeed, instead of working with all the primes, it is enough to work p-component per p-component. We establish this by showing how to determine the value of the divisor function $d(n)$ through this process. We have $d = \mathbb{1} \star \mathbb{1}$, and the p-component $\mathcal{D}_p(\mathbb{1}, s)$ of the Dirichlet series of the function $\mathbb{1}$ is $1 + p^{-s} + p^{-2s} + \ldots = 1/(1 - p^{-s})$. Whence, the p-component $\mathcal{D}_p(d, s)$ of the Dirichlet series of the function d is $1/(1 - p^{-s})^2$. Let us recall that

$$\frac{1}{(1 - X)^2} = \sum_{k \geq 0}(k + 1)X^k,$$

from which we deduce that $d(p^k) = k + 1$. We conclude that

$$d(n) = \prod_{1 \leq i \leq r}(\alpha_i + 1) \quad \text{when} \quad n = \prod_{1 \leq i \leq r} p_i^{\alpha_i},$$

where the p_i's are distinct. The readers will find in the next section as well as in Chap. 10 several practical cases where this technique is being used.

Pari/GP proposes the function `direuler` to get the Dirchlet series of an Euler product. On combining our goods, here is another manner to get the successive values of the Möbius function under Pari/GP:

```
d0 = direuler(p = 2, 50, 1, 50);
d1 = direuler(p = 2, 50, 1/(1-X), 50);
dmoebius = dirdiv(d0, d1)
```

Exercise 2-2.

1 ⋄ Compute the Dirichlet series of $\varphi(n)/n$ and deduce that $\frac{\varphi(n)}{n} = \prod_{p|n}(1 - 1/p)$

2 ⋄ Let $d_3(n)$ be the number of triples of positive integers (a, b, c) such that $abc = n$. Show that $d_3(n) = \prod_{p^k \| n} \frac{(k+1)(k+2)}{2}$.

See Exer. 4-12 for a generalization to r-tuples of divisors.

The unitary convolution has also an easy interpretation in terms of Dirichlet series.

Theorem 2.3

For any two multiplicative functions f and g, we have

$$\mathcal{D}(f \,\square\, g, s) = \prod_{p \geq 2}\left(1 + \sum_{k \geq 1}(f(p^k) + g(p^k))p^{-ks}\right),$$

where the product is over all primes p and the sum is over all positive integers k.

Proof By Theorem 1.3, we know that $f \,\square\, g$ is a multiplicative function. We readily compute that

$$(f \,\square\, g)(p^k) = f(p^k) + g(p^k)$$

when $k \geq 1$, whence Theorem 2.2 applies.   □

We mention for the readers curious about the algebraic structure of the ring of formal Dirichlet series that Cashwell & Everett [3] showed that it is factorial.

2.2 Further Examples

Suppose one would like to compare $f(k) = \mu(k)k/\varphi(k)$ to the Liouville function $\lambda(k)$. The local factor of the Dirichlet series of f is

$$1 + \sum_{k=1}^{\infty} f(p^k)p^{-ks} = 1 - \frac{p}{p-1}p^{-s}.$$

Similarly, the local factor of the Dirichlet series for the Liouville function is

$$1 - p^{-s} + p^{-2s} - \ldots = 1/(1 + p^{-s}) \, .$$

If we divide the first expression above by the second one, we get

$$\left(1 - \frac{p}{p-1} p^{-s}\right)(1 + p^{-s}) = 1 - \frac{1}{p-1} p^{-s} - \frac{p}{p-1} p^{-2s} \, .$$

We next convert this identity into a convolution identity by considering a multiplicative function g that is given on prime powers by

$$g(p) = -\frac{1}{p-1} \, , \quad g(p^2) = -\frac{p}{p-1} \, , \quad g(p^k) = 0, \; \forall k \geq 3 \, .$$

By construction, we have $\mathcal{D}(g, s) = \mathcal{D}(f, s)/\mathcal{D}(\lambda, s)$ thus $\mathcal{D}(f, s) = \mathcal{D}(g, s)\,\mathcal{D}(\lambda, s) = \mathcal{D}(g \star \lambda, s)$, which translates into $f = g \star \lambda$. The expression for g above does not look very useful, so we define the two auxiliary multiplicative functions by

$$g'(p) = -\frac{1}{p-1} \, , \quad g'(p^k) = 0, \; \forall k \geq 2 \, ,$$

(i.e. $g(a) = \mu(a)/\varphi(a)$) and

$$g''(p) = 0 \, , \quad g''(p^2) = -\frac{p}{p-1} \, , \quad g''(p^k) = 0, \; \forall k \geq 3 \, .$$

On appealing to Theorem 2.3, we discover that $g = g' \square g''$, which may be rewritten as

$$g(m) = \sum_{\substack{ab^2=m \\ (a,b)=1}} \frac{\mu(a)\,\mu(b)}{\varphi(a)\,\varphi(b)} b \, .$$

This expression is the one used, for instance, in [6].

We continue our journey with some exercises.

Exercise 2-3.

1 ⋄ Find the Dirichlet series of $d(n^2)$.

2 ⋄ Find the Dirichlet series of $d(n)^2$.

3 ⋄ Using the previous two exercises show that $d(n)^2 = \sum_{m|n} d(m^2)$.

Exercise 2-4. Show that $d(n^2) = \sum_{d|n} 2^{\omega(d)}$.

Exercise 2-5.

1 ⋄ Determine the convolution inverse of the Liouville λ-function.

2 ⋄ Show that $\sum_{m|n} \lambda(n/m)\mu(m) = \lambda(n)2^{\omega(n)}$ and that $\sum_{m|n} \lambda(m)\mu(m) = 2^{\omega(n)}$.

Exercise 2-6.

1 ◇ Using the Dirichlet series, compute the inverse $g(n)$ of the multiplicative function Id that to every integer n associates n.

2 ◇ Prove that $\sigma(n)^2 = n \sum_{d|n} \sigma(d^2)/d$.

Exercise 2-7. Show that $\sum_{m|n} d(m)^3 = \left(\sum_{m|n} d(m) \right)^2$.

Here are now two related identities; the first one is due to E. Busche in [2] and the second one to S. Ramanujan around 1916 (see [5, Theorem 1.12 and around] for more details).

Exercise 2-8. (Busche-Ramanujan Identities). Prove the two identities $\sigma_k(mn) = \sum_{d|(m,n)} \mu(d) d^k \sigma_k(m/d) \sigma_k(n/d)$ and $\sigma_k(m)\sigma_k(n) = \sum_{d|(m,n)} d^k \sigma_k(mn/d^2)$.

Exercise 2-9. Prove that $\varphi \star d = \sigma$.

Exercise 2-10. Prove that, for every positive integer n, we have the identity $\sum_{m|n} \varphi(m)\sigma(n/m) = nd(n)$.

Let us close this section with an exercise on *powerful numbers* as defined by S.W. Golomb in [4].

Exercise 2-11. An integer n is said to be *powerful* if, whenever a prime p divides n, then so does p^2.

1 ◇ Prove that any such integer can be written uniquely in the form $n = a^2 b^3$, where b is square-free and deduce the Dirichlet series of their characteristic function.

2 ◇ Use Pari/GP and the functions direuler and dirmul to get the list of the gaps between the powerful numbers below 1000.

3 ◇ Golomb conjectured that the difference between any two consecutive powerful numbers could never be equal to 6. What do you think about this conjecture? What about the gap 14?

Powerful numbers are sometimes called *squarefull integers*. Their characteristic function is denoted by ispowerful in Pari/GP.

Further Reading

The calculus of Dirichlet series is developed in many books. Of particular interest on this aspect is the monograph [1] by P.T. Bateman and H. Diamond.

References

[1]   P.T. Bateman and H.G. Diamond. *Analytic number theory*. An introductory course. World Scientific Publishing Co. Pte. Ltd., Hackensack, NJ, 2004, pp. xiv+360 (cit. on p. 22).

[2]   E. Busche. "Über Gitterpunkte in der Ebene". In: *J. Reine Angew. Math.* 131 (1906), pp. 113–135. https://doi.org/10.1515/crll.1906.131.113 (cit. on p. 21).

[3]   E.D. Cashwell and C.J. Jr. Everett. "The ring of number-theoretic functions". In: *Pacific J. Math.* 9 (1959), pp. 975–985. http://projecteuclid.org/euclid.pjm/1103038878 (cit. on p. 20).

[4]   S.W. Golomb. "Powerful numbers". In: *Amer. Math. Monthly* 77 (1970), pp. 848–855. https://doi.org/10.2307/2317020 (cit. on p. 22).

[5]   P. J. McCarthy. *Introduction to arithmetical functions*. Universitext. Springer-Verlag, New York, 1986, pp. vii+365. https://doi.org/10.1007/978-1-4613-8620-9 (cit. on p. 21).

[6]   Akhilesh P and O. Ramaré. "Explicit averages of non-negative multiplicative functions: going beyond the main term". In: *Colloq. Math.* 147.2 (2017), pp. 275–313. https://doi.org/10.4064/cm6080-4-2016 (cit. on p. 21).

Chapter 3
Analytical Dirichlet Series

We use the Dirichlet series of a complex argument. When n is a positive integer and $s = \sigma + it$ is a complex number having real part σ and imaginary part t, we can write $n^{-s} = (\cos(t \log n) - i \sin(t \log n))/n^{\sigma}$. In particular, $|1/n^s| = 1/n^{\sigma}$. If the readers are unfamiliar with these notions, they may well consider only real values of s, as this will change neither the theory nor the proofs we present. However, as experience shows, having complex variables has clear advantages, as will become apparent in the second walk.

3.1 Abscissa of Absolute Convergence

To any arithmetical function f, we associate its *Dirichlet series* given for any complex number s by

$$D(f, s) = \sum_{n \geq 1} f(n)/n^s , \qquad (3.1)$$

provided this series converges absolutely. This is the analytic counterpart of the formal Dirichlet series introduced in Chap. 2.

Series of this form are called *Dirichlet series* in honour of J.P.G.L. Dirichlet, who used them in 1837/39 (see [6, 7] and Chap. 15) to prove that prime numbers are equidistributed over arithmetic progressions $a, a + d, a + 2d, \ldots$ with a and d coprime. Dirichlet considered only real arguments s. The final extension to the complex domain, where the most important information lies, was one of the major innovations made by B. Riemann in his epoch-making eight pages long memoir [18].

Lemma 3.1

Let f be an arithmetical function such that its Dirichlet series converges absolutely for a certain complex number s. Then for any real number $r \geq \Re s$ the series $D(f, r)$ converges absolutely. Moreover, for any complex number s' such that $\Re s' \geq \Re s$ the series $D(f, s')$ converges absolutely as well.

O. Ramaré, *Excursions in Multiplicative Number Theory*, Birkhäuser Advanced Texts Basler Lehrbücher, https://doi.org/10.1007/978-3-030-73169-4_3

Proof We have

$$D(f,r) = \sum_{n \geq 1} \frac{f(n)}{n^s} \frac{n^s}{n^r} \ .$$

Since r is real and $r \geq \Re s$ therefore $|n^s|/n^r = n^{\Re s - r} < 1$ and $D(|f|, r) \leq D(|f|, \Re s)$. Hence, the Dirichlet series of f converges absolutely for any $r \geq \Re s$. The complex case is completely analogous. □

This property brings us to the concept of abscissa of absolute convergence.

Definition 3.1

We define *the abscissa of absolute convergence* of a function f as the infimum of the real numbers s such that the Dirichlet series $D(f, s)$ converges absolutely. If $D(f, s)$ converges absolutely for all s, we say that the abscissa of absolute convergence is $-\infty$.

A series may not converge at its abscissa of convergence. According to a theorem of Landau, nicely presented by F. Dress in [8], this situation indeed never arises when this abscissa is finite and f is non-negative. The case $f(n) = e^{-n}$ shows that the abscissa of convergence can be equal to $-\infty$ though the function f is non-negative without a compact support.

To each arithmetical function f, we may associate its Dirichlet series, and this series will have some finite abscissa of convergence when $f(n)$ grows at most polynomially at infinity. We show now that the values taken by this Dirichlet series of f indeed determines the function f.

Lemma 3.2

Let f and g be two arithmetical functions, whose respective Dirichlet series converge absolutely for some real number s. Assume further that $D(f, r) = D(g, r)$ for every $r > s$. Then $f = g$.

Proof We define $h_1 = f - g$ and note that $D(h_1, r) = 0$ for every $r > s$. Since this series converges at $r = s + 1$, we infer that $h_2(n) = h_1(n)/n^{r+1}$ is bounded in absolute value and satisfies $D(h_2, r) = 0$ for every $r > -1$. We have to show that $h_2 = 0$. Assume it is not the case and call n_0 the smallest integer n such that $h_2(n) \neq 0$. A comparison to an integral gives us directly that for any $r > 1$, we have

$$|n_0^r D(h_2, r) - h_2(n_0)| \leq \max_n |h_2(n)| \sum_{n \geq n_0 + 1} \frac{n_0^r}{n^r}$$

$$\leq \max_n |h_2(n)| n_0^r \int_{n_0}^\infty \frac{dt}{t^r} \leq n_0 \max_n |h_2(n)|/(r-1) \ .$$

Clearly, the ultimate upper bound above goes to 0 when r goes to infinity. Since $D(h_2, r) = 0$, this tells us that $h_2(n_0) = 0$, contradicting our hypothesis. □

The readers may want to improve this proof in two ways: first by replacing the reduction ad absurdum by a recursion. Secondly, a small change in the argument gives us $n_0^r D(h_2, r) - h_2(n_0) = O((1 + n_0^{-1})^{-r})$, while we only obtained $O(1/r)$.

Exercise 3-1. Assume that the Dirichlet series $D(f, s_0)$ converges *pointwise* at some complex number s_0. Let s be some other complex number such that $\Re s > \Re s_0$. Show that the series $D(f, s)$ also converges *pointwise*.

Exercise 3-2. Assume that the Dirichlet series $D(f, s_0)$ converges *pointwise* at some complex number s_0. Show that the Dirichlet series $D(f, s)$ converges absolutely in the domain $\Re s > \Re s_0 + 1$.

Exercise 3-3. Show that $-D(f \cdot \log, s)$ is the derivative of $D(f, s)$, provided that it converges absolutely. Give an example of a non-negative function f for which $D(f, 1)$ converges, but $D(f \cdot \log, 1)$ does not.
Note that the function $n \mapsto f(n) \log n$ has been called ∂f just after Theorem 1.1.

We encourage the readers to prove the important fact that the abscissa of absolute convergence of $D(f \cdot \log, s)$ and of $D(f, s)$ are the same. This apparent contradiction with the result of the above exercise comes from the minimum appearing in the definition of the abscissa of absolute convergence.

3.2 An Introduction to the Riemann Zeta-Function

The Riemann ζ-function is the simplest Dirichlet series. It is defined, when $\Re s > 1$, by

$$\zeta(s) = \sum_{n \geq 1} n^{-s} .$$

Though it is the most famous of all Dirichlet series, many of its properties are largely unknown, many are guessed and few are proven. It has been named after B. Riemann though it had already been considered and used by L. Euler. In his 1859-memoir, Riemann started a revolution in prime number theory and sketched results that would take researchers decades to prove rigorously (after complex function theory was put on a firm footing). We will continue the study of this function in Chap. 19, but the readers should look at Exer. 3-18.

Exercise 3-4. Show that the series defining the Riemann ζ-function is indeed absolutely convergent when $\Re s > 1$.

Exercise 3-5. Show that $\zeta(s) = \dfrac{s}{s-1} - s \displaystyle\int_1^\infty \{t\} \dfrac{dt}{t^{s+1}}$ and deduce a formula for $\zeta(1 + u)$ as u goes to 0.

Exercise 3-6. Show that for $\sigma > 1$, we have $1/(\sigma - 1) < \zeta(\sigma) < \sigma/(\sigma - 1)$.

Both Sage and Pari/GP propose primitives to get numerical values of the Riemann zeta-function. In Pari/GP, it is enough to use zeta(1/2), while here is how to do so in Sage:

```
C = ComplexIntervalField(500); C(zeta(1/2))
def f(t):
  return(float(abs(zeta(1/2+I*t))))

plot(f, (0,30))
```

The above script gives a graph of the $|\zeta(1/2 + it)|$ for t ranging over $[0, 30]$. We detect on it the first two zeros of $\zeta(s)$ at $\frac{1}{2} + i14.13472\ldots$ and at $\frac{1}{2} + i21.02203\ldots$.

Exercise 3-7. Show that for real $\sigma > 1$ we have $\zeta(\sigma) = (\sigma - 1)^{-1} + \gamma + O(\sigma - 1)$

Exercise 3-8. Define $L(s) = \sum_{n \geq 1}(-1)^n n^{-s}$.

1 ◇ Show that this series is convergent when $\Re s > 0$, then that $L(1) < 0$ and finally that $L'(1) < 0$.

2 ◇ Show that $-\zeta'(\sigma)/\zeta(\sigma) < 1/(\sigma - 1)$ for real $\sigma > 1$.

The next exercise gives an elementary improvement due to H. Delange in [5] of the above inequality for $-\zeta'(\sigma)/\zeta(\sigma)$.

Exercise 3-9.

1 ◇ Prove that, when $s > 1$ is a real number, we have

$$(s - 1)\zeta(s) = s - s(s - 1)\int_1^\infty \{t\}dt/t^{s+1} .$$

Deduce that $(s - 1)\zeta(s) > s$.

2 ◇ Show that $\zeta(s) + (s - 1)\zeta'(s) > 1 - (2s - 1)\int_1^\infty \{t\}dt/t^{s+1}$.

3 ◇ Show that $\int_1^\infty \dfrac{\{t\}}{t^{s+1}}dt < \dfrac{1}{2s}$.

4 ◇ Show that $-\dfrac{\zeta'(s)}{\zeta(s)} < \dfrac{1}{s-1} - \dfrac{1}{2s^2}$ when $s > 1$.

Getting a plot to illustrate the inequality of the previous exercise is more tricky as we have to get access to $\zeta'(s)$. We shall go much further on this topic in Chap. 19 but mention here some magic words in Sage that lead to the graph of Fig. 3.1. This way of doing will generalize to very intricate functions, a glimpse of which is given in Sect. 15.2. However, explaining the `Dokchitser` package of T. Dokchitser is out of the scope of this book!

```
myprec = 200
RR = RealField(myprec)
myzeta = Dokchitser(conductor=1, gammaV=[0], weight=1,
```

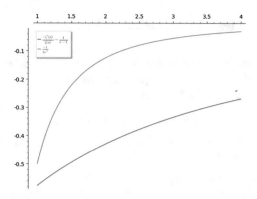

Fig. 3.1: $-(\zeta'/\zeta)(s) - 1/(s-1)$ below in blue versus $-1/(2s^2)$ above in green

```
                eps=1, poles=[1], residues=[-1],

                prec=myprec, init='1')
def ref(s):
    return (-myzeta.derivative(s)/myzeta(s)-RR(1/(s-1)))
def comp(s):
    return RR(-1/2/s^2)


smin = 1.0001
smax = 4
rp = plot(ref, (smin, smax), rgbcolor=(0, 0, 0.5),


    legend_label=r"$\frac{-\zeta'(s)}{\zeta(s)}-\frac{1}{s-1}$",
    legend_color=(0, 0, 0.5))
cp = plot(comp, (smin, smax), rgbcolor=(0, 0.5, 0),
            legend_label=r"$\frac{-1}{2s^2}$",
            legend_color=(0, 0.5, 0))

(rp+cp).show(figsize = 5, dpi = 300)
save(rp+cp,'CompDelange.png')
```

Sage provides also the simpler primitive `zetaderiv`. Here is how to compute values of $\zeta'(1.2)$ in Pari/GP:

```
myzeta = lfuncreate(1);
print("zeta'(1.2) : ", lfun(myzeta, 1.2, 1));
```

3.3 Dirichlet Series and Convolution Product

The two internal laws on multiplicative functions have a simple and elegant translation in terms of Dirichlet series. First, given any two arithmetical functions f and g, whose Dirichlet series converge absolutely at s we have

$$D(f + g, s) = D(f, s) + D(g, s) .$$

Secondly, the convolution product ($\star$) is translated to the usual product as the following theorem reveals.

Theorem 3.1

Let f and g be two arithmetical functions, whose Dirichlet series converge absolutely at some point s. Then the same is true for $D(f \star g, s)$ and we have

$$D(f \star g, s) = D(f, s)D(g, s) .$$

Proof Both series converge absolutely, so one can shift the terms as deemed necessary. The proof then reduces to the one of Theorem 2.1. □

This theorem shows that the operator D that associates to any arithmetical function of polynomial growth its Dirichlet series, *trivializes* the convolution product. In exactly the same way, the Fourier transform trivializes the convolution product. No such link exists in general between $D(f \square g, s)$ and $D(f, s)$ and $D(g, s)$, but we shall see in Sect. 3.5 that such a formula indeed exists in the case of multiplicative functions.

The previous theorem implies that the abscissa of convergence of the Dirichlet series associated to $f \star g$ is bounded above by the maximum of the abscissas of the Dirichlet series associated to both f and g. This upper bound is often an equality when both abscissas are not equal and no factor vanishes identically!

Exercise 3-10. Show that $D(\varphi, s) = \zeta(s - 1)/\zeta(s)$, $D(\lambda, s) = \zeta(2s)/\zeta(s)$, and $D(\mu^2, s) = \zeta(s)/\zeta(2s)$. Here λ is the Liouville function defined by $\lambda(n) = (-1)^{\Omega(n)}$, where $\Omega(n)$ is the number of prime factors of n counted with multiplicities.

Exercise 3-11. Show that the Dirichlet series associated with μ is $1/\zeta(s)$ and deduce that the abscissa of convergence of a product can be strictly smaller than the abscissa of each factor (consider $\mathbb{1} \star \mu$).

Exercise 3-12. Express the Dirichlet series of the function that associates $2^{\omega(n)}$ to n in terms of the Riemann zeta-function. Here $\omega(n)$ is the number of prime factors of n counted without multiplicities.

Exercise 3-13. Express the Dirichlet series of the function that associates $2^{\omega(n)}\lambda(n)$ to n in terms of the Riemann zeta-function.

Exercise 3-14. In this exercise, we define $\kappa(n) = v_1 v_2 \cdots v_k$, where $n = p_1^{v_1} p_2^{v_2} \cdots p_k^{v_k}$. Establish the next two identities:

$$1 \diamond \sum_{n \geq 1} \kappa(n)/n^s = \zeta(s)\zeta(2s)\zeta(3s)/\zeta(6s),$$

$$2 \diamond \sum_{n \geq 1} 3^{\omega(n)} \kappa(n)/n^s = \zeta^3(s)/\zeta(3s).$$

Link the function κ with the characteristic function of powerful numbers defined in Exer. 2-11.

Exercise 3-15. (Inspired from [17] by S. Ramanujan). This exercise proposes to find the Dirichlet series of different sets of integers.

$1 \diamond$ Following Ramanujan, we call an integer *irregular* if it has an odd number of prime factors. Show that $\displaystyle\sum_{n \text{ irregular}} 1/n^s = \frac{\zeta(s)^2 - \zeta(2s)}{2\zeta(s)}$.

$2 \diamond$ Let us restrict this set to square-free irregular numbers. Show that its Dirichlet series satisfies $\displaystyle\sum_{n \text{ irregular}} \mu^2(n)/n^s = \frac{\zeta(s)^2 - 1}{2\zeta(s)\zeta(2s)}$.

$3 \diamond$ Let us finally consider the set S of the integers that are divisible some square integer distinct from 1. Show that $\displaystyle\sum_{n \in S} 1/n^s = \zeta(s)\frac{\zeta(2s) - 1}{\zeta(2s)}$.

3.4 Dirichlet Series and Multiplicativity

The connection between the Riemann zeta-function and the primes can be seen via *Euler's product formula* proven in Exer. 3-18

$$\zeta(s) = \prod_p \left(1 - \frac{1}{p^s}\right)^{-1},$$

which is valid for all complex s such that $\Re s > 1$. This is a consequence of the fundamental theorem of arithmetic.

In fact, the identity above is not an isolated curiosity, but a general feature of Dirichlet series associated to multiplicative functions.

Theorem 3.2

Assume that the Dirichlet series of the multiplicative function f converges absolutely for some s. Then $D(f, s)$ can be expressed as an Euler product

$$D(f, s) = \prod_{p \geq 2} \left(1 + \sum_{k \geq 1} \frac{f(p^k)}{p^{ks}}\right). \tag{3.2}$$

This harmless result is the source of a lot of confusion and misunderstanding, so we shall trod cautiously ahead and start with some comments. Notice that its formal version is contained in Theorem 2.2.

The factors on the right-hand side of (3.2) are called the *Euler factors at the prime p*.

Convergence and convergent products

We now discuss the meaning of (3.2). The sequence $1/2$, $(1/2) \times (1/2)$, $(1/2) \times (1/2) \times (1/2)$, having $(1/2)^n$ as its generic term, is convergent, with limit 0. Nothing exciting about that. Some people say that a (formal) product $\prod_{n \geq 1} a_n$ is convergent if and only if

1. The sequence of its partial products converge towards a limit, say P.
2. $P \neq 0$.

Cutting Eq. (3.2) at a given prime naturally gives rise to partial products, it then remains to explicitly check condition 2. To add to the confusion, some authors say that a product *converges strictly* whenever this additional condition is verified. This implies that their definition of a convergent product is restricted only to condition 1 above. Some other authors say that the product is convergent whenever the series of the logarithm is convergent. But since we want to use complex numbers, the usage of logarithms is unclear.

Convergence and absolutely convergent product

We shall employ the following terminology.

Definition 3.2

The product $\prod_{n \geq 1} a_n$ is said to be absolutely convergent in the sense of Godement if and only if $\sum_{n \geq 1} |a_n - 1| < \infty$.

It is clear that we have put a lot of emphasis on our definition by using the adverb *absolutely* while also completing the expression by *in the sense of Godement*, but we hope to stimulate the reader to adopt a good mathematical habit in this way. The main interest of our definition comes from the next theorem.

Theorem 3.3

Assume that $\sum_{n \geq 1} |u_n| = M < \infty$. Then

1. $P_n = \prod_{1 \leq j \leq n} (1 + u_j) \to P \in \mathbb{C}$.
2. If $1 + u_j \neq 0$ for every j, then $P \neq 0$.

Furthermore, we have $|P| \leq e^M$.

Proof We start this proof by noticing that $P_n - P_{n-1} = u_n P_{n-1}$ and thus

$$|P_n - P_{n-1}| \leq |u_n||P_{n-1}| \leq |u_n| \prod_{1 \leq j \leq n-1} e^{|u_j|} \leq |u_n|e^M .$$

We conclude from this inequality that $\sum_n (P_n - P_{n-1})$ is absolutely convergent and thus the sequence P_n converges towards, say, $P \in \mathbb{C}$.

To show that $P \neq 0$, we shall exhibit Q such that $PQ = 1$ and $Q = \prod_{j \geq 1}(1 + v_j)$ with $\sum_{j \geq 1} |v_j| < \infty$. We want to adjust v_j so that $(1 + u_j)(1 + v_j) = 1$ for every j. It turns out that this is possible because $1 + u_j \neq 0$. We get

$$v_j = \frac{1}{1 + u_j} - 1 = \frac{-u_j}{1 + u_j} , \quad \text{hence } |v_j| \sim |u_j| ,$$

and the readers will swiftly complete the proof. $\square$

The readers may see that in this definition in the inequalities $1 + x \leq e^x$ and $|P_{n-1}| \leq e^M$ logarithms are hidden. They remain implicit, with the consequence that our definition has no difficulty with complex valued terms.

The classical book [14] of K. Knopp treats series of *real numbers* and relies on logarithms, a notion that is more delicate to handle in the context of complex numbers.

A novel statement and its proof

> **Theorem 3.4**
>
> Assume the Dirichlet series of the multiplicative function f converges absolutely for some s. Then $D(f, s)$ can be expressed as an Euler product
>
> $$D(f, s) = \prod_{p \geq 2}\left(1 + \sum_{k \geq 1} \frac{f(p^k)}{p^{ks}}\right). \tag{3.2}$$
>
> where the product converges absolutely in the sense of Godement.

When f is a multiplicative function, so is $|f|$, and we can apply our theorem to this function. We deduce that the product

$$\prod_{p \geq 2}\left(1 + \sum_{k \geq 1} \frac{|f(p^k)|}{p^{ks}}\right)$$

is also absolutely convergent in the sense of Godement.

Proof The product under consideration can be written as $\prod_p(1 + u_p)$ with

$$u_p = \sum_{k \geq 1} \frac{f(p^k)}{p^{ks}} . \tag{3.3}$$

We immediately see that

$$\sum_p |u_p| \le \sum_p \sum_{k \ge 1} \frac{|f(p^k)|}{|p^{ks}|} \le \sum_{n \ge 1} \frac{|f(n)|}{|n^s|}$$

and Theorem 3.3 applies.   □

According to our definition, a product may be convergent and have value 0, in which case one of its factors vanishes. Here is an example. Define the multiplicative function f_5 by

$$\begin{cases} f_5(2) = 2, \quad f_5(4) = -24, \quad \forall k \ge 3, \quad f_5(2^k) = 0, \\ \forall p \ge 3, \forall k \ge 1, \quad f_5(p^k) = 1 \end{cases}$$

We readily check that $\sum_{n \ge 1} f_5(n)/n^s$ is absolutely convergent when $\Re s > 1$ and that we have

$$\sum_{n \ge 1} \frac{f_5(n)}{n^s} = \left(1 + \frac{2}{2^s} - \frac{24}{2^{2s}}\right) \prod_{p \ge 3} \left(1 - \frac{1}{p^s}\right)^{-1}.$$

This series vanishes at $s = 2$.

Exercise 3-16. Show that the function $n \mapsto (-1)^{n+1}$ is multiplicative and find its Dirichlet series.

The curious readers will find interesting material in the paper [13] of J.-P. Kahane & É. Saias. Theorem 9.2 on exchange of product signs is also of interest.

3.5 Dirichlet Series, Multiplicativity and Unitarian Convolution

The central result of this section is the following.

Theorem 3.5

Let s be a complex number. Let f and g be two multiplicative functions, whose Dirichlet series are absolutely convergent at s. Then the same is true for $D(f \square g, s)$ and we have

$$D(f \square g, s) = \prod_{p \ge 2} \left(1 + \sum_{k \ge 1} \frac{f(p^k)}{p^{ks}} + \sum_{k \ge 1} \frac{g(p^k)}{p^{ks}}\right),$$

where the Euler product converges absolutely in the sense of Godement.

Recall that the unitary convolution product is defined in Sect. 1.6 and that we have already proved a formal version of this result in Theorem 2.3.

Proof Theorem 1.3 shows that $f \square g$ is a multiplicative function and Theorem 3.1 shows that the series $D(|f| \star |g|, s)$ converges absolutely. Since one can easily check that

$$(|f| \square |g|)(n) \le (|f| \star |g|)(n)$$

for any integer n, we deduce that the series $D(|f| \square |g|, s)$ converges absolutely, the same conclusion is true for $D(f \square g, s)$. Now we can use Theorem 3.4 to show that the last series can be written as an Euler product that converges absolutely in the sense of Godement. It remains to identify the factors of the Euler product, which is easy thanks to $(f \square g)(p^k) = f(p^k) + g(p^k)$. $\qquad\square$

3.6 A Popular Variation

Let f be an arithmetic multiplicative function that is bounded in absolute value by 1. In all practical situations we can restrict to that case by considering a function $f(n)/n^a$, which is also multiplicative.

Let $y \geq 1$ be a real parameter. Consider a multiplicative function f_y given by

$$\forall p \leq y, \forall \alpha \geq 1, f_y(p^a) = f(p^a), \quad \forall p > y, \forall \alpha \geq 1, f_y(p^a) = 0, \qquad (3.4)$$

(p being a prime number) and, symmetrically,

$$\forall p \leq y, \forall \alpha \geq 1, f^y(p^a) = 0, \quad \forall p > y, \forall \alpha \geq 1, f^y(p^a) = f(p^a). \qquad (3.5)$$

Here is the central property of these functions.

Lemma 3.3

Given any multiplicative function f, we have the decomposition

$$f = f_y \star f^y \qquad (3.6)$$

Proof Indeed, the summands of

$$\sum_{d_1 d_2 = n} f_y(d_1) f^y(d_2)$$

are all zero, save one of them, since an integer n is uniquely written as $n = \ell m$, where all the prime factors of ℓ are less or equal than y and all the prime factors of m are larger than y. Thus, the sum above can be simply reduced to $f_y(\ell) f^y(m)$, which is $f(n)$. $\qquad\square$

Write

$$D_y^\flat(f, s) = D(f_y, s), \quad D_y^\sharp(f, s) = D(f^y, s) \qquad (3.7)$$

so that $D(f, y) = D_y^\flat(f, s) D_y^\sharp(f, s)$. The Dirichlet series $D_y^\flat(f, s)$ for $\Re s > 1$ can be written as an Euler product:

$$D_y^\flat(f, s) = \prod_{p \leq y} \left(1 + \frac{f(p)}{p^s} + \frac{f(p^2)}{p^{2s}} + \frac{f(p^3)}{p^{3s}} + \dots\right).$$

The series $D_y^\sharp(f, s)$ becomes small when y goes to infinity. For $\sigma = \Re s$, we have

$$|D_y^\sharp(f, s) - 1| \le \sum_{n > y} 1/n^\sigma \le y^{-\sigma} + \int_y^\infty dt/t^\sigma \le \frac{\sigma}{(\sigma - 1)y^{\sigma - 1}}.$$

On letting $y \to \infty$, we obtain the following for $D(f, s)$ in the form of Euler product:

$$D(f, s) = \prod_{p \ge 2} \left(1 + \frac{f(p)}{p^s} + \frac{f(p^2)}{p^{2s}} + \frac{f(p^3)}{p^{3s}} + \cdots \right), \quad (\Re s > 1), \qquad (3.8)$$

where every factor is called *local* at p.

Historical background

The decomposition (3.6) was used a lot, especially by H. Daboussi in [2] (see also [3] and [4]) and allowed, in the cited work, to give an alternative proof of the prime number theorem by elementary methods.

   This decomposition is the key part of the recent work [9], where the authors developed the following philosophy: the behaviour of the function f_y can be well described by its Euler product, while the prevailing behaviour of f^y can be understood via functional equations (we do not discuss them in this book). One can try to reduce understanding the multiplicative function f, to understanding the factors f_y, which is the essence of the probabilistic model developed by J. Kubilius.

   Exercise 3-17. Let a be non-negative real. Write $\lambda_a(n) = \sum_{d|n} d^a \lambda(d)$, where $\lambda(d)$ is the Liouville function. Show that $\displaystyle\sum_{n \ge 1} \frac{\lambda_a(n)}{n^s} = \frac{\zeta(s)\zeta(2s - 2a)}{\zeta(s - a)}$ and

$$\sum_{n \ge 1} \frac{\lambda(n)\lambda_a(n)}{n^s} = \frac{\zeta(2s)\zeta(s - a)}{\zeta(s)} \text{ for } \Re s > 1 + a.$$

   Exercise 3-18. Prove the Euler Identity: for $\Re s > 1$ we have $\zeta(s) = \displaystyle\prod_{p \ge 2} \left(1 - \frac{1}{p^s}\right)^{-1}$. where the Euler product converges absolutely in the sense of Godement.

3.7 Some Remarks Without Proof

Dirichlet series were introduced in [6] by J.P.G.L. Dirichlet in 1837 to show that there are infinitely many primes of the type $a + nq$ with a and q being coprime. This a topic is discussed in Chap. 15. A student and later a friend of Dirichlet, R. Dedekind established many properties of Dirichlet series, see [7].

   The next main step is due to E. Cahen, and is made in his PhD thesis [1] that is notorious for ... the inexact character of several of its statements and proofs! This memoir triggered however further investigations and the year 1915 saw the publication of the short book [10] by G.H. Hardy & M. Riesz that remains a classical reference on the subject.

We next comment on the following two points:

1. How are the average order and the abscissa of convergence linked?
2. Does a decomposition $D(f, s) = D(h, s)D(g, s)$ enable us to conclude that the abscissa of absolute convergence of $D(f, s)$ is the maximum between the one of $D(h, s)$ and the one of $D(g, s)$?

A one-sided answer to the first question is given by the simple identity

$$D(f, s) = s \int_1^\infty \left(\sum_{n \le t} f(n) \right) dt/t^{s+1} , \tag{3.9}$$

which is valid for $\Re s$ strictly greater than the abscissa of absolute convergence of $D(f, s)$. See also Theorem 20.5.

Proof Here is a formal verification of this identity. We have

$$s \int_1^\infty \left(\sum_{n \le t} |f(n)| \right) dt/t^{s+1} = \sum_{n \ge 1} |f(n)| \, s \int_n^\infty dt/t^{s+1} = \sum_{n \ge 1} |f(n)|/n^s .$$

On applying this identity to $|f|$, we see that knowing the average order of $|f|$ would enable one to get an upper bound for the abscissa of absolute convergence of $D(f, s)$. The reverse is false, simply because it is possible that $|f|$ does not admit any average order. These two notions are however linked as shown by the next result first exposed by E. Cahen in [1].

> **Theorem 3.6**
>
> If the abscissa of absolute convergence σ_0 of $D(f, s)$ is positive, then it is given by
>
> $$\sigma_0 = \limsup_{N \to \infty} \frac{\log \sum_{n \le N} |f(n)|}{\log N} .$$

A similar theorem exists to establish the *abscissa of convergence* (but we have not established its existence in this book!). It is also possible to handle the case when $\sigma_0 \le 0$, although the formula would be different this time. The educated readers will notice that the above expression is the exact analogue of Hadamard's formula for the radius of convergence of an entire series.

Let us comment on the second problem. We start with a decomposition $D(f, s) = D(h, s)D(g, s)$, where we know that the abscissa of absolute convergence of $D(h, s)$ is strictly smaller than the one of $D(g, s)$, say σ_0. Can we conclude that σ_0 is the abscissa of absolute convergence, say σ_0', of $D(f, s)$? By Theorem 3.1, we have $\sigma_0' \le \sigma_0$, but can it be smaller? A trivial case arises when $h = 0$, so let us exclude this situation. Even so, we do not know how to answer to this general question. Let us however add the (harmless in practice) hypothesis: for each $\delta > 0$, the modulus of $D(h, s)$ is bounded below when s ranges the half-plane $\Re s \ge \sigma_0 + \delta$. For instance in the convolution method, we get $D(h, s)$ in the form of an absolutely convergent Dirichlet product in the sense of Godement and it is easy to ensure that it is bounded below away from 0. A theorem of E. Hewitt & J.H. Williamson from [11] then answers our question.

Theorem 3.7

Let $D(h, s)$ be a Dirichlet series that is absolutely convergent when $\Re s \geq \sigma$ and is bounded below in modulus by a positive constant. Then $1/D(h, s)$ is again an absolutely convergent Dirichlet series in the domain $\Re s \geq \sigma$.

With this result, we can write $D(g, s) = D(h, s)^{-1}D(f, s)$ and conclude that $\sigma'_0 \geq \sigma_0$, thus reaching $\sigma_0 = \sigma'_0$.

Several investigations compared the different abscissas of convergence, *pointwise*, *uniform* or *absolute* of the three components of the equality $D(f, s) = D(h, s)D(g, s)$. In [12], J.-P. Kahane & H. Queffélec extended these studies to the case of several factors and many of their results are optimal.

Exercise 3-19. Show that $\mathbb{1} \star \lambda$ is the characteristic function of the squares and using that, find the Dirichlet series of λ. Here λ is the Liouville function defined by $\lambda(n) = (-1)^{\Omega(n)}$, where $\Omega(n)$ is the number of prime factors of n counted with multiplicities.

Exercise 3-20. Recall that a function f is called *completely multiplicative* if and only if $f(mn) = f(m)f(n)$ for all pairs of positive integers m and n.

1 ◇ Determine all the completely multiplicative functions f such that $\mathbb{1} \star f$ is also completely multiplicative.

2 ◇ Determine the inverse of a completely multiplicative function f.

Further Reading

We stayed in the domain of *absolute convergence* and ignored more refined convergence notions. We refer the readers to the book [16] by H. Queffélec and Martine Queffélec for such questions. More on the decomposition $f = f_y \star f^y$ can be found in the book [15] of D. Koukoulopoulos, see for instance Chap. 22 therein.

References

[1]   E. Cahen. "Sur la fonction $\zeta(s)$ de Riemann et sur des fonctions analogues". French. In: (1894). www.numdam.org/item?id=ASENS_1894_3_11__75_0 (cit. on pp. 34, 35).

[2]   H. Daboussi. "Sur le théorème des nombres premiers". In: *C. R. Acad. Sci., Paris, Sér. I* 298 (1984), pp. 161–164 (cit. on p. 34).

[3]   H. Daboussi. "Effective estimates of exponential sums over primes". In: *Analytic number theory. Vol. 1. Proceedings of a conference in honor of Heini Halberstam, May 16-20, 1995, Urbana, IL, USA. Boston, MA: Birkhaeuser. Prog. Math.* 138 (1996), pp. 231–244 (cit. on p. 34).

[4]   H. Daboussi and J. Rivat. "Explicit upper bounds for exponential sums over primes". In: *Math. Comp.* 70.233 (2001), pp. 431–447 (cit. on p. 34).

[5] H. Delange. "Une remarque sur la dérivée logarithmique de la fonction zêta de Riemann". In: *Colloq. Math.* 53.2 (1987), pp. 333–335 (cit. on p. 26).

[6] J.P.G.L. Dirichlet. "Beweis des Satzes, das jede unbegrenzte arithmetische Progression, deren erstes Glied und Differenz ganze Zahlen ohne gemeinschaftlichen Factor sind, unendlich viele Primzahlen enthält". In: *Abhandlungen der Königlichen Preussischen Akademie der Wissenschaften zu Berlin* (1837) (cit. on pp. 23, 34).

[7] J.P.G.L. Dirichlet. *Lectures on Number Theory, edited by R. Dedekind. Second edition. (Vorlesungen über Zahlentheorie, herausgegeben von R. Dedekind. Zweite Auflage.)* German. Première édition en 1863. Braunschweig. Vieweg, 1871 (cit. on pp. 23, 34).

[8] F. Dress. "Théorèmes d'oscillations et fonction de Möbius". In: *Sémin. Théor. Nombres, Univ. Bordeaux I* Exp. No 33 (1983/84). resolver.sub.uni-goettingen.de/purl? GDZPPN002545454, 33pp (cit. on p. 24).

[9] A.J. Granville and K. Soundararajan. "The spectrum of multiplicative functions". In: *Ann. of Math. (2)* 153.2 (2001), pp. 407–470. https://doi.org/10.2307/2661346 (cit. on p. 34).

[10] G.H. Hardy and M. Riesz. *The general theory of Dirichlet's series.* Cambridge Tracts in Mathematics and Mathematical Physics, No. 18. Première édition en 1915. Stechert-Hafner, Inc., New York, 1964, pp. vii+78 (cit. on p. 34).

[11] E. Hewitt and J.H. Williamson. "Note on absolutely convergent Dirichlet series". In: *Proc. Amer. Math. Soc.* 8 (1957), pp. 863–868 (cit. on p. 35).

[12] J.-P. Kahane and H. Queffélec. "Ordre, convergence et sommabilité de produits de séries de Dirichlet". In: *Ann. Inst. Fourier (Grenoble)* 47.2 (1997). www.numdam.org/item?id=AIF_ 1997__47_2_485_0, pp. 485–529 (cit. on p. 35).

[13] J.-P. Kahane and É. Saias. "Fonctions complètement multiplicatives de somme nulle". In: *Expo. Math.* 35.4 (2017), pp. 364–389. https://doi.org/10.1016/j.exmath.2017.05.002 (cit. on p. 32).

[14] K. Knopp. *Infinite sequences and series.* Translated by Frederick Bagemihl. Dover Publications, Inc., New York, 1956, pp. v+186 (cit. on p. 31).

[15] D. Koukoulopoulos. *The distribution of prime numbers.* Vol. 203. Graduate Studies in Mathematics. American Mathematical Society, Providence, RI, ©2019, pp. xii + 356. https:// doi.org/10.1090/gsm/203 (cit. on p. 36).

[16] H. Queffélec and M. Queffélec. *Diophantine Approximation and Dirichlet Series.* Vol. 2. Harish-Chandra Research Institute Lecture Notes. Hindustan Book Agency, 2013, p. 244 (cit. on p. 36).

[17] S. Ramanujan. "Irregular numbers". In: *J. Indian Math. Soc.* 5 (1913). Coll. Papers 20-21., pp. 105–108 (cit. on p. 29).

[18] B. Riemann. "Ueber die Anzahl der Primzahlen unter einer gegebenen Grösse". In: *Monatsberichte der Königlich Preussischen Akademie der Wissenschaften zu Berlin* (1859). url: www.claymath.org/sites/default/files/zeta.pdf (cit. on p. 23).

Chapter 4
Growth of Arithmetical Functions

In this chapter, we prove pointwise upper bounds for the values of arithmetic functions. This question is crucial to evaluate the abscissa of convergence of a series.

4.1 The Order of the Divisor Function

We consider first the divisor function $d(n) = \sum_{d|n} 1$. For any prime number p, we have $d(p) = 2$, while $d(n) = 1$ if and only if $n = 1$.

Exercise 4-1. Show that $d(n) = 2$ if and only if n is a prime number.

Exercise 4-2. Show that $d(n) = 7$ if and only if n is equal to p^6 for some prime number p.

We recall the explicit expression for the divisor function:

$$d(n) = \prod_{p|n} (\alpha_p + 1), \tag{4.1}$$

where $n = \prod_{p|n} p^{\alpha_p}$, $\alpha_p \geq 1$ for all $p|n$ and is zero otherwise. This expression implies in particular that the function $d(n)$ is unbounded. Here is a script that gives the list of its first values.

```
{WriteList( lowerlimit, upperlimit ) =
  for( n = lowerlimit, upperlimit,
    print("d(", n, ") = ", numdiv(n))); }
WriteList( 10, 20);
```

Exercise 4-3. What is the smallest integer n such that $d(n) \geq 10$? What is the smallest integer n such that $d(n) \geq 11$?

© The Author(s), under exclusive license to Springer Nature Switzerland AG 2022
O. Ramaré, *Excursions in Multiplicative Number Theory*, Birkhäuser Advanced Texts
Basler Lehrbücher, https://doi.org/10.1007/978-3-030-73169-4_4

Exercise 4-4.

1 ◇ Prove that $2^{\omega(n)} \le d(n) \le 2^{\Omega(n)}$, where $\omega(n)$ is the number of prime factors of n *without* multiplicity, while $\Omega(n)$ is the number of prime factors of n *with* multiplicity.

2 ◇ Use the arithmetic/geometric inequality to improve the above upper bound to
$$d(n) \le \left(1 + \frac{\Omega(n)}{\omega(n)}\right)^{\omega(n)}.$$

We have borrowed the last question from P. Letendre's PhD thesis. See also Exer. 25-2
One may wonder how fast the function $d(n)$ may go to infinity.

> **Theorem 4.1**
>
> For every integer $n \ge 2$, we have $\log d(n) \le 4 \log(2n)/\log \log(2n)$.

The constant 4 is *not* optimal.

Proof The inequality is clearly true for $n = 2$. Now assume that $n \ge 3$ and write
$n = \prod_{p|n} p^{\alpha_p}$, $\alpha_p \ge 1$. Let P be a positive real number depending on n, and that is to be set later. We have
$$\log d(n) = \sum_{p^{\alpha_p} \| n} \log(\alpha_p + 1) \le \sum_{\substack{p^{\alpha_p} \| n \\ p < P}} \log(\alpha_p + 1) + \sum_{\substack{p^{\alpha_p} \| n \\ p \ge P}} \alpha_p ,$$

where in the last inequality, we used the crude bound $\log(1 + x) \le x$ valid for $x \ge 0$.
Similarly, we can write
$$\log n = \sum_{p^{\alpha_p} \| n} \alpha_p \log p = \sum_{\substack{p^{\alpha_p} \| n \\ p < P}} \alpha_p \log p + \sum_{\substack{p^{\alpha_p} \| n \\ p \ge P}} \alpha_p \log p \ge \log P \sum_{\substack{p^{\alpha_p} \| n \\ p \ge P}} \alpha_p .$$

Therefore, the second sum satisfies $\displaystyle\sum_{\substack{p^{\alpha_p} \| n \\ p \ge P}} \alpha_p \le \frac{\log n}{\log P}.$

On using $1 + \alpha_p \le \log(2n)$, for the first sum, we simply get
$$\sum_{\substack{p^{\alpha_p} \| n \\ p < P}} \log(\alpha_p + 1) \le \log \log(2n) \sum_{p < P} 1 \le P \log \log(2n) .$$

Combining the two bounds above, we get
$$\log d(n) \le P \log \log(2n) + \log(2n)/\log P .$$

Let $y = \log(2n) \ge \log 6$ and choose $P = y/(\log y)^2$. The upper bound becomes

$$\log d(n) \le \frac{y}{\log y} + \frac{y}{\log y - 2 \log \log y} \le \frac{y}{\log y}\left(1 + \frac{\log y}{\log y - 2 \log \log y}\right).$$

Consider the function g of $z = \log y$ given by

$$g : [\log \log 6, +\infty) \to \mathbb{R}$$

$$z \mapsto \frac{z}{z - 2 \log z}.$$

It is clear from the graph of $g(z)$ that this function is increasing and then decreasing and, moreover, that it satisfies $g(z) \le e/(e-2) < 3.8$. Showing it properly only involves computing the derivative

$$g'(z) = -\frac{2(\log z - 1)}{(z - 2 \log z)^2}.$$

We conclude that g reaches its maximum at $z = e$. This ends the proof. $\qquad\square$

Exercise 4-5. Let $\omega(n)$ be the number of distinct prime factors of n. Prove that, when $n \ge 2$, we have $\omega(n) \le 6 \log(2n)/\log \log(2n)$.

Please note that this exercise did *not* use any bounds on the number of primes.

Corollary 4.1

For any $\varepsilon > 0$, there is a constant $C(\varepsilon)$ such that, for every $n \ge 1$, we have $d(n) \le C(\varepsilon)n^\varepsilon$.

Exer. 1-10 proposes another proof of this corollary.

Proof By Theorem 4.1, we have $d(n) \le \exp\big(4 \log(2n)/\log \log(2n)\big)$. We remark that the function $n \mapsto 1/\log \log(2n)$ tends to 0 when n tends to infinity. Thus, there exists a real number $N(\varepsilon)$ such that $1/\log \log(2n) \le \varepsilon/4$ for $n \ge N(\varepsilon)$. We may actually take $N(\varepsilon) = \exp(\exp(4/\varepsilon))$ and get that, for all $n \ge N(\varepsilon)$, we have $d(n) \le \exp(\varepsilon \log(2n)) = 2^\varepsilon n^\varepsilon$.

Consider now

$$C(\varepsilon) = \max\left(2^\varepsilon, \max_{n < N(\varepsilon)} d(n)/n^\varepsilon\right).$$

Surely, we have $d(n) \le C(\varepsilon)n^\varepsilon$, as it holds for $n < N(\varepsilon)$ and for $n \ge N(\varepsilon)$, hence all the time. $\qquad\square$

From a numerical perspective, this corollary is not very useful. Actually, if we follow the proof with $\varepsilon = 1/10$, we reach $N(1/10) = \exp(\exp(40))$, which means that we would have to compute

$$\max_{n < e^{e^{40}}} \left(d(n)/n^{1/10}\right).$$

The astrophysicists tell us that there are less than 10^{85} atoms in the universe, while $\exp(\exp(40)) \ge 10^{10^{17}}$. The number that we need to compute is thus more than astronomically large! See [6] by J.-L. Nicolas and G. Robin for better numerical bounds, and [3] by A. Derbal for corresponding results for the number $d^*(n)$ of unitary divisors.

Exercise 4-6. Show that there exists a sequence (n_k) tending to infinity such that $\inf \dfrac{\log d(n_k) \log \log n_k}{\log n_k} > 0$. What is the best possible lower bound?

Exercise 4-7. Show that $\sigma(n)/n \le 1 + \log n$, where $\sigma(n)$ is the sum of the divisors of n. By invoking Theorem 12.3, show that $\sigma(n)/n \ll \log \log(3n)$.

Exercise 4-8. Use Pari/GP or Sage to find the smallest integer n such that $\sigma(n)/n \ge 2$ and $\sigma(n+1)/(n+1) \ge 2$.

4.2 Extensions and Exercises

Exercise 4-9. Show that for every $\varepsilon > 0$, there is a strictly positive constant $C(\varepsilon)$ such that, for any integer n, we have $C(\varepsilon)n^{1-\varepsilon} \le \varphi(n) \le C(\varepsilon)n^{1+\varepsilon}$. Find the abscissa of absolute convergence of $\sum_{n \ge 1} 1/(\varphi(n)n^s)$. What can be said about the abscissa of absolute convergence of $\sum_{n \ge 1} \mu^2(n)/(\varphi(n)n^s)$?

Exercise 4-10. Again applying Theorem 12.3, show that there exists a constant $c_1 > 0$ such that, for any integer q, we have $q/\varphi(q) \le c_1 \log \log 3q$.

Exercise 4-11. Let $\mathcal{D}$ be a set of integers without prime factors $\le D$, where D is some parameter ≥ 1.

1 ◇ Show that $\displaystyle\sum_{d \in \mathcal{D}} 1/d \ll \log(2D)$ by using Theorem 12.3.

2 ◇ Show that for all $q \ge 1$, we have $d(m)^q \le \displaystyle\sum_{k\ell=m} d(k)^{q-1} d(\ell)^{q-1}$.

3 ◇ Show that for any integer $q \ge 1$, we have $\displaystyle\sum_{m \in \mathcal{D}} d(m)^q/m \ll (\log(2D))^{2^q}$.

4 ◇ Let D be a fixed parameter ≥ 1 and $d_D(n)$ be the number of divisors of n that are $\le D$. Let $q \ge 1$ be an integer. Show that $\displaystyle\sum_{n \le X} d_D(n)^q \ll X(\log(2D))^{2^q}$.

Exercise 4-12. Denote by $d_r(n)$ the number of r-tuples of positive integers $(n_1, n_2, \ldots, n_r)$ such that $n_1 n_2 \cdots n_r = n$.

1 ◇ Compute $d_r(p^a)$ where p is a prime and a is a positive integer.

2 ◇ Show that $\displaystyle\sum_{n \le N} d_r(n)^2/n \le (\log N + 1)^{r^2}$. The readers may start by showing that $d_r(n_1 n_2 \cdots n_r) \le d_r(n_1)d_r(n_2)\cdots d_r(n_r)$ for any r-tuple $(n_1, n_2, \ldots, n_r)$.

3 ◇ Let $\ell \ge 1$ be a real number such that r^ℓ is an integer. Prove that $d_r(n)^\ell \le d_{r^\ell}(n)$ with equality when n is square-free.

Exercise 4-13. (From K.K. Mardjanichvili[a] in [5]). Let the function d_r be the same as in Exer. 4-12. The paper [5] proves the general bound

$$\sum_{n \le x} d_r(n)^\ell \le x \frac{r^\ell}{r! \frac{r^{\ell-1}}{r-1}} \left(\log x + r^\ell - 1 \right)^{r^\ell - 1}, \tag{4.2}$$

which is valid for any real number $x \ge 1$, and any integers $r \ge 2$ and $\ell \ge 1$. The next questions reproduce Mardjanichvili's approach in case $\ell = 1$ and $\ell = 2$.

1⋄ Show by recursion on $r \ge 2$ that $\displaystyle\sum_{n \le x} d_r(n) \le x \frac{(\log x + r - 1)^{r-1}}{(r-1)!}$ for every real number $x \ge 1$. This proves (4.2) when $\ell = 1$.

2⋄ Show that $g([m_1, m_2]) \le \displaystyle\sum_{\substack{d | m_1, \\ d | m_2}} g(m_1 m_2/d)$, for any non-negative function g, with m_1 and m_2 any integers having $[m_1, m_2]$ as their least common multiple.

3⋄ By using $d_r = \mathbb{1} \star d_{r-1}$, prove that

$$\sum_{n \le x} d_r(n)^2 \le x \sum_{m_1 \le x} \frac{d_{r-1}(m_1) d(m_1)}{m_1} \sum_{n_2 \le x} \frac{d_{r-1}(n_2)}{n_2}.$$

4⋄ Proceed by recursion on r to prove (4.2) when $\ell = 2$.

5⋄ Show that the last inequality proved in Exer. 4-12 combined with (4.2) with $\ell = 1$ improves (4.2) in

$$\sum_{n \le x} d_r(n)^\ell \le \frac{x}{(r^\ell - 1)!} \left(\log x + r^\ell - 1 \right)^{r^\ell - 1}, \tag{4.3}$$

which is valid for any real number $x \ge 1$, any integer $r \ge 2$ and any real number $\ell \ge 1$ that is such that r^ℓ is an integer.

See [1] by O. Bordelles for an application. We continue our journey with a modern installment of an idea of J. van der Corput in [7].

Exercise 4-14. (From B. Landreau in [4]). Show that for every integer $k \ge 1$ and every real number $s > 0$, we have $d^s(n) \le k^{k(k-1)s} \displaystyle\sum_{\substack{\delta | n \\ \delta \le n^{1/k}}} d^{ks}(\delta)$.

See Exer. 22-3 for some examples of use. Similar inequalities have been considered by several authors. The paper [2] by Z. Brady is the latest at the time of writing. Proposition 1 of this paper contains a result of A. Granville, a simplification of which is the topic of the next exercise.

---

[a] Due to transliteration evolution, this name may be spelled differently, in particular the "j" may become "zh", while the initial of the first name may become "C" in some references.

Exercise 4-15. Let f and g be two non-negative multiplicative functions with $f(m)/(\mathbb{1} \star f)(m) \leq g(m)/(\mathbb{1} \star g)(m)$ for all m. Select a square-free integer n and a subset $\mathscr{D}$ of its set of divisors. We assume that, when $d \in \mathscr{D}$, every divisor of d lies also in $\mathscr{D}$ (i.e. $\mathscr{D}$ *is divisor-closed*). Our aim is to show that

$$\frac{\sum_{d\in\mathscr{D}} f(d)}{\sum_{d|n} f(d)} \geq \frac{\sum_{d\in\mathscr{D}} g(d)}{\sum_{d|n} g(d)} \ . \tag{4.4}$$

1 ◇ Let $\mathcal{D}_2(n) = \{(d_1, d_2)/d_1|d_2|n\}$. Assume there exists a non-negative function r_n on $\mathcal{D}_2(n)$ such that, for every divisor d of n, we have

$$\sum_{d|\ell} r_n(d, \ell) = \frac{f(d)}{(\mathbb{1} \star f)(n)} \ , \quad \sum_{\ell|d} r_n(\ell, d) = \frac{g(d)}{(\mathbb{1} \star g)(n)} \ . \tag{4.5}$$

Show that (4.4) holds in that case.

2 ◇ Let n_1 and n_2 be two coprime integers. Show that, if one can build functions r_{n_1} and r_{n_2} as in the previous question, respectively, for $n = n_1$ and $n = n_2$, then one can build one for $n = n_1 n_2$.

3 ◇ Find a non-negative solution to (4.5) when n is a prime and establish (4.4) when n is square-free.

4 ◇ Find a non-negative solution to (4.5) when n is a prime square and establish (4.4) when n is *cube-free*, i.e. when no prime to the power 3 divides it.

5 ◇ Show that the condition "$f(m)/(\mathbb{1} \star f)(m) \leq g(m)/(\mathbb{1} \star g)(m)$ for all m" is equivalent, when f and g are multiplicative, to: for every prime p and every positive integer k, we have $f(p) + \cdots + f(p^k) \leq g(p) + \cdots + g(p^k)$.

Inequality (4.4) can be proved for every integer n by following the same idea.

References

[1]   O. Bordellès. "An inequality for the class number". In: *JIPAM. J. Inequal. Pure Appl. Math.* 7.3 (2006), Article 87, 8 pp. (electronic) (cit. on p. 43).

[2]   Z. Brady, "Divisor function inequalities, entropy, and the chance of being below average". In: *Math. Proc. Cambridge Philos. Soc.* 163.3 (2017), pp. 547–560. https://doi.org/10.1017/0305004117000147 (cit. on p. 43).

[3]   A. Derbal. "Ordre maximum d'une fonction liée aux diviseurs d'un nombre entier". In: *Integers* 12 (2012), Paper No. A44, 15 (cit. on p. 41).

[4]   B. Landreau. "A new proof of a theorem of van der Corput". In: *Bull. Lond. Math. Soc.* 21.4 (1989), pp. 366–368 (cit. on p. 43).

[5]   K.K. Mardjanichvili. "Estimation d'une somme arithmétique". In: *Comptes Rendus Acad. Sciences URSS* N. s. 22 (1939), pp. 387–389 (cit. on p. 42).

[6]   J.-L. Nicolas and G. Robin. "Majorations explicites pour le nombre de diviseurs de N". In: *Canad. Math. Bull.* 26.4 (1983), pp. 485–492. https://doi.org/10.4153/CMB-1983-078-5 (cit. on p. 41).

[7]   J.G. van der Corput. "Une inégalité relative au nombre des diviseurs". In: *Nederl. Akad. Wetensch., Proc.* 42 (1939), pp. 547–553 (cit. on p. 43).

Chapter 5
An "Algebraical" Multiplicative Function

This chapter deals with a family of multiplicative functions appearing in algebraic problems. Their Dirichlet series are the object of intensive research.

5.1 The Integers Modulo q

Let q be a positive integer. We define $\mathbb{Z}/q\mathbb{Z}$ as being the finite set of the subsets

$$a + q\mathbb{Z} = \{a + kq, \ k \in \mathbb{Z}\} \ .$$

We may also use the notation $\bar{a}$ for this set, and call it *the class of a modulo q*. Alternatively, we say that *a is a representative of* $a + q\mathbb{Z}$. There are q such distinct subsets, it is enough to select $a \in \{0, \ldots, q-1\}$ to get all of them. This last notation is sometimes less explicit as the modulus, i.e. q, is not specified. Note that we have

$$\bar{b} = \bar{a} \iff b \in a + q\mathbb{Z} \iff q|b - a \ . \tag{5.1}$$

There is yet another notation we have to introduce, concerning equality. Instead of writing $a + q\mathbb{Z} = b + q\mathbb{Z}$, we write

$$a \equiv b \ \mathrm{mod} \ q \ . \tag{5.2}$$

It is sometimes convenient to confuse classes (i.e. subsets) and representatives and one can find equations like

$$x \equiv b \ \mathrm{mod} \ q$$

when in fact $x = a + q\mathbb{Z}$. We shall try to remain rather strict in our notation. The set $\mathbb{Z}/q\mathbb{Z}$ is not only a set but it can also be equipped with an addition and a multiplication, giving it a *ring* structure. We detail the multiplication for $q = b$ being a prime number in the next section.

Here is a final note to be comprehensive about our notation: in inlined text, we prefer the more legible "$a \equiv b \ (\mathrm{mod} \ q)$", while in subscript, we use the shortened form "$a \equiv b \ [q]$".

© The Author(s), under exclusive license to Springer Nature Switzerland AG 2022 47
O. Ramaré, *Excursions in Multiplicative Number Theory*, Birkhäuser Advanced Texts
Basler Lehrbücher, https://doi.org/10.1007/978-3-030-73169-4_5

Exercise 5-1. Show that the equation $5x^2 = 7y^3 + 2$ has no integer solutions by looking at it modulo 7.

5.2 The Multiplicative Group Modulo b

Let b be a fixed prime number ≥ 3. Consider

$$(\mathbb{Z}/b\mathbb{Z})^{\times} = \mathbb{Z}/b\mathbb{Z} \setminus \{0\} = \{\overline{1}, \overline{2}, \ldots, \overline{b-1}\} \,. \tag{5.3}$$

The set $(\mathbb{Z}/b\mathbb{Z})^{\times}$ is the set of *invertible* (in a sense clarified below) classes* modulo b. A class $\overline{n}$ belongs to $(\mathbb{Z}/b\mathbb{Z})^{\times}$ if and only if $\gcd(n, b) = 1$. Since b is a prime number, saying that n is coprime to b is exactly the same as saying that b does not divide n. The set $(\mathbb{Z}/b\mathbb{Z})^{\times}$ endowed with the multiplication forms a *group*. In particular, every element x has an inverse y, i.e. an element such that $xy = yx = 1$. We prove this important feature in two different ways.

Proof (First proof) Let $x = n + q\mathbb{Z}$ be an element of $(\mathbb{Z}/b\mathbb{Z})^{\times}$. Consider the sequence x, x^2, x^3, etc. Since this sequence varies over a finite set, there are two distinct positive integers ℓ and k such that $x^{\ell} = x^k$. Suppose that, for instance, $\ell < k$. The class x^{ℓ} (resp. x^k) is represented by n^{ℓ} (resp. n^k). Whence, we have $n^{\ell} \equiv n^k \pmod{b}$, i.e. $n^{\ell}(n^{k-\ell} - 1) \equiv 0 \pmod{b}$, which in turn means that $b | n^{k-\ell}(n^{\ell} - 1)$. Then Lemma 1.1 tells us that b divides either $n^{k-\ell}$ or $n^{\ell} - 1$. Since b does not divide n, it cannot divide $n^{k-\ell}$. Therefore b divides $n^{\ell} - 1$. On working modulo b, we obtain $n^{\ell} \equiv 1 \pmod{b}$, i.e. $x^{\ell} = 1$. This shows as well that $x \cdot x^{\ell-1} = 1$, whence $x^{\ell-1}$ is the inverse of x. $\square$

This classical proof comes from the general theory of finite groups. It is the key element of Lagrange's theorem, which states that the order of a subgroup divides the order of the full group. Actually, J.-L. Lagrange did not prove precisely this theorem, since the notion of the group did not exist before, say, 1830 with the work of É. Galois. He proved it in the special case of permutation groups, without using of course the notion of group.

Proof (Second proof) Consider an element x of $(\mathbb{Z}/b\mathbb{Z})^{\times}$, which is the class of an integer n coprime to b. By the É. Bézout[†] identity there exist two integers a and b, such that $an + bx = 1$. On reducing the latter modulo b, we conclude that $an \equiv 1 \pmod{b}$. The class $\overline{a}$ gives us the inverse of x. $\square$

---

* One may also call them *reduced residue classes*. The difficulty with the word *reduced* is that it somehow supposes a bigger thing that has been ... reduced! This is the case of fractions: 2/6 that can be reduced to 1/3. To wade their way around this linguistic hazard, some authors explain that, when you remove from all the classes the ones that have a representative not coprime to the modulus b, *what is left* are the reduced classes.

[†] This identity has the name of Bézout, while, in fact, it dates back to 1624, when it was established by C.-G. Bachet de Méziriac in his book "Problèmes plaisans et délectables qui se font par les nombres".

We could go faster by establishing directly Fermat's little theorem, named after P. Fermat who proved it in 1640.

Theorem 5.1 (Fermat's little theorem)

For every integer n and prime b we have $n^b \equiv n \pmod{b}$.

Two other proofs of this theorem are proposed in Exers. 5-2 and 9-5.

Proof There are many proofs of this result. We present the unpublished argument of Leibniz of 1683 that has also been later given by Euler in [5]. We work with the binomial coefficients $\binom{b}{k}$ as in Chap. 14. The essential observation is that b divides this coefficient if $1 \le k \le b-1$ (we leave it for the readers to check). Hence for every integer m, we have $(m + 1)^b \equiv m^b + 1 \pmod{b}$. This allows to show the theorem via recursion on m: clearly, we have $0^b \equiv 0 \pmod{b}$ and if $m^b \equiv m \pmod{b}$, then $(m + 1)^b \equiv m + 1 \pmod{b}$. $\square$

Consequently, when n is coprime to b, then $n^{b-1} \equiv 1 \pmod{b}$.

Exercise 5-2. Show that the map $h : (\mathbb{Z}/b\mathbb{Z})^\times \to (\mathbb{Z}/b\mathbb{Z})^\times$ permutes the ele-
$$x \mapsto nx$$
ments of $(\mathbb{Z}/b\mathbb{Z})^\times$. Deduce again that $n^b \equiv n \pmod{b}$.

Exercise 5-3. Let n be an integer that is neither divisible by the prime b nor congruent to 1 modulo b. Assume further that $n^3 \equiv 1 \pmod{b}$. Show that $3 | b - 1$

Exercise 5-4.

1 ◇ Show that every integer belongs to $(0 + 2\mathbb{Z}) \cup (0 + 3\mathbb{Z}) \cup (1 + 4\mathbb{Z}) \cup (5 + 6\mathbb{Z}) \cup (7 + 12\mathbb{Z})$, where $A \cup B$ denotes the *union* of the sets A and B.

2 ◇ Show that the same holds true for $(0 + 2\mathbb{Z}) \cup (0 + 3\mathbb{Z}) \cup (1 + 4\mathbb{Z}) \cup (3 + 8\mathbb{Z}) \cup (7 + 12\mathbb{Z}) \cup (23 + 24\mathbb{Z})$.

A collection of congruences such that every integer belongs to at least one of them is called *a covering system*. Many problems are open in this area, see for instance, [15] by P.P. Nielsen. The second covering system in the above exercise comes from [4] by P. Erdős and will be used in Exer. 29-8.

5.3 The Subgroup of Squares

Consider the subgroup $(\mathbb{Z}/b\mathbb{Z})^{\times(2)}$ of $(\mathbb{Z}/b\mathbb{Z})^\times$ given by

$$(\mathbb{Z}/b\mathbb{Z})^{\times(2)} = \{x^2, \ x \in (\mathbb{Z}/b\mathbb{Z})^\times\} \ . \tag{5.4}$$

Since the product of two squares is also a square, the set $(\mathbb{Z}/b\mathbb{Z})^{\times(2)}$ is closed under multiplication. Moreover, the inverse of a square is also a square[*].

Can we have $(\mathbb{Z}/b\mathbb{Z})^{\times(2)} = (\mathbb{Z}/b\mathbb{Z})^{\times}$? This question is meaningful since there are many finite groups where every element is a square. Therefore, we have to use some specific information about our group.

Let us show that it is not so by using Fermat's little theorem. Let $\ell \geq 1$ be an integer such that $2^{\ell}|b-1$ and $(b-1)/2^{\ell}$ is odd. Suppose that every element $(\mathbb{Z}/b\mathbb{Z})^{\times}$ is a square, i.e. any x is of the form y_1^2. But then y_1 is also a square and we can write $y_1 = y_2^2$. This implies that for a certain y we have $x = y^{2^{\ell}}$. Fermat's little theorem tells us that $x^{(b-1)/2^{\ell}} = y^{b-1} = 1$. Take $x = \overline{-1}$ (the class of -1). Since $(b-1)/2^{\ell}$ is odd, we have $x^{(b-1)/2^{\ell}} = -1$. We do not have $-1 \equiv 1 \pmod{b}$ if and only if $b = 2$, but by our assumption $b > 2$. This contradicts the hypothesis: consequently, there is an invertible class that is not a square.

Here is a more conceptual approach which can be ignored at first reading. Look at the map

$$h : (\mathbb{Z}/b\mathbb{Z})^{\times} \to (\mathbb{Z}/b\mathbb{Z})^{\times(2)}$$
$$x \mapsto x^2.$$

This is a group homomorphism, since $h(xy) = (xy)^2 = x^2y^2 = h(x)h(y)$. What can be said about its kernel? The equation $x^2 = 1$ can be rewritten as $n^2 \equiv 1 \pmod{b}$ if n is a representative of the class x. Then $b|(n^2 - 1) = (n+1)(n-1)$. By Lemma 1.1, we have that either $b|n-1$ or $b|n+1$, i.e. $x = 1$ or $x = -1$. Therefore the kernel of h is $\{\pm 1\}$. Thus, all the elements of $(\mathbb{Z}/b\mathbb{Z})^{\times(2)}$ have two pre-images, whence

$$\left|(\mathbb{Z}/b\mathbb{Z})^{\times(2)}\right| = (b-1)/2 . \tag{5.5}$$

We conclude that there is an element y_0, which is not a square and such that every invertible class may be written either in the form x^2 or in the form $y_0 x^2$. Let us prove this last claim: consider $H = \{y_0 x^2, x \in (\mathbb{Z}/b\mathbb{Z})^{\times}\}$. This set is of the same cardinality as $(\mathbb{Z}/b\mathbb{Z})^{\times(2)}$, i.e. $(b-1)/2$ and it has no intersection with $(\mathbb{Z}/b\mathbb{Z})^{\times(2)}$. Indeed, an element of the intersection could be written as $y_0 x^2$ and z^2, whence $y_0 = (zx^{-1})^2$ contradicting the fact that y_0 is not a square. Thus the sum of the cardinalities of H and $(\mathbb{Z}/b\mathbb{Z})^{\times(2)}$ is equal to $2 \cdot (b-1)/2 = b-1$, which implies that $H \cup (\mathbb{Z}/b\mathbb{Z})^{\times(2)}$ covers $(\mathbb{Z}/b\mathbb{Z})^{\times}$ as claimed.

5.4 The Legendre Symbol

We are finally ready to define the Legendre symbol (due to A.-M. Legendre in 1783). Define

---

[*] This is quite obvious: if $x \cdot x^{-1} = 1$, then $x^2 \cdot (x^{-1})^2 = 1$.

$$\left(\frac{n}{b}\right) = \begin{cases} 0, & \text{if } b|n, \\ 1, & \text{if } n \text{ is a square modulo b,} \\ -1, & \text{otherwise.} \end{cases} \tag{5.6}$$

The following theorem is the main result concerning this *symbol**.

Theorem 5.2

The Legendre symbol $n \mapsto \left(\frac{n}{b}\right)$ is a completely multiplicative function.

Recall that a function f is *completely multiplicative* if $f(mn) = f(m)f(n)$ for every integers m and n (not necessarily coprime as for multiplicative functions).

Proof Let m and n be two integers. If m or n is divisible by b, then b divides the product mn. Either $\left(\frac{m}{b}\right)$ or $\left(\frac{n}{b}\right)$ is zero in which case $\left(\frac{mn}{b}\right) = 0 = \left(\frac{m}{b}\right)\left(\frac{n}{b}\right)$. Now let m and n be two integers coprime to b.

Firstly, let m and n be congruent to a square modulo b, say $m \equiv u^2 \pmod{b}$ and $n \equiv v^2 \pmod{b}$. Then $mn \equiv (uv)^2 \pmod{b}$ and $\left(\frac{mn}{b}\right) = 1 = 1 \times 1 = \left(\frac{m}{b}\right)\left(\frac{n}{b}\right)$.

Secondly, when m is congruent to a square modulo b, but n is not, then we can write $m \equiv u^2 \pmod{b}$ and $n \equiv w_0 v^2 \pmod{b}$, where w_0 is a representative of the class y_0 that appeared in the previous section (just after (5.5)): it is a chosen non-square. Consequently, $mn \equiv w_0(uv)^2 \pmod{b}$ is not a square modulo b and $\left(\frac{mn}{b}\right) = -1 = 1 \times (-1) = \left(\frac{m}{b}\right)\left(\frac{n}{b}\right)$ as expected.

Similarly, when n is congruent to a square modulo b, but m is not, the relation $\left(\frac{mn}{b}\right) = \left(\frac{m}{b}\right)\left(\frac{n}{b}\right)$ gives us the result.

Finally, when neither m nor n are squares modulo b, then we can write $m \equiv w_0 u^2 \pmod{b}$ and $n \equiv w_0 v^2 \pmod{b}$. Hence $mn \equiv (w_0 uv)^2 \pmod{b}$ is a square, i.e. $\left(\frac{mn}{b}\right) = 1 = (-1) \times (-1) = \left(\frac{m}{b}\right)\left(\frac{n}{b}\right)$ as needed. □

The values of the Legendre symbol are directly accessible in Pari/GP:

```
for(n = 1, 20, print("(", n, "/ 11) = ", kronecker(n, 11)));
```

The name **kronecker** comes from the mathematician L. Kronecker who introduced a generalization of the Legendre symbol in [12].

Exercise 5-5.

1 ◇ Let b be a prime. Prove Wilson's theorem, i.e. that $(b - 1)! \equiv -1 \pmod{b}$ by pairing each class with its inverse.

2 ◇ Let $b \neq 2$ be a prime. Show that $\left\{\left(\frac{b-1}{2}\right)!\right\}^2 \equiv (-1)^{\frac{b+1}{2}} \pmod{b}$ and deduce that, if $(-1)^{\frac{b-1}{2}} = 1$, then -1 is a square modulo b. Prove that if $(-1)^{\frac{b-1}{2}} = -1$, then -1 is not a square modulo b. Conclude that

$$\left(\frac{-1}{b}\right) = (-1)^{\frac{b-1}{2}}. \tag{5.7}$$

---

* In fact, it is a function that associates to n either 0, 1 or -1.

3 ◇ Let $b \neq 2$ be a prime. Show that, in general, for any integer a prime to b, we have $\left(\dfrac{a}{b}\right) \equiv a^{\frac{b-1}{2}} \pmod{b}$.

The above proof of Wilson's theorem is due to Gauss in [6, art. 77], while its usage to compute the Legendre symbol of some arbitrary integer a is due to Dirichlet in [3]. We refer the readers who wonder about the value of $\left(\dfrac{b-1}{2}\right)! \pmod{b}$ to the elegant Chap. 26 of the monograph [18] by P. Pollack.

Exercise 5-6. Deduce from (5.7) that the prime factors of $(n!)^2 + 1$ are all congruent to 1 modulo 4.

Exercise 5-7.

1 ◇ Let b be an odd prime congruent to 1 modulo 4. By investigating the decomposition $2 \cdot 4 \cdot 6 \cdots 2\dfrac{b-1}{2} = \left(2 \cdots 2\dfrac{b-1}{4}\right) \cdot \left(2\dfrac{b+3}{4} \cdots 2\dfrac{b-1}{2}\right)$ compute $2^{\frac{b-1}{2}}$ modulo b and by using Exer. 5-5, deduce the value of $\left(\dfrac{2}{b}\right)$.

2 ◇ Adapt the above proof to the case when $b \equiv 3 \pmod{4}$.

In general, the readers will discover that

$$\left(\frac{2}{b}\right) = (-1)^{\frac{b^2-1}{8}} = \begin{cases} 1 & \text{when } b \equiv 1 \text{ or } 7 \text{ mod } 8, \\ -1 & \text{when } b \equiv 3 \text{ or } 5 \text{ mod } 8. \end{cases} \tag{5.8}$$

Here is now a way to compute the Legendre symbol at 3.

Exercise 5-8.

1 ◇ Let b be an odd prime congruent to 1 modulo 3. By investigating the decomposition

$$3 \cdot 6 \cdot 9 \cdots 3\frac{b-1}{2} = \left(3 \cdots 3\frac{b-1}{6}\right) \cdot \left(3\frac{b+5}{6} \cdots 3\frac{2b-2}{6}\right) \cdot \left(3\frac{2b+4}{6} \cdots 3\frac{b-1}{2}\right)$$

compute $3^{\frac{b-1}{2}}$ modulo b and by using Exer. 5-5, deduce the value of $\left(\dfrac{3}{b}\right)$.

2 ◇ Adapt the previous proof to the case when $p \equiv 2 \pmod{3}$.

3 ◇ Show that all the prime factors of $(n!)^2 + 3$ are congruent to 1 modulo 3, save the prime factor 3.

The above proof is clearly ad hoc. The general way to compute $\left(\dfrac{q}{p}\right)$ for primes p and q is via the Quadratic Reciprocity Law but we do not cover this here.

Theorem 5.3

We have $\displaystyle\sum_{1 \le n \le b} \left(\frac{n}{b}\right) = 0$.

Proof In fact, there are exactly $(b-1)/2$ integers modulo b which are congruent to squares and are not zero modulo b, hence

$$\sum_{1 \le n \le b} \left(\frac{n}{b}\right) = \frac{b-1}{2} + (-1)\frac{b-1}{2} + 0 = 0 \, .$$

This is what we wanted to show. $\qquad\square$

Here is a useful consequence of the above theorem.

Lemma 5.1

Let I be a subset of $\{1, \ldots, b\}$. We have

$$\left| \sum_{n \in I} \left(\frac{n}{b}\right) \right| \le b - 1 \, .$$

The same bound holds true for any finite interval instead of I.

Proof Since the values $\left(\frac{n}{b}\right)$ are bounded by 1 in absolute value, we can bound $\left| \sum_{n \in I} \left(\frac{n}{b}\right) \right|$ by $b - 1$ when I is any subset of $\{1, \ldots, b\}$. When I is a finite interval, we note that the sum of the values of $\left(\frac{n}{b}\right)$ on any b consecutive integers vanishes, reducing the problem to the first case. $\qquad\square$

Much more is believed to be true when I is restricted to be an interval. Here is a script that compares the value of the sum above with $\sqrt{b}$.

```
{GetMax(myf, Verbose = 1) =
  my(mymax = 0, sqrtmyf = sqrt(myf), localsum = 0);
  for(i = 1, myf - 2,
    localsum = kronecker(i, myf);
    for( j = i + 1, myf-1,
      localsum += kronecker(j, myf);
      mymax = max(mymax, abs(localsum/sqrtmyf))));
  if(Verbose == 1,
    print("|sum_{n in I} (n / ", myf,
          ")| / sqrt(", myf, ") <= ", mymax),);
  return(mymax);}


GetMax(211);
mm = 0; forprime(p = 10, 2000, mm = max(mm, GetMax(p,0))); mm
%% Returns 1.819...
```

Exercise 5-9. Let b be an odd prime.

$1 \diamond$ When k is invertible modulo b, we denote by $\overline{k}$ its inverse, i.e. the class of integers ℓ such that $k\ell \equiv 1 \pmod{b}$. Show that the map $k \mapsto \overline{k}$ is one-to-one on the multiplicative group modulo b.

$2 \diamond$ Prove that $\displaystyle\sum_{1 \le k \le b-2} \left(\frac{k(k+1)}{b} \right) = -1$.

$3 \diamond$ Prove that the number N of consecutive quadratic residues modulo b, i.e. the number of integers $k \le p-2$ such that both k and $k+1$ are congruent to a square modulo b, is given by $N = \dfrac{p - 4 - (-1)^{(p-1)/2}}{4}$.

We investigated the number of pairs $(k, k+1)$ such that both k and $k+1$ are quadratic residues modulo b in the previous exercise; the readers may wonder about strings, like $(k, k+1, k+2, k+3)$. It is indeed possible to extend the above study, but the methods required are up to now difficult.

5.5 The L-Function of the Legendre Symbol

The Dirichlet series for Legendre symbol is written as

$$L\left(s, \left(\frac{\cdot}{b}\right)\right) = \sum_{n \ge 1} \left(\frac{n}{b}\right) n^{-s} . \tag{5.9}$$

We have

$$L\left(s, \left(\frac{\cdot}{b}\right)\right) = \prod_{p \ge 2} \left(1 - \frac{\left(\frac{p}{b}\right)}{p^s}\right)^{-1} , \tag{5.10}$$

where the product is taken over all the primes p. This is the reason to use the letter b for the modulus, since it is convenient to have the variable p in the product above. Pari/GP allows one to compute values of L-series directly, and the readers should look for the functions `lfun` and `Ldata`.

There has been considerable efforts to show that $L\left(1, \left(\frac{\cdot}{b}\right)\right)$ is non-zero and bounded from below. Getting a good lower bound for $L\left(1, \left(\frac{\cdot}{b}\right)\right)$ is one of the most important problems in Analytic Number Theory. We introduce the readers to the subject by proving a simple bound.

> **Theorem 5.4**
>
> We have $L\left(1, \left(\frac{\cdot}{b}\right)\right) \geq \frac{\pi}{16b}$.

The argument we propose is a modification of an idea of A. Gel'fond from [7], which is maybe more clearly exposed in [8]. We refer to [16] by J. Oesterlé for a much better explicit lower estimate, expanding on some work of D. Goldfeld that is presented in a very readable way in [9].

Proof It is typographically simpler to set $\chi(n) = \left(\frac{n}{b}\right)$. The fundamental remark is that $\mathbb{1} \star \chi \geq 0$. Indeed, by multiplicativity, it is enough to check this on powers of primes. Let p be a prime and p^k an arbitrary prime power. We have

$$(\mathbb{1} \star \chi)(p^k) = \sum_{d \mid p^k} \chi(d) = \sum_{0 \leq \ell \leq k} \chi(p^\ell) = \sum_{0 \leq \ell \leq k} \chi(p)^\ell .$$

Now, when $\chi(p) = 0$, i.e. $p = b$, then the final value is 1 (since $\chi(p^0) = 1$). When $\chi(p) = 1$, the final value is $k + 1$ and when $\chi(p) = -1$, the final value is 0 when k is odd, and 1 when k is even. So, not only is $(\mathbb{1} \star \chi)(n)$ non-negative, we also know that it is ≥ 1 when n is a square. We next define

$$S(\alpha) = \sum_{n \geq 1} (\mathbb{1} \star \chi)(n) e^{-n\alpha}$$

for some positive parameter α. A comparison with an integral gives us

$$1 + S(\alpha) \geq \sum_{m \geq 0} e^{-m^2 \alpha} \geq \int_0^\infty e^{-\alpha t^2} dt = \frac{\Gamma(1/2)}{2\sqrt{\alpha}} = \frac{\sqrt{\pi}}{2\sqrt{\alpha}} .$$

On the other hand, we can expand $(\mathbb{1} \star \chi)(n) = \sum_{d \mid n} \chi(d)$ and get

$$S(\alpha) = \sum_{d \geq 1} \frac{\chi(d)}{e^{\alpha d} - 1} = \frac{L(1, \chi)}{\alpha} - \sum_{d \geq 1} \chi(d) g(\alpha d) ,$$

by using the non-negative non-increasing function $g(x) = 1/x - 1/(e^x - 1)$. We find that, by Lemma 5.1,

$$\sum_{d \geq 1} \chi(d) g(\alpha d) = -\sum_{d \geq 1} \chi(d) \int_{\alpha d}^\infty g'(t) dt$$

$$= -\int_0^\infty \sum_{d \leq t/\alpha} \chi(d) g'(t) dt \geq (b - 1) \int_0^\infty g'(t) dt = -(b - 1)/2 ,$$

since $\lim g(x) = 1/2$ as x tends to 0 from above. By comparing the upper and lower bound for $S(\alpha)$, we obtain

$$L(1, \chi) \geq \frac{\sqrt{\pi \alpha}}{2} - \alpha - \frac{\alpha(b-1)}{2} \geq \frac{\sqrt{\pi \alpha}}{2} - \alpha b .$$

The theorem follows on selecting $\alpha = \pi/(4b)^2$.   □

Since there exists some integer $n \leq b$ for which $(\frac{n}{b}) = -1$, there exists forcibly a prime p for which $(\frac{p}{b}) = -1$. But what about primes such that $(\frac{p}{b}) = 1$? The next exercise gives a preliminary answer to this question. It will be continued in Exer. 15-17.

Exercise 5-10. Let $\chi(n) = (\frac{n}{b})$. Assume that the positive integer x is so that $\chi(p) = -1$ for every prime less than x, save for b for which we have $\chi(b) = 0$.

1 ◇ Show that $S(x) = \sum_{n \leq x}(\mathbb{1} \star \chi)(n) \leq \sqrt{x}$.

2 ◇ Show that $L(1, \chi) = \sum_{d \leq x} \frac{\chi(d)}{d} + O^*(b/x)$.

3 ◇ Let m be some positive integer. Use summation by parts to prove that
$$\left| \sum_{d/[x/d]=m} \chi(d)\{x/d\} \right| \leq b.$$

4 ◇ Show that $\left| \sum_{d \leq x} \chi(d)\{x/d\} \right| \leq 2\sqrt{bx}$.

5 ◇ Show that $S(x) = xL(1, \chi) + O^*(b) + O^*(2\sqrt{bx})$.

6 ◇ Conclude that $x \leq 32b^2/\pi$.

We have adapted the proof of J. Pintz taken from [17]. The evaluation that takes place in Question 4 and 5 of the previous exercise follows a method developed by A. Axer in [2]; in essence, we remark that the fractional parts $\{x/d\}$ behave in a monotonous way in long intervals when d is not too small, and that summation by parts then gives an excellent result.

Exercise 5-11. We establish Thue's Lemma (from [19]): let $a \geq 1$ be an integer and $p \geq 3$ be a prime number such that $p \nmid a$. The equation $au \equiv v \pmod{p}$ has one solution $(u, v) \in \mathbb{Z}^2$ such that $1 \leq |u| < \sqrt{p}$, $1 \leq |v| < \sqrt{p}$.

1 ◇ Consider $S = \{0, \ldots, [\sqrt{p}]\}$ and the map
$$f : S^2 \to \{0, \ldots, p-1\},$$
$$(u, v) \mapsto f(u, v) \equiv au - v \pmod{p}.$$

Show that f is not injective.

2 ◇ Let (u_1, v_1) and (u_2, v_2) be two distinct pairs such that $f(u_1, v_1) = f(u_2, v_2)$. Write $u = u_1 - u_2$ and $v = v_1 - v_2$. Show that $au \equiv v \pmod{p}$, $|u| < \sqrt{p}$ and $|v| < \sqrt{p}$ and, finally, that neither u nor v is zero.

A different form of Thue's Lemma appears in L. Aubry's paper [1]. In both cases, it is not useful to restrict ones attention to the modulus p to be prime. We refer the readers to Chap. 7 of the book [13] of Leveque. The next two exercises exhibit a beautiful usage of Thue's Lemma.

Exercise 5-12. Let p be a prime congruent to 1 modulo 4. Show that p is the sum of two (integer) squares.

Sage has the function `two_squares` that returns a possible writing of n as a sum of two squares and raises a `ValueError` otherwise.

Exercise 5-13.

1 ◇ Let $p \geq 3$ be a prime for which -2 is a square residue. By using the strategy of Exer. 5-12, show that there exist two integers x and y such that $x^2 + 2y^2 = kp$ where $k \in \{1, 2\}$. Refine this estimate by showing that in fact $k = 1$.

2 ◇ Let $p \geq 5$ be a prime for which -3 is a square residue. Use the previous strategy to show that there exist two integers x and y such that $x^2 + 3y^2 = kp$ with $k \in \{1, 2\}$. Prove that, if $k = 2$, then x and y are odd, and by looking at the quantity $0 = x^2 + 3y^2 - 2p$ modulo 4, obtain a contradiction.

3 ◇ Let $p \geq 5$ be a prime for which 3 is a quadratic residue. Prove that there exist two integers x and y for which $x^2 - 3y^2 = \pm p$.

4 ◇ Let $p \geq 3$ be a prime for which 2 is a quadratic residue. Prove that there exist two integers x and y for which $x^2 - 2y^2 = \pm p$. If $x^2 - 2y^2 = -p$, find two integers a and b such that $(x + ay)^2 - 2(x + by)^2 = p$.

Exercise 5-14. Let D be an integer. Show that the set of integers n expressible in the form $a^2 + Db^2$, where a and b are integers, is closed under multiplication.

Exercise 5-15.

1 ◇ By using (5.7) together with (5.8), Exers. 5-13 and 5-14, show that every positive integer whose prime factors are congruent to 1 or 3 modulo 8 is expressible in the form $a^2 + 2b^2$, where both a and b are integers.

2 ◇ Find a similar theorem for integers expressible in the form $a^2 - 2b^2$ and $a^2 + 3b^2$.

The readers will find in Chap. VI of [14] by T. Nagell more representations of primes as sums of two squares; they are given in Theorem 100 and in the exercises at the end of this chapter. The sequence of integers of the form $a^2 + 2b^2$ is referenced A002479 in [10], where some more information is available, while the sequence of integers of the form $a^2 + 3b^2$, the so-called *loeschian numbers* is referenced A003136 therein.

An important tool for analysing the Legendre symbol is the so-called *Gauss's Lemma*. It may be found with applications as Lemma 13.4 in the book [11] by J.-M. de Koninck and F. Luca.

References

[1] L. Aubry. "Un théorème d'"arithmétique". French. In: *Mathesis* 3 (1913), pp. 33–35 (cit. on p. 54).

[2] A. Axer. Über einige Grenzwertsätze". In: *Wien. Ber.* 120 (1911), pp. 1253–1298 (cit. on p. 54).

[3] J.P.G.L. Dirichlet. "Démonstrations nouvelles de quelques théorèmes relatifs aux nombres". In: *J. Reine Angew. Math.* 3 (1828), pp. 390–393 (cit. on p. 50).

[4] P. Erdös. "On integers of the form $2^k + p$ and some related problems". In: *Summa Brasil. Math.* 2 (1950), pp. 113–123 (cit. on p. 47).

[5] L. Euler. "Theorematum quorundam ad numeros primos spectantium demonstratio". In: *Comment. Acad. Sci. Petrop.* 8 (1741), pp. 141–146 (cit. on p. 47).

[6] J.C.F. Gauss. *Disquitiones arithmeticae*. Leipzig: Gerhard Fleischer, 1801, p. 668 (cit. on p. 50).

[7] A.O. Gel'fond. "On the arithmetic equivalent of analyticity of the Dirichlet L-series on the line Re$s = 1$". In: *Izv. Akad. Nauk SSSR. Ser. Mat.* 20 (1956), pp. 145–166 (cit. on p. 52).

[8] A.O. Gelfond and Y.V. Linnik. *Elementary methods in analytic number theory*. Translated by Amiel Feinstein. Revised and edited by L. J. Mordell. Rand McNally & Co., Chicago, Ill., 1965, pp. xii+242 (cit. on p. 52).

[9] D. Goldfeld. "Gauss's class number problem for imaginary quadratic fields". In: *Bull. Amer. Math. Soc.* (1) 13 (1985), pp. 23–37 (cit. on p. 52).

[10] OEIS Foundation Inc. *The On-Line Encyclopedia of Integer Sequence*. oeis.org/. 2019 (cit. on p. 55).

[11] J.-M. de Koninck and F. Luca. *Analytic number theory*. Vol. 134. Graduate Studies in Mathematics. Exploring the anatomy of integers. American Mathematical Society, Providence, RI, 2012, pp. xviii+414. https://doi.org/10.1090/gsm/134 (cit. on p. 55).

[12] L. Kronecker. "Zur Theorie der elliptischen Funktionen". In: *Sitzungsberichte der Königlich Preussischen Akademie der Wissenschaften zu Berlin* 5 (1885), pp. 761–784 (cit. on p. 49).

[13] W.J. LeVeque. *Fundamentals of number theory*. Reprint of the 1977 original. Dover Publications, Inc., Mineola, NY, 1996, pp. viii+280 (cit. on p. 54).

[14] Trygve Nagell. *Introduction to number theory*. Second edition. Chelsea Publishing Co., New York, 1964, p. 309 (cit. on p. 55).

[15] P.P. Nielsen. "A covering system whose smallest modulus is 40". In: *J. Number Theory* 129.3 (2009), pp. 640–666. https://doi.org/10.1016/j.jnt.2008.09.016 (cit. on p. 47).

[16] J. Oesterlé. "Nombres de classes des corps quadratiques imaginaires". In: *Astérisque* 121/122 (1985), pp. 309–323 (cit. on p. 52).

[17] J. Pintz. "Elementary methods in the theory of L-functions, VI. On the least prime quadratic residue (modρ)". In: *Acta Arith.* 32.2 (1977), pp. 173–178 (cit. on p. 54).

[18] P. Pollack. *A conversational introduction to algebraic number theory*. Vol. 84. Student Mathematical Library. Arithmetic beyond $\mathbb{Z}$. American Mathematical Society, Providence, RI, 2017, pp. ix + 316 (cit. on p. 50).

[19] A. Thue. "Et par antydninger til en taltheoretisk metod". In: *Kra. Vidensk. Selsk. Forh.* 7 (1902), pp. 57–75 (cit. on p. 54).

Chapter 6
Möbius Inversions

This chapter is combinatorial and rather rewarding: we are going to present three procedures that yield interesting formulas in a rather magical way.

Even though the function called the *Möbius function* already appeared in the works of Euler in 1749 and in "Disquisitiones arithmeticae" of Gauss, it was A.F. Möebius who started its systematical study in 1832 with the celebrated "inversion formula" presented below. We would like to point out that the notion of Möbius function was later successfully generalized for lattices by G.-C. Rota in [14].

6.1 Pointwise Inversion

We have already seen this inversion in Theorem 1.1: it simply says that the Möbius function is the convolution inverse of the function $\mathbb{1}$. In a more surprising manner, we can write the following: given two arithmetic functions f and g, if for all $n \geq 1$, we have $f(n) = \sum_{d|n} g(d)$, then the function g can be described in terms of f via $g(n) = \sum_{d|n} \mu(n/d)f(d)$. Algebraically, it can be written as: if $f = \mathbb{1} \star g$, then $g = \mu \star f$. Actually, on "convolving" (which is not the same as multiplying) the terms of the left and on the right sides of the identity $f = \mathbb{1} \star g$ by μ, we obtain

$$\mu \star f = \mu \star (\mathbb{1} \star g) = (\mu \star \mathbb{1}) \star g = \delta_{n=1} \star g = g \,,$$

which is what we wanted to show.

Exercise 6-1. Let f and g be two functions such that, for all $n \geq 1$, we have $f(n) = \sum_{\substack{d|n \\ (d,n/d)=1}} g(d)$. Then the function g can be described in terms of f via

$$g(n) = \sum_{\substack{d|n \\ (d,n/d)=1}} (-1)^{\omega(n/d)} f(d).$$

© The Author(s), under exclusive license to Springer Nature Switzerland AG 2022   59
O. Ramaré, *Excursions in Multiplicative Number Theory*, Birkhäuser Advanced Texts Basler Lehrbücher, https://doi.org/10.1007/978-3-030-73169-4_6

Here is a very classical way for establishing the value of the Euler φ-function.

Exercise 6-2.

1 ⋄ By reducing the fractions of $\{a/n, 1 \leq a \leq n\}$, show that $\sum_{d|n} \varphi(d) = n$.

2 ⋄ Deduce the value of $\varphi(d)$.

We use below an idea we borrow from [1] by R.P. Brent and J. van de Lune, though the identity may already be found earlier, for instance as Exer. 71 in Part VIII of the book [11] by G. Pólya and G. Szegő.

Exercise 6-3.

1 ⋄ Show that the *Lambert series* of the Möbius function, $L(z) = \sum_{n \geq 1} \dfrac{\mu(n)z^n}{1 - z^n}$, satisfies $L(z) = z$ when $|z| < 1$

2 ⋄ Show that we have $L^b(z) = \sum_{n \geq 1} \dfrac{\mu(n)z^n}{1 + z^n} = z - 2z^2$ when $|z| < 1$.

We shall see here and there some *Lambert series* appear. They were introduced as early as 1771 by J.H. Lambert in § 875, p. 507 of the second volume of [8], a reference we owe to K. Knopp in [7].

6.2 Functional Möbius Inversion

We now present a functional variant of the Möbius inversion formula.

Theorem 6.1

Let f be a function of $[1, \infty)$. Consider the function F of $x \in [1, \infty)$ given by

$$F(x) = \sum_{n \leq x} f(x/n) .$$

Then for $x \in [1, \infty)$, we have $f(x) = \sum_{n \leq x} \mu(n)F(x/n)$.

Proof We proceed directly via the definition:

$$\sum_{n \leq x} \mu(n) \sum_{m \leq x/n} f(x/(nm)) = \sum_{\ell \leq x} f(x/\ell) \sum_{nm = \ell} \mu(n) = f(x) .$$

The Iseki & Tatuzawa formula of Theorem 24.1 gives another inversion formula of the same spirit.

Exercise 6-4.

1 ⋄ Given any two arithmetical functions f and g, show that

$$\sum_{n \le x} g(n) \sum_{m \le x/n} f(m) = \sum_{\ell \le x} (f \star g)(\ell)$$

for every real number $x \ge 1$.

2 ⋄ Show that $M(x) + M(x/2) + M(x/3) + M(x/4) + \cdots = 1$ for every real number $x \ge 1$, where $M(z) = \sum_{d \le z} \mu(d)$.

See also Exer. 7-6.

Exercise 6-5.

1 ⋄ Using $\sum_{n \le x} 1 = [x]$, where $[x]$ stands for the integer part of x, show the identity due to E. Meissel in [9, Eq. (6)][a]: $\sum_{n \le x} \mu(n)[x/n] = 1$.

2 ⋄ Deduce that $\sum_{n \le x} \mu(n)\{x/n\} = -1 + x \sum_{n \le x} \dfrac{\mu(n)}{n}$.

3 ⋄ Prove the J.P. Gram inequality[b]: $\left| \sum_{n \le x} \dfrac{\mu(n)}{n} \right| \le 1$.

4 ⋄ Deduce that, for $\epsilon \ge 0$, we have $\left| \sum_{n \le x} \dfrac{\mu(n)}{n^{1+\epsilon}} \right| \le 1 + \epsilon$.

5 ⋄ Using a similar process, show that for every real number $x \ge 0$ and every integer $q > 0$, we have $\left| \sum_{\substack{n \le x \\ \gcd(n,q)=1}} \dfrac{\mu(n)}{n} \right| \le 1$.

This last result is one of these results that is often rediscovered: in the form given here, it is [6, Lemma 10.2], but it can in fact be found 60 years earlier, though in a slightly weaker form, in [3, Lemma 1] by Davenport and later, in a slightly stronger form, in [15] by T. Tao.

Exercise 6-6. (From [17, Lemma 2] by von Mangoldt). Find an upper bound for
$\left| \sum_{n \le x} \dfrac{\mu(n)}{n} \log \dfrac{x}{n} - 1 \right|$ by using $\sum_{nm \le x} \mu(n)/(mn)$.

---

[a] Thanks to the DigiZeitschriften project hosted by the university of Göttingen, we can have an access to this text online, though some knowledge of Latin is required. The classical reference book [4] on history of numbers of L.E. Dickson may serve as a first guide, and for instance, the paper [9] is mentioned in Chap. XIX of this series of three books.

[b] Gram won the Gold Medal of the Royal Danish Academy of Sciences in 1884 for the paper he published that contains inter alia this inequality, see [5, p 196-197].

See also Exer. 11-1 for another proof. It can be shown, as a consequence of the Prime Number Theorem (see Sect. 23), that $\sum_{n \le x} \frac{\mu(n)}{n} \log \frac{x}{n}$ is indeed asymptotic to 1.

Exercise 6-7. (Meissel's identity). Define $m(x) = \sum_{d \le x} \mu(d)/d$ and $M(x) = \sum_{d \le x} \mu(d)$. Show that for $x \ge 1$ we have

$$m(x) = \frac{M(x)}{x} + \frac{1}{x} \int_1^x \left\{ \frac{x}{t} \right\} \frac{M(t)dt}{t} + \frac{\log x}{x}.$$

Exercise 6-8.

1 ⋄ Compute the convolution inverse of $g(n) = \mathbb{1}_{(n,2)=1}$.

2 ⋄ Let f be a function of $[1, \infty)$. Consider the function F of $x \in [1, \infty)$ given by

$$F(x) = \sum_{n \le (x-1)/2} f\left(\frac{x}{2n+1} \right).$$

Show that for $x \in [1, \infty)$ we have $f(x) = \sum_{n \le (x-1)/2} \mu(2n+1)F\left(\frac{x}{2n+1} \right).$

3 ⋄ Given an arithmetical function g such that $g(1) \ne 0$, formulate a general inversion formula starting from the expression $F(x) = \sum_{n \le x} g(n)f(x/n).$

This will be used in Exer. 17-7.

6.3 An Identity Factory

The next theorem is inspired by the work of F. Daval, see [2].

Theorem 6.2

Let f and g be two arithmetical functions. We define $S_f(t) = \sum_{n \le t} f(n)$ and $S_{f \star g}(t)$ similarly. Let $h : (0, 1] \to \mathbb{C}$ be Lebesgue-integrable over every segment $\subset (0, 1]$ and let H be a C^1-function over $[1, \infty)$. When $x \ge 1$, we have

$$\sum_{n \le x} f(n)H(x/n) - H(1)S_f(x) = \int_1^x S_{f \star g}\left(\frac{x}{y} \right) \frac{h(1/y)}{y} dy$$

$$+ \int_1^x S_f\left(\frac{x}{t} \right) \left(H'(t) - \frac{1}{t} \sum_{n \le t} g(n)h(n/t) \right) dt.$$

After reading Sect. 7.2 on absolute continuity, the readers will easily relax the condition on H. It is enough to require that it is absolutely continuous on every finite interval $\subset [1, \infty)$.

Proof On the one hand, we get immediately

$$\int_1^x S_f\left(\frac{x}{t}\right)H'(t)dt = \sum_{n \le x} f(n)H(x/n) - H(1)S_f(x),$$

while on the other hand, we have

$$\int_1^x S_f\left(\frac{x}{t}\right)\frac{1}{t}\sum_{n \le t} g(n)h(n/t)dt = \sum_{n \le x} g(n)\int_n^x S_f\left(\frac{x}{t}\right)h(n/t)\frac{dt}{t}$$

$$= \sum_{n \le x} g(n)\int_1^{x/n} S_f\left(\frac{x/n}{u}\right)h(1/u)\frac{du}{u}$$

$$= \int_1^x \sum_{n \le x/u} g(n)S_f\left(\frac{x/u}{n}\right)h(1/u)\frac{du}{u}.$$

The proof follows on invoking the result of Exer. 6-4. □

Already the case $H(t) = t$ and $g = 1$ yields an interesting statement, which was one of the initial results of F. Daval.

Corollary 6.1

We define $M(z) = \sum_{d \le z} \mu(d)$ and $m(z) = \sum_{d \le z} \mu(d)/d$. Let $h : (0,1] \to \mathbb{C}$ be Lebesgue-integrable over every segment $\subset (0, 1]$. When $x \ge 1$, we have

$$m(x) - \frac{M(x)}{x} = \frac{1}{x}\int_1^x M(x/t)\left(1 - \frac{1}{t}\sum_{n \le t} h(n/t)\right)dt + \frac{1}{x}\int_{1/x}^1 \frac{h(y)}{y}dy.$$

Though not required, it is better to normalize h by $\int_0^1 h(u)du = 1$. Like many identities, once it is written down, it is not very difficult to establish. In the paper [13, Theorem 7.4], it is proven that one recovers all (regular enough) identities of this shape linking $m(x)$ and $M(x)$ so that the above is not only a curiosity that will be superseded by a further stream of identities. On selecting $h = 1$, we recover the Meissel identity proved in Exer. 6-7, and on selecting $h(t) = 2t$, the identity of R.A. MacLeod, also proved in Exer. 7-6. The functional transform that, to a function h, associates the function that maps any positive t to $\int_0^1 h(u)du - \sum_{n \le t} h(n/t)/t$ is closely related to a transform introduced by Ch.H. Müntz in [10]. An easier reference is surely [16, Sect. 2.11]. The readers will find information on this transform in the paper [18] by S. Yakubovich.

Exercise 6-9. Let $g(n) = (-1)^n$.

1 ⋄ Express the convolution product $\mu \star g$ in a simpler manner.

2 ⋄ Similarly, express the convolution product $\Lambda \star g$ in a simpler manner where Λ is the von Mangoldt function.

3 ⋄ The hypotheses being the one of Cor. 6.1, show that

$$m(x) - \frac{M(x)}{x} = -\frac{1}{x} \int_1^x M(x/t)\left(1 - \frac{1}{t}\sum_{n\leq t}(-1)^n h(n/t)\right)dt$$

$$+ \frac{1}{x}\int_{1/x}^{1/2} \frac{h(y)}{y}dy - \frac{1}{x}\int_{1/2}^1 \frac{h(y)}{y}dy .$$

Exercise 6-10. (Daval in [2]). Let $k : (0, 1] \to \mathbb{C}$ be Lebesgue-integrable over every segment $\subset (0, 1]$. Prove that, under the notation of Cor. 6.1, we have, for $x \geq 1$,

$$m(x) - \frac{M(x)}{x} = \frac{1}{x}\int_1^x m(x/t)\left(1 - \sum_{n\leq t}k(n/t)\right)\frac{dt}{t^2} + \int_{1/x}^1 k(y)dy .$$

Notice that, in this identity, it is better to choose k such that $\int_0^1 k(u)du = 0$.

Theorem 6.2 can be put to work in more difficult situations as shown by the next identity.

Corollary 6.2

With the same notation of Cor. 6.1, we have, for $x \geq 1$,

$$\sum_{n\leq x} \frac{\mu(n)}{n} \log \frac{x}{n} + \gamma\left(\sum_{n\leq x} \frac{\mu(n)}{n} - \frac{M(x)}{x}\right) = 1 - \frac{1}{x}$$

$$+ \frac{1}{x}\int_1^x M(x/t)\left(\log t + \gamma + \frac{1}{t} - \sum_{n\leq t}\frac{1}{n}\right)dt ,$$

where γ is Euler's constant.

Proof We select $H(t) = t \log t - \log t + \gamma t$ and $h(t) = 1/t$ in Theorem 6.2. We readily get

$$\sum_{n\leq x} \frac{\mu(n)}{n} \log \frac{x}{n} - \sum_{n\leq x} \frac{\mu(n)}{x} \log \frac{x}{n} + \gamma\left(\sum_{n\leq x} \frac{\mu(n)}{n} - \frac{M(x)}{x}\right) = 1 - \frac{1}{x}$$

$$+ \frac{1}{x}\int_1^x M(x/t)\left(\log t + \gamma - \sum_{n\leq t}\frac{1}{n}\right)dt ,$$

in which we plug the easily proved identity

$$\sum_{n\leq x} \mu(n) \log \frac{x}{n} = \int_1^x M(x/t) \frac{dt}{t} .$$

Such identities and several others have been applied in [12]. Here is a novel corollary.

Corollary 6.3

For every $x \geq 1$, we have

$$\sum_{n\leq x} \frac{\lambda(n)}{n} - \frac{1}{x} \sum_{n\leq x} \lambda(n) = \frac{2}{\sqrt{x}} - \frac{1}{x} \int_1^x (\sqrt{y} - [\sqrt{y}]) \frac{dy}{y}$$

$$+ \frac{1}{x} \int_1^x \sum_{n\leq x/t} \lambda(n) \{t\} \frac{dt}{t} .$$

Proof This is obtained from Theorem 6.2 with the choice $H(t) = t$, $h(t) = 1$, $f(n) = \lambda(n)$ and $g = \mathbb{1}$. □

Exercise 6-11.

1 ◇ Prove that we have $\displaystyle\sum_{\substack{n\leq x, \\ (n,7)=1}} \frac{\mu(n)}{n} \log \frac{x}{n} \ll 1$ for every $x \geq 1$.

2 ◇ Prove that there exists a constant C' such that, for every $x \geq 1$, we have

$$\left| \sum_{\substack{n\leq x, \\ (n,7)=1}} \frac{\mu(n)}{n} \left(\log \frac{x}{n}\right)^2 - \frac{6}{7} \log x \right| \leq C'.$$

3 ◇ Deduce from the above that $\displaystyle\sum_{\substack{n\leq x, \\ (n,7)=1}} \frac{\mu(n)}{n} \left(\log \frac{x}{n}\right)^2 \geq 0$.

4 ◇ Deduce that $\displaystyle\sum_{\substack{n\leq x, \\ (n,7)=1}} \frac{\mu(n)}{n} \left(\log \frac{x}{n}\right)^k \geq 0$ for every integer $k \geq 2$.

The reader may look at [12] and [13], and more precisely at Corollary 1.11 therein, for more on this question. The result of the last question is a remark of Priyamvad Srivastav in his PhD thesis.

6.4 Another Moebius Inversion Formula

In the context of the Selberg sieve, we encounter another inversion.

Theorem 6.3

Let $(\lambda_\ell)_{\ell \geq 1}$ be a sequence such that the series $\sum_\ell d(\ell)|\lambda_\ell|$ converges, where $d(\ell)$ is the number of divisors of ℓ. For any integer $q \geq 1$, consider

$$F(q) = \sum_{\substack{\ell \geq 1 \\ q \mid \ell}} \lambda_\ell \ .$$

Then the series $\sum_q |F(q)|$ converges, and for any integer $\ell > 0$, we have

$$\lambda_\ell = \sum_{\substack{q \geq 1 \\ \ell \mid q}} \mu(q/\ell) F(q) \ .$$

We usually apply this theorem to finite series.

Proof We consider $\sum_{q \geq 1, \ell \mid q} \mu(q/\ell) F(q)$: it immediately follows that

$$\sum_{\substack{q \geq 1 \\ \ell \mid q}} \mu(q/\ell) \sum_{\substack{d \geq 1 \\ q \mid d}} \lambda_d = \sum_{\substack{d \geq 1 \\ \ell \mid d}} \lambda_d \sum_{\substack{q \mid d \\ q \mid \ell}} \mu(q/\ell) = \lambda_\ell \ ,$$

as expected. $\square$

References

[1] R.P. Brent and J. van de Lune. "A note on Pólya's observation concerning Liouville's function". In: *Leven met getallen : liber amicorum ter gelegenheid van de pensionering van Herman te Riele*. Ed. by J.A.J. van Vonderen (Coby). CWI (cit. on p. 58).

[2] F. Daval. "Identités intégrales et estimations explicites associées pour les fonctions somma-toires liées á la fonction de Möbius et autres fonctions arithmétiques". PhD thesis. Université de Lille, Mathematics, 2019 (cit. on pp. 60, 62).

[3] H. Davenport. "On some infinite series involving arithmetical functions". In: *Quart. J. Math., Oxf. Ser.* 8 (1937), pp. 8–13 (cit. on p. 59).

[4] L.E. Dickson. *Theory of numbers*. Chelsea Publishing Company, 1971 (cit. on p. 59).

[5] J.P. Gram. *Undersøgelser angaaende Maengden af Primtal under en given Graense. Résumé en français*. Danish. Kjöbenhavn. Skrift. (6) II. 185-308 (1884). 1884 (cit. on p. 59).

[6] A.J. Granville and O. Ramaré. "Explicit bounds on exponential sums and the scarcity of squarefree binomial coefficients". In: *Mathematika* 43.1 (1996), pp. 73–107 (cit. on p. 59).

[7] K. Knopp. "Über *Lambertsche* Reihen." In: *J. Reine Angew. Math.* 142 (1913), pp. 283–315 (cit. on p. 58).

[8] J.H. Lambert. *Anlage zur Architectonik oder Theorie des Einfachen und Ersten in des philosophischen und mathematischen Erkenntnis*. 2 vol. Riga, 1771 (cit. on p. 58).

[9] D.F.E. Meissel. "Observationes quaedam in theoria numerorum". Latin. In: *J. Reine Angew. Math.* 48 (1854), pp. 301–316. https://doi.org/10.1515/crll.1854.48.301 (cit. on p. 59).

[10]   Ch.H. Müntz. "Beziehungen der Riemannschen ζ-Funktion zu willkürlichen reellen Funktionen". German. In: *Mat. Tidsskr. B* 1922 (1922), pp. 39–47 (cit. on p. 61).

[11]   G. Pólya and G. Szegő. *Problems and theorems in analysis. II.* Classics in Mathematics. Theory of functions, zeros, polynomials, determinants, number theory, geometry, Translated from the German by C. E. Billigheimer, Reprint of the 1976 English translation. Springer-Verlag, Berlin, 1998, pp. xii+392. https://doi.org/10.1007/978-3-642-61905-2_7 (cit. on p. 58).

[12]   O. Ramaré "Explicit estimates on several summatory functions involving the Moebius function". In: *Math. Comp.* 84.293 (2015), pp. 1359–1387 (cit. on pp. 62, 63).

[13]   O. Ramaré. "Explicit average orders: news and problems". In: *Number theory week 2017.* Vol. 118. Banach Center Publ. Polish Acad. Sci. Inst. Math., Warsaw, 2019, pp. 153–176 (cit. on pp. 61, 63).

[14]   G.-C. Rota. "On the foundations of combinatorial theory. I. Theory of Möbius functions". In: *Z. Wahrscheinlichkeitstheorie und Verw. Gebiete* 2 (1964), 340–368 (1964). https://doi.org/10.1007/BF00531932 (cit. on p. 57).

[15]   T. Tao. "A remark on partial sums involving the Möbius function". English. In: *Bull. Aust. Math. Soc.* 81.2 (2010), pp. 343–349. https://doi.org/10.1017/S0004972709000884 (cit. on p. 59).

[16]   E.C. Titchmarsh. *The theory of the Riemann zeta-function.* Second. Edited and with a preface by D.R. Heath-Brown. The Clarendon Press, Oxford University Press, New York, 1986, pp. x+412 (cit. on p. 61).

[17]   H.C.F. von Mangoldt. "Beweis der Gleichung $\sum_{k=1}^{\infty} \frac{\mu(k)}{k} = 0$" . German. In: *Berl. Ber.* 1897 (1897), pp. 835–852 (cit. on p. 59).

[18]   S. Yakubovich. "New summation and transformation formulas of the Poisson, Müntz, Möbius and Voronoi type". In: *Integral Transforms Spec. Funct.* 26.10 (2015), pp. 768–795. https://doi.org/10.1080/10652469.2015.1051483 (cit. on p. 61).

Chapter 7
Handling a Smooth Factor

We have seen that many techniques in analytic number theory were developed to evaluate sums of the type

$$\sum_{n \le x} a(n) F(n),$$

where $a(n)$ is some arithmetical function and F is a smooth factor, like $\log n$. This chapter examines the smooth part.

7.1 Summing Smooth Functions

Let us start with the case when a is $\mathbb{1}$ and F is C^1, meaning that it is differentiable with a continuous derivative. Another hypothesis is in fact hidden in what follows, which is that this derivative should be of controlled growth. As a first example, we prove that

$$\sum_{n \le x} \frac{1}{n} = \log x + \gamma + O(1/x), \quad (x \ge 1) \tag{7.1}$$

where

$$\gamma = 1 - \int_1^\infty \frac{\{t\} dt}{t^2} = 0.5777\cdots \tag{7.2}$$

is the *Euler constant* (sometimes called *Euler–Mascheroni constant*). We begin from the basic estimate

$$\sum_{n \le x} 1 = x + O^*(1), \quad (x \ge 1). \tag{7.3}$$

Proof *(of* (7.1)*)* We start by using (7.3) naively. First notice that

$$\frac{1}{n} = \frac{1}{x} + \int_n^x \frac{dt}{t^2}. \tag{7.4}$$

© The Author(s), under exclusive license to Springer Nature Switzerland AG 2022
O. Ramaré, *Excursions in Multiplicative Number Theory*, Birkhäuser Advanced Texts
Basler Lehrbücher, https://doi.org/10.1007/978-3-030-73169-4_7

On using this, we get

$$\sum_{n \le x} \frac{1}{n} = \frac{1}{x} \sum_{n \le x} 1 + \sum_{n \le x} \int_n^x \frac{dt}{t^2} = \frac{1}{x} \sum_{n \le x} 1 + \int_1^x \sum_{n \le t} 1 \frac{dt}{t^2}$$

$$= \frac{1}{x} \sum_{n \le x} 1 + \int_1^x (t + O^*(1)) \frac{dt}{t^2} = \log x + O^*(1) . \qquad (7.5)$$

The exchange of the summation sign $\sum_n$ and the integral $\int dt$ is legal since our sums are finite or absolutely convergent, but the readers may have difficulties with the domains over which the variables should range. We surely have

$$\sum_{n \le x} \int_n^x \frac{dt}{t^2} = \int_{\cdots}^{\cdots} \sum_{\cdots \le n \le \cdots} 1 \frac{dt}{t^2}$$

but which limits should we put? Notice that, in the former quantity, the summation conditions are $1 \le n \le t \le x$. The same domain is described with the two inequalities $1 \le t \le x$ and $1 \le n \le t$, which gives us what we have announced.

So far, so good, but we have only reached $\sum_{n \le x} 1/n = \log x + O(1)$! Here is how to do better. First note that

$$\sum_{n \le t} 1 = [t] = t - \{t\} , \qquad (7.6)$$

where $[t]$ denotes the integer part of t and $\{t\}$ its fractional part. Our loss stems from the crude approximation $\{t\} = O^*(1)$, which we do not use anymore. Instead, we infer from the above that

$$\sum_{n \le x} \frac{1}{n} = \frac{x + O^*(1)}{x} - \int_1^x \{t\} \frac{dt}{t^2} + \log x$$

$$= \log x + 1 + O^*(1/x) - \int_1^\infty \{t\} \frac{dt}{t^2} + \int_x^\infty \{t\} \frac{dt}{t^2}$$

$$= \log x + \gamma + O^*(2/x) ,$$

which is a more precise version of what we have announced. Equation (10.5) will be even more precise. □

The technique used is *summation by parts*, also called *Abel summation* when used in a slightly broader context. Within the formalism of the Stieltjes integral, it is indeed an integration by parts as we find in the more usual calculus. The same technique leads to the next estimate.

Lemma 7.1

We have

$$\sum_{n \le x} \frac{\log n}{n} = \tfrac{1}{2} (\log x)^2 + \gamma_1 + O(\log(2x)/x) , \quad (x \ge 1)$$

where γ_1 is a constant called the first *Laurent–Stieltjes Constant* given by $\gamma_1 =$
$$\int_1^\infty \{t\} \frac{\log t - 1}{t^2} dt = -0.07281\ 58454\ 83676\ 72486\ldots$$

In general, the Laurent–Stieltjes constants are implicitly defined by

$$\zeta(s) = \frac{1}{s-1} + \sum_{r \geq 0} \frac{(-1)^r}{r!} \gamma_r (s-1)^r,$$

which is the Laurent expansion of the Riemann zeta-function around $s = 1$.

Remark 7.1

In the three previous estimates, we were careful to write "$x \geq 1$" despite the fact that these estimates are of interest only when x is large, or so we think! However, applications show that we need estimates with a large domain of validity (the proof of Lemma 10.1 below will be an extreme example). In particular, there exists a constant C such that, for any $x \geq 1$, we have

$$\left| \sum_{n \leq x} \frac{\log n}{n} - \tfrac{1}{2} \log^2 x - \gamma_1 \right| \leq C \log(2x)/x .$$

This would become **false** if we had replaced $\log(2x)/x$ by $(\log x)/x$, simply because of the case $x = 1$. Such questions are usually trivial to handle when establishing the estimates but may lead to disastrous mistakes when a sloppy estimate is used later. The readers should always remain extremely vigilant concerning this point. We close this remark by signalling that there is nothing magical in our "$\log(2x)$" and we could as well have put "$1 + \log x$" or "$\log(3x)$".

Proof We simply write

$$\frac{\log n}{n} = \frac{\log x}{x} + \int_n^x \frac{\log t - 1}{t^2} dt .$$

This gives us

$$\sum_{n \leq x} \frac{\log n}{n} = [x] \frac{\log x}{x} + \sum_{n \leq x} \int_n^x \frac{\log t - 1}{t^2} dt$$

$$= [x] \frac{\log x}{x} + \int_1^x \left(\sum_{n \leq t} 1 \right) \frac{\log t - 1}{t^2} dt .$$

We continue by using (7.6):

$$\sum_{n \leq x} \frac{\log n}{n} = \int_1^x \frac{\log t - 1}{t} dt + \log x - \int_1^\infty \{t\} \frac{\log t - 1}{t^2} dt + O\left(\frac{\log(2x)}{x} \right),$$

from which Lemma 7.1 follows readily.   □

The readers will find different versions of this in the literature, usually more intricate than the one above, relying either on Abel summation or on Stieltjes integration. If we want to gain precision in the error term, we may appeal to the Euler–MacLaurin summation formula, but as the readers will see by analysing the example we treated, there is no way one can avoid fractional parts in the development. Note however that:

- We do not know how to evaluate $\sum_{n \le x} n^{it}$ with enough precision when t is large with respect to x.
- The error term in (7.3) (i.e. the fractional part) is much more important than it looks. In our proofs, we want very often to show that the resulting error term is very small but, if it simply did not exist, then we would have $\zeta(s)$ "=" $s/(s-1)$ (see Exer. 3-5). This shows that this error term is responsible for the Euler-product of the Riemann zeta-function as well as for its *functional equation**.

Before ending this part with exercises, we mention that Lemmas 8.1, 10.2 and Eq. (10.5) contain further evaluations of some smooth sums.

Exercise 7-1. Establish the formula $\gamma = \int_1^\infty \frac{\{x\}dx}{[x]x}$.

Exercise 7-2.

1 ⋄ Establish the formula $\gamma = 1 + \sum_{n \ge 1} \left(\frac{1}{n+1} + \log \frac{n}{n+1} \right)$.

2 ⋄ The constant γ_1 being defined in Lemma 7.1, establish similarly that

$$\gamma_1 = \sum_{n \ge 2} \left(\left(\frac{1}{n} + \log \frac{n-1}{n} \right) \log n + \frac{1}{2} \log^2 \frac{n-1}{n} \right).$$

These formulae are used in Exers. 9-14 and 9-15 to compute accurately the decimals of γ and γ_1.

Exercise 7-3. Show that, for any integer $k \ge 1$, we have $\sum_{n \le x} \left(\log \frac{x}{n} \right)^k \ll x$.

The next exercise explains how to handle a coprimality condition on the summation variable. This will be used again and again.

Exercise 7-4.

1 ⋄ Show that for any $x \ge 1$, we have $\displaystyle\sum_{\substack{n \le x, \\ (n,2)=1}} \frac{1}{n} = \frac{1}{2} \log x + \frac{\gamma - \log 2}{2} + O(1/x)$.

2 ⋄ Show that, for every integer $q \ge 1$, there exists a constant $C(q)$ such that, for any real number $x \ge 1$, we have $\displaystyle\sum_{\substack{n \le x, \\ (n,q)=1}} \frac{1}{n} = \frac{\varphi(q)}{q} \log x + C(q) + O(\sigma(q)/x)$

---

* Which is not presented in this book.

where the constant in the O-symbol does not depend on q and where $\sigma(q)$ is the sum of the divisors of q.

Exercise 7-5.

1⋄ Let $B_2(x) = x^2 - x + \frac{1}{6}$ be the second Bernoulli polynomial. Check that $B_2(X + 1) - B_2(X) = 2X$ and that $B_2'(X) = 2X - 1$.

2⋄ Show that $2 \sum_{n \le x} n = x^2 + O^*(3x)$ when $x \ge 0$.

3⋄ Show that $\sum_{n \ge 1} n e^{-n/x} = x^2 + O(3x/2)$.

Concerning the second question, a better approximation is given in Lemma 8.1.

7.2 Some Notes on Integrals and Derivatives

The readers have noticed that the summation by parts version we use, relies only on the formula

$$F(b) - F(a) = \int_a^b F'(t)dt . \tag{7.7}$$

The limitation to continuously differentiable functions (i.e. C^1-functions as above) is too restrictive in practice, where we want to admit some functions that are still continuous, but have less regularity like $|\sin x|$. One way of handling the problem is to refer to Stieltjes integrals. This would work for functions of bounded variation, but we then have to take care about endpoints which diminishes the ease of use. Another way is to appeal to *absolutely continuous functions*.

It is expedient to proceed in an axiomatic manner.

Definition 7.1

A complex-valued function F on an interval I is said to be *absolutely continuous over I* if and only if, for every $\varepsilon > 0$, there exists $\delta > 0$ such that, for any finite sequence of pairwise disjoint sub-intervals (x_k, y_k) of I with $\sum_k (y_k - x_k) < \delta$, we have $\sum_k |f(y_k) - f(x_k)| < \varepsilon$.

The set of absolutely continuous functions over a segment (i.e. a compact interval $[a, b]$) is closed under addition, multiplication and contains the Lipschitz-continuous functions and hence also the C^1-functions. We could have limited our class to these Lipschitz-continuous functions, but this would not simplify the presentation by much. However, in the case of absolutely continuous functions, we have the following elegant theorem due to H. Lebesgue, sometimes called *the fundamental theorem of Lebesgue integral calculus*.

> ### Theorem 7.1
>
> The following conditions on a complex-valued function F on a compact interval $[a, b]$ are equivalent:
>
> - F is absolutely continuous;
> - F has a derivative F' almost everywhere, the derivative is Lebesgue integrable, and
>
> $$F(x) = F(a) + \int_a^x F'(t)\, dt$$
>
> for every x in $[a, b]$;
> - There exists a Lebesgue integrable function G on $[a, b]$ such that
>
> $$F(x) = F(a) + \int_a^x G(t)\, dt$$
>
> for every x in $[a, b]$.

This theorem answers our question fully. In general, the functions we encounter are absolutely continuous over every segment $\subset [0, \infty)$ (which is not equivalent to being absolutely continuous on $[0, \infty)$).

The next exercise is inspired by identities developed by R.A. MacLeod in [3], as exposed by M. Balazard in [2] (see also [1]).

Exercise 7-6. Let $B_2(X) = X^2 - X + 1/6$ be the second Bernoulli polynomial as in Exer. 7-5. Check again that $B_2(X + 1) - B_2(X) = 2X$. For any positive real number x define:

$$g_2(x) = \frac{B_2(x) - B_2(\{x\})}{x}.$$

1 $\diamond$ Show that $\displaystyle\sum_{n \leq x} 2(1 - n/x) = g_2(x)$.

2 $\diamond$ Prove that, for any real number $x \geq 1$, we have $\displaystyle\sum_{n \leq x} \mu(n) g_2(x/n) = 2(1 - 1/x)$

3 $\diamond$ Show that, whenever g is absolutely continuous over every segment in $[1, \infty)$ that vanishes at 1, we have $\displaystyle\sum_{n \geq 1} \mu(n) g(x/n) = \int_1^x M(x/t) g'(t)\, dt$, where $M(t) = \sum_{n \leq t} \mu(n)$.

4 $\diamond$ Show that, for every real number $t \geq 0$, we have $|t(2\{t\} - 1) - \{t\}^2 + \{t\}| \leq t$

5 $\diamond$ Show that $x \mapsto \epsilon(x) = (\{x\}^2 - \{x\})/x$ is an absolutely continuous function over every segment in $[1, \infty)$ and deduce from the above that, for every $x \geq 1$,

$$x \sum_{n \leq x} \frac{\mu(n)}{n} - M(x) = 2 - \frac{2}{x} + \int_1^x M(x/t) \epsilon'(t)\, dt \text{ and that } |\epsilon'(t)| \leq 1/t.$$

6 $\diamond$ Recover the above identity by using Theorem 6.1 with $h(t) = 2t$.

Exercise 7-7. Show that, for integers $N \geq 1$ and $d \geq 1$, we have

$$\frac{1}{d} \sum_{\substack{n,m \leq N \\ d \mid n-m}} 1 = \frac{N^2}{d^2} + \{N/d\} - \{N/d\}^2.$$

References

[1] M. Balazard. "Remarques élémentaires sur la fonction de Moebius". In: *Hal preprint* (2011). hal-00732694, pp. 1–17 (cit. on p. 72).

[2] M. Balazard. "Elementary Remarks on Möbius' Function". In: *Proceedings of the Steklov Intitute of Mathematics* 276 (2012), pp. 33–39 (cit. on p. 72).

[3] R.A. MacLeod. "A curious identity for the Möbius function". In: *Utilitas Math.* 46 (1994), pp. 91–95 (cit. on p. 72).

Part II
The Convolution Walk

Chapter 8
The Convolution Method

Let us present the main features of the *convolution method*. The motivating problem is to determine the average order of a given multiplicative function f. To do so, we compare f with a "model" function g whose average order is known. The model for the function f_0 defined in (0.1) is the function Id that associates n to n. How do we show that g is a model for f? We proceed by proving that there exists a function h such that $f = h \star g$, where h is "smaller" than f. Then the average order of f will be inferred from the one of g.

The argument in itself is very short and takes up only Sect. 8.1, as we have prepared the ground in the previous chapters.

8.1 Proof of Theorem $\mathscr{A}$

Here is the first thing to check.

Claim (STEP 1) The function f_0 is multiplicative.

Our main tool to prove the multiplicativity is Lemma 1.1. Let m and n be any two coprime integers. By definition, we have

$$f_0(mn) = \prod_{p \mid mn} (p - 2) .$$

By Lemma 1.1, the primes p that divide the product mn split in two groups: the ones that divide m and the ones that divide n. These two groups are distinct since every prime in their intersection would divide the gcd of m and n, which is 1 by assumption. Therefore,

$$f_0(mn) = \prod_{p \mid m} (p - 2) \prod_{p \mid n} (p - 2) = f_0(m) f_0(n) ,$$

© The Author(s), under exclusive license to Springer Nature Switzerland AG 2022
O. Ramaré, *Excursions in Multiplicative Number Theory*, Birkhäuser Advanced Texts
Basler Lehrbücher, https://doi.org/10.1007/978-3-030-73169-4_8

which is what was to be proved. What about the condition $f_0(1) = 1$? In the expression $\prod_{p|1}(p-2)$, the set of primes is empty, and, by convention, a product over an empty set is equal to 1.

Claim (STEP 2) The Dirichlet series $D(f_0, s)$ can be expressed as an Euler product by $D(f_0, s) = \prod_{p \geq 2}\left(1 + \dfrac{p-2}{p^s - 1}\right)$.

This Euler product is absolutely convergent in the sense of Godement. By Theorem 3.4, the Dirichlet series of f_0 is given by

$$D(f_0, s) = \prod_{p \geq 2}\left(1 + \sum_{k \geq 1} \frac{\prod_{\ell | p^k}(\ell - 2)}{p^{ks}}\right),$$

where ℓ ranges prime numbers. Since ℓ and p are primes, we have $\ell = p$. We are left with

$$\sum_{k \geq 1} \frac{p-2}{p^{ks}} = \frac{p-2}{p^s}\left(1 - \frac{1}{p^s}\right)^{-1} = \frac{p-2}{p^s - 1},$$

as announced.

Claim (STEP 3) The series $D(f_0, s)$ can be decomposed in $D(f_0, s) = H(s)\zeta(s-1)$, where

$$H(s) = \prod_{p \geq 2}\left(1 - \frac{2p^{s-1} + p - 3}{(p^s - 1)p^{s-1}}\right). \tag{8.1}$$

This product is absolutely convergent in the sense of Godement when $\Re s > 3/2$.

Indeed, the product defining $D(f_0, s)$ *looks like* $\prod_{p \geq 2}\left(1 + 1/(p^{s-1} - 1)\right)$, which in turn looks like $\prod_{p \geq 2}\left(1 - \frac{1}{p^{s-1}}\right)^{-1} = \zeta(s-1)$. Let us proceed to extract the factor $\zeta(s-1)$ from our product. We notice that

$$D(f_0, s) = \prod_{p \geq 2}\left(1 + \frac{p-2}{p^s - 1}\right) = \prod_{p \geq 2}\left(1 - \frac{2p^{s-1} + p - 3}{(p^s - 1)p^{s-1}}\right)\left(\frac{1}{1 - 1/p^{s-1}}\right)$$

$$= H(s)\zeta(s-1) \tag{8.2}$$

as announced. Though this is still far away, we mention that the readers will be able to show in Exer. 17-5 that $H(s)$ admits a meromorphic continuation to the half-plane $\Re s > 1$. The product that defines $H(s)$ is absolutely convergent in the sense of Godement for s such that the series $\sum \dfrac{2p^{s-1} + p - 3}{(p^s - 1)p^{s-1}}$ converges absolutely. This takes place at least, by extending this sum to every integer, when $\Re s > 3/2$. The abscissa of absolute convergence of $\zeta(s-1)$ being equal to 2, it is *probably* the one of $D(f_0, s)$, so that the series H converges effectively in a larger domain.

If we manage to realize the series H as a reasonable Dirichlet series, then the corresponding function will indeed be much "smaller" than f_0, in the sense that the absolute convergence abscissa of its Dirichlet series will be smaller than the one for f_0. Let us specify that this notion of size is only here to guide our computations, and it does not need to be made rigorous: a heuristic control is enough. However, we show a posteriori that the abscissa of convergence (and here of absolute convergence since f_0 is non-negative) is effectively equal to 2. Indeed, assuming that Theorem $\mathscr{A}$ is proved, a summation by parts yields

$$
\begin{aligned}
\sum_{1 \leq n \leq N} f_0(n)/n^s &= \sum_{1 \leq n \leq N} f_0(n)\left(\frac{1}{N^s} + s \int_n^N \frac{dt}{t^{s+1}}\right) \\
&= \frac{1}{N^s} \sum_{1 \leq n \leq N} f_0(n) + s \int_n^N \sum_{1 \leq n \leq t} f_0(n) \frac{dt}{t^{s+1}} \\
&= \tfrac{1}{2}\mathscr{C}_0 N^{2-s} + s\mathscr{C}_0 \int_1^N \frac{dt}{t^{s-1}} + O(N^{1+\sigma-s}) + O\left(\int_1^N dt/t^{s-\sigma}\right).
\end{aligned}
$$
(8.3)

By selecting $\sigma = 0.6$ for instance, we see that the series defining $D(f_0, s)$ converges when $\Re s > 2$ and diverges when $\Re s < 2$.

We now have to transform the two series we have obtained, namely, $H(s)$ and $G(s) = \zeta(s-1)$, into Dirichlet series. The case of G is easily disposed with, as $G(s) = D(\mathrm{Id}, s)$. Let us now consider H.

Claim (STEP 4) The multiplicative function h defined on prime powers by

$$
\begin{cases}
h(p) = -2, \\
h(p^k) = -(p^2 - 3p + 2) & \text{for } k \geq 2,
\end{cases}
$$
(8.4)

satisfies $H(s) = \sum_{n \geq 1} h(n)/n^s$.

We need to construct a multiplicative function h such that $H(s) = D(h, s)$. It is enough to do so for each Euler factor, which means that we want that

$$
1 + \sum_{k \geq 1} \frac{h(p^k)}{p^{ks}} = 1 - \frac{2p^{s-1} + p - 3}{(p^s - 1)p^{s-1}}.
$$
(8.5)

We set $z = 1/p^s$ and proceed as follows:

$$
-\frac{2/(pz) + p - 3}{(1/z - 1)1/(pz)} = \frac{2z + p^2 z^2 - 3pz^2}{z - 1} = -2\sum_{k \geq 1} z^k - (p^2 - 3p)\sum_{k \geq 2} z^k.
$$

On noting that the coefficient of z^k equals $h(p^k)$, we have checked that the multiplicative function h we defined in (8.4) solves our problem.

Claim (STEP 5) The convolution identity $f_0 = \mathrm{Id} \star h$ holds true.

Indeed, we have $D(f_0, s) = H(s)\zeta(s - 1)$, and we know how to develop $H(s)$ and $\zeta(s - 1)$ in Dirichlet series. We are thus looking at a product of two Dirichlet series, which corresponds to the Dirichlet series of the convolution product. By Lemma 3.2, we can identify termwise and conclude that $f_0 = \mathrm{Id} \star h$.

Here is a more computational proof. Firstly, we notice that $\mathrm{Id} \star h$ is still a multiplicative function; we thus only have to check that $f_0(p^k)$ equals $(\mathrm{Id} \star h)(p^k)$ on prime powers p^k. Now, let p be a prime number and $k \geq 1$ be a natural number. We have

$$(\mathrm{Id} \star h)(p^k) = \sum_{\ell/p^k} \mathrm{Id}(p^k/\ell)h(\ell) = \sum_{\ell/p^k} \frac{p^k}{\ell}h(\ell)$$

$$= p^k\left(1 - \frac{2}{p} - \sum_{2 \leq t \leq k} \frac{p^2 - 3p + 2}{p^t}\right) = f_0(p^k)$$

as required. Let us write this convolution equation in a more deployed form for the sequel:

$$f_0(n) = \sum_{\ell m = n} h(\ell)\,\mathrm{Id}(m) = \sum_{\ell m = n} h(\ell)m \ .$$

We next need a lemma that we copy from [3, Lemma 4.3].

Lemma 8.1

We have $2 \displaystyle\sum_{q \leq x} q = x^2 + O^*(x^c)$ for any real numbers $x \geq 0$ and $c \in [1, 2]$.

Proof When $x < 1$, we readily check that $x^c/2 \geq x^2/2$ for every $c \in [1, 2]$. This proves that

$$\sum_{q \leq x} q = x^2/2 + O^*(x^c/2) \quad (\forall x \in [x^*, x^* + 1)) \tag{8.6}$$

for every $c \in [1, 2]$ and $x^* = 0$.

Now let Q be a positive integer and define $N = \sum_{q \leq Q} q = Q(Q + 1)/2$. Notice that $Q \leq \sqrt{2N} < Q + 1$. We show successively the next two points:

- When $Q \leq x < \sqrt{2N}$, the inequality $|N - x^2/2| \leq x^c/2$ is valid for every $c \in [1, 2]$. Indeed, it is equivalent to $N \leq x^2/2 + x^c/2$, which is implied by $N \leq Q^2/2 + Q/2 \leq x^2/2 + x^c/2$.
- When $\sqrt{2N} \leq x < Q + 1$, the inequality $|N - x^2/2| \leq x^c/2$ is this time equivalent to $x^c - x^2 + 2N \geq 0$. The derivative of this function of x is $cx^{c-1} - 2x$ which is negative since $c \leq 2$. Therefore, it remains to check that $(Q+1)^c - (Q+1)^2 + 2N \geq 0$ i.e. $2N \geq Q^2 + Q$ which is true.

This establishes (8.6) for $x^* = Q$, for any Q. $\qquad\square$

After these preparations, let us start the computation of the average order of f_0.

Claim (STEP 6) For every $x > 0$, we have $\sum_{n \leq x} f_0(n) = \frac{1}{2}\mathscr{C}_0 x^2 + O^*(3.5\,x^{1.6})$.

By the convolution identity we proved in Step 5, we have

$$\sum_{n \leq x} f_0(n) = \sum_{\ell m \leq x} h(\ell)m = \sum_{\ell \leq x} h(\ell) \sum_{m \leq x/\ell} m . \tag{8.7}$$

We can extend the sum over ℓ to the sum over every integer as the second sum $\sum_{m \leq x/\ell} m$ vanishes when $\ell > x$. We resort to the approximation given by Lemma 8.1 which is indeed valid for every $x/\ell > 0$ and not only for $x/\ell \geq 1$. This yields

$$\sum_{n \leq x} f_0(n) = \frac{x^2}{2} \sum_{\ell \geq 1} \frac{h(\ell)}{\ell^2} + O^*\left(\frac{1}{2}x^c \sum_{\ell \geq 1} \frac{|h(\ell)|}{\ell^c}\right).$$

Since $H(s)$ converges absolutely for $s > 3/2$, the sum $\sum_{\ell \geq 1} |h(\ell)|/\ell^c$ is finite for every $c > 3/2$. We choose $c = 1.6$ and use Pari/GP to check that $\sum_{\ell \leq 1} |h(\ell)|/\ell^{1.6} \leq 7$.

```
Hbar(p) = 1 + 2/p^1.6+(p^2-3*p+2)/p^(1.6)/(p^(1.6)-1);
prodeuler(p = 2, 100000, Hbar(p))
```

To complete the validation of Theorem $\mathscr{A}$, we only need to compute accurately the first 20 decimals of $\mathscr{C}_0$. This is explained in Exer. 9-8. The proof of Theorem $\mathscr{A}$ is now complete.

Exercise 8-1. Give an asymptotic for $\sum_{\substack{n \leq x, \\ (n,6)=1}} f_0(n)$.

For the same price ...

The readers may follow the same steps as the ones above to determine the average order of the Euler φ-function.

Exercise 8-2. Define the *Jordan totient function* by $J_k(n) = \sum_{d|n} \mu(d)\left(\frac{n}{d}\right)^k$ for $k \geq 2$. Show that $\sum_{n \leq x} J_k(n) = c_k x^{k+1} + O(x^k)$, where $1/c_k = (k+1)\zeta(k+1)$.

Exercise 8-3. Consider the function $f = \mu \star \varphi$. Show that it is multiplicative and compute its average order. Though this is outside the scope of this book, $f(n)$ counts the number of primitive Dirichlet characters modulo n.

8.2 An Exercise on Summation by Parts

We illustrate further the summation by parts technique by deducing the following consequence of Theorem $\mathscr{A}$, which we formulate in two different ways.

Theorem 8.1

We have $\sum_{n \leq x} f_0(n)/n = \mathscr{C}_0 x + O(x^\sigma)$ for every real number σ in $(1/2, 1]$ and where $\mathscr{C}_0$ is defined in Theorem $\mathscr{A}$.

Here is a more usual manner to state the preceding result.

Theorem 8.2

For every $\varepsilon > 0$, we have $\sum_{n \leq x} f_0(n)/n = \mathscr{C}_0 x + O(x^{1/2+\varepsilon})$.

Proof A closer look at Step 6 shows that we can prove a modified form of Theorem $\mathscr{A}$, namely

$$\sum_{n \leq x} f_0(n) = \tfrac{1}{2}\mathscr{C}_0 x^2 + O(x^{\sigma'}) \tag{8.8}$$

for every $\sigma' > 3/2$, where the constant in the O-symbol may depend on σ'. We select $\sigma' = \sigma + 1$. Once we have that in hand, we use (7.4) to infer that

$$\sum_{n \leq x} f_0(n)/n = \sum_{n \leq x} f_0(n)\left(\frac{1}{x} + \int_n^x \frac{dt}{t^2}\right) = \frac{1}{x}\sum_{n \leq x} f_0(n) + \int_1^x \sum_{n \leq t} f_0(n)\frac{dt}{t^2}$$

$$= \tfrac{1}{2}\mathscr{C}_0 x + O(x^\sigma) + \tfrac{1}{2}\mathscr{C}_0 \int_1^x dt + O\left(\int_1^x t^{\sigma-1}dt\right),$$

which is indeed what we have announced.   □

8.3 A Warning and some Limitations

The process we use in the convolution method is to "extract" some simple factor and hope to get a more convergent remaining series. It may be expected to work in every situation but it is not the case, and here is a telling example.

Consider $k(n) = \prod_{p|n} p$ the *square-free kernel* of n. This function appears rather naturally in different problems. The average order of $1/k(n)$ has been obtained by de Bruijn in [1], by using a result from [2] by Hardy and Ramanujan. The Dirichlet series of $1/k(n)$ simply reads

$$D(1/k, s) = \prod_{p \geq 2}\left(1 + \frac{1}{p(p^s - 1)}\right). \tag{8.9}$$

One proves (see [4] by Robert and Tenenbaum to get the latest information on this issue) that

$$\sum_{n \leq x} 1/k(n) = \exp\left((1 + o(1))\sqrt{8(\log x)/\log \log x}\right) . \tag{8.10}$$

In particular, it is *not* of the asymptotical shape $C \log x$. This implies that the convolution method (which would compare it to a simpler function, say $1/n$) will *not* work here. By using the Prime Number Theorem (this is Theorem 23.1), the readers may show that $D(1/k, s) \sim 1/(s \log(1/s))$ as $s > 0$ goes to 0.

Another interesting example one stumbles upon reads simply:

$$D(s) = \prod_{p \geq 2}\left(1 + \frac{1}{p^s} - \frac{1}{p^{2s-1}}\right) . \tag{8.11}$$

Here, however, the convolution method gives some grasp on this series in a two steps approach. First, we may write

$$D(s) = \zeta(s) \prod_{p \geq 2}\left(1 - \frac{1}{p^{2s}} - \frac{1}{p^{2s-1}} + \frac{1}{p^{3s}}\right),$$

which we continue by using

$$\left(1 + \frac{p}{p^{2s}}\right)\left(1 - \frac{1}{p^{2s}} - \frac{p}{p^{2s}} + \frac{1}{p^{3s}}\right) = 1 - \frac{p^2}{p^{4s}} - \frac{1}{p^{2s}} - \frac{p}{p^{4s}} + \frac{1}{p^{3s}} + \frac{p}{p^{5s}} .$$

This last Euler factor defines a Dirichlet series that is absolutely convergent when $\Re s > 3/4$. We thus find that

$$D(s) = \zeta(s)\frac{\zeta(4s - 2)}{\zeta(2s - 1)} \prod_{p \geq 2}\left(1 - \frac{p^2}{p^{4s}} - \frac{1}{p^{2s}} - \frac{p}{p^{4s}} + \frac{1}{p^{3s}} + \frac{p}{p^{5s}}\right), \tag{8.12}$$

getting an analytic continuation until $\Re s = 3/4$. Whether and how far one can continue this procedure is an open question. It is shown in Exer. 17-4 that continuation to the domain $\Re s > 1/2$ is possible. The question of continuation to a larger domain (or showing that it is impossible) is open. We end our detour by mentioning that such questions are also addressed in the context of enumerative algebra; on this issue, see [5] by M. du Sautoy and L. Woodward.

It is interesting to take another viewpoint and see on the pointwise values what this decomposition entails. On using the unitary convolution, we see that (8.11) is the Dirichlet series of the function

$$\sum_{\substack{ab^2=n \\ (a,b)=1}} b\mu(b) ,$$

while $\zeta(s)/\zeta(2s - 1)$ is the Dirichlet series of the function $\sum_{ab^2=n} b\mu(b)$ where the coprimality condition simply disappeared.

Exercise 8-4. Let $\omega(n)$ be the number of distinct prime factors of n.

1 ◇ Let g be the multiplicative function uniquely defined by

$$g(p) = \frac{2}{p-1}, \quad g(p^2) = \frac{1+p}{1-p}, \quad \text{and} \quad g(p^k) = 0, \quad (k \geq 3).$$

Show that $G = \sum_{d=1}^{\infty} \frac{g(d) \log d}{d}$ converges and that, asymptotically,

$$\sum_{n \leq x} \frac{2^{\omega(n)}}{\varphi(n)} = \frac{15}{2\pi^2} (\log x)^2 + \left(\frac{30\gamma}{\pi^2} - G \right) \log x + O(1).$$

2 ◇ Let g be the multiplicative function uniquely defined as follows:

$$g(p) = \frac{2}{p-1}, \quad g(p^2) = -\frac{3p+1}{p-1} \quad \text{and} \quad g(p^k) = 0, \quad (k \geq 3).$$

By a variation of the previous question establish that $G = \sum_{d=1}^{\infty} \frac{g(d) \log d}{d}$ con-

verges and that we have $\sum_{n \leq x} \frac{2^{\omega(n)}}{\varphi(n)} \mu^2(n) = C(\log x)^2 + (4C\gamma - G) \log x + O(1)$,

where $C = \frac{1}{2} \prod_{p} \left(1 - \frac{p+1}{p^2(p-1)} \right)$.

3 ◇ Using the notation of the previous question, write $\sum_{n \geq 1} g(d)/d$ as an Euler product. Write also $D'(g, 1)/D(g, 1)$ as a sum over primes and, anticipating on the next chapter, deduce an approximate value for the constant G via Theorem 9.8.

References

[1] N.G. de Bruijn. "On the number of integers $\leq x$ whose prime factors divide n". In: *Illinois J. Math.* 6 (1962), pp. 137–141. http://projecteuclid.org/euclid.ijm/1255631814 (cit. on p. 80).

[2] G.H. Hardy and S.A. Ramanujan. "Asymptotic formulæ for the distribution of integers of various types". In: *Proc. London Math. Soc.* (2) (1917), pp. 112–132 (cit. on p. 80).

[3] O. Ramaré. "An explicit density estimate for Dirichlet L-series". In: *Math. Comp.* 85.297 (2016), pp. 335–356 (cit. on p. 78).

[4] O. Robert and G. Tenenbaum. "Sur la répartition du noyau d'un entier". In: *Indag. Math. (N.S.)* 24.4 (2013), pp. 802–914. https://doi.org/10.1016/j.indag.2013.07.007 (cit. on p. 80).

[5] M. du Sautoy and L. Woodward. *Zeta functions of groups and rings.* Vol. 1925. Lecture Notes in Mathematics. Springer-Verlag, Berlin, 2008, pp. xii+208. https://doi.org/10.1007/978-3-540-74776-5 (cit. on p. 81).

Chapter 9
Euler Products and Euler Sums

Many constants in number theory are Euler products of the form $\prod_{p>P} h(1/p)$, where h is an analytic function satisfying $h(z) = 1 + O(z^2)$ around $z = 0$. Examples are, e.g. the *twin prime constant*

$$T = \prod_{p \geq 3} \left(1 - \frac{1}{(p-1)^2}\right),\qquad(9.1)$$

and the *Artin constant*

$$A = \prod_{p} \left(1 - \frac{1}{p(p-1)}\right).\qquad(9.2)$$

In these cases and actually in several Euler products occurring in nature, so to say, the function h can be written as $h = F/G$, where F and G are two polynomials with *integer* coefficients satisfying $F(0) = G(0) = 1$.

The obvious way to evaluate Euler products numerically is by computing the partial product $\prod_{P<p\leq x} h(1/p)$ for a large value of x but this leads to a very limited accuracy. To do better a completely different approach exists, where the leading idea is to write the constant as a product of Riemann zeta values. As these latter can be computed with high accuracy, so can the constant. This process has long been known (e.g. Mertens in 1874 was aware of it) and is folklore. Some authors, however, dealt with this problem in a more systematic way [3, 9, 10] and particularly H. Cohen in [2]. The little theory we develop in this chapter provides a control of the error terms involved.

We assume in this chapter that we have at our disposal efficient means for computing $\zeta(s)$ and $\zeta'(s)$ when s is a real number > 1 (and most often an integer). These are sums over *integers* and the problematic we address is to reduce a product or a sum over *primes* to products or sums defined in terms of *integers*.

The next exercise is a specific problem taken from a recent paper.

© The Author(s), under exclusive license to Springer Nature Switzerland AG 2022
O. Ramaré, *Excursions in Multiplicative Number Theory*, Birkhäuser Advanced Texts
Basler Lehrbücher, https://doi.org/10.1007/978-3-030-73169-4_9

Exercise 9-1. Define $B = \sum_{b \geq 1} \dfrac{\mu^2(b)d^2(b)}{\sqrt{b}\varphi(b)}$. Our aim is to find an approximate value of B.

1 ⋄ Show that

$$\left(1 - \frac{1}{p^{3/2}}\right)^4 \left(1 + \frac{4}{\sqrt{p}(p-1)}\right)$$
$$= 1 + \frac{4p^5 - 10p^{9/2} - 6p^{7/2} + 20p^3 + 4p^2 - 15p^{3/2} - \sqrt{p} + 4}{p^{13/2}(p-1)}.$$

2 ⋄ Show that, when $p \geq 7$, we have

$$0 \leq \frac{4p^5 - 10p^{9/2} - 6p^{7/2} + 20p^3 + 4p^2 - 15p^{3/2} - \sqrt{p} + 4}{p^{13/2}(p-1)} \leq \frac{4}{p^{5/2}}.$$

3 ⋄ Show that, for any $P \geq 7$, the constant $B/\zeta(3/2)^4$ is equal to

$$\prod_{2 \leq p \leq P} \left(1 + \frac{4p^5 - 10p^{9/2} - 6p^{7/2} + 20p^3 + 4p^2 - 15p^{3/2} - \sqrt{p} + 4}{p^{13/2}(p-1)}\right) \times I,$$

where $I \in [1, \exp(8/(3P^{3/2}))]$. In particular, $B \leq 28.8$.

Even if we are willing to use 10^6 initial primes, the formula obtained in Exer. 9-1 ensures only an accuracy of roughly 10^{-8} for B, while we obtain rapidly the first hundred digits of B in Sect. 9.5. We first have to cover some ground.

9.1 Exchanging Sum and Product Signs

The next two theorems will be required later on.

Theorem 9.1

Let $(a_{m,n})_{m,n \geq 1}$ be a sequence of complex numbers. We assume that, for each m, the series $\sum_{n \geq 1} a_{m,n}$ converges absolutely towards some A_m, and that $\sum_{m \geq 1} |A_m| < \infty$. Then, for each n, the series $\sum_m a_{m,n}$ is absolutely convergent, say towards B_n, the series $\sum_m A_m$ and $\sum_n B_n$ are also absolutely convergent and we have $\sum_m A_m = \sum_n B_n$.

In short, when $\sum_m \sum_n |a_{m,n}| < \infty$, we have $\sum_{m \geq 1} \left(\sum_{n \geq 1} a_{m,n}\right) = \sum_{n \geq 1} \left(\sum_{m \geq 1} a_{m,n}\right)$ and the intervening series are all absolutely convergent. Here is the multiplicative analogue.

Theorem 9.2

Let $(a_{m,n})_{m,n\geq 1}$ be a sequence of complex numbers. We assume that, for each m, the product $\prod_{n\geq 1}(1 + a_{m,n})$ converges absolutely in the sense of Godement towards some $1 + A_m^*$, and that $\sum_{m\geq 1}|A_m^*| < \infty$. Then, for each n, the product $\prod_m(1 + a_{m,n})$ is absolutely convergent in the sense of Godement, say towards $1 + B_n^*$, the products $\prod_m(1 + A_m^*)$ and $\prod_n(1 + B_n^*)$ are also absolutely convergent in the sense of Godement and we have $\prod_m(1 + A_m^*) = \prod_n(1 + B_n^*)$.

9.2 The Witt Expansion

We first establish the existence of the Witt expansion.

Theorem 9.3

Let F be a function on the disc $|z| < r$ for some $r \in (0, 1]$. Assume $F(0) = 1$ and that F admits there a Taylor development of order $L \geq 1$ with integer coefficients, for some parameter L, with remainder $O(z^{L+1})$. Then there exist unique integers $e_1, e_2, \ldots, e_L$ such that

$$F(z) = \prod_{1\leq \ell \leq L} (1 - z^\ell)^{-e_\ell} F_1(z) \tag{9.3}$$

in the disc $|z| < r$, where $F_1(z) = 1 + O(z^{L+1})$.

How does the mathematician E. Witt enter the scene? In the paper [15] on Lie algebras (whatever they are), Witt produced in Eq. (11) therein a decomposition that is the prototype of the above expansion. This decomposition, known as the *cyclotomic identity* or the *necklace identity* (see Exer. 9-2), may well have been discovered earlier. D. Shanks used in Eq. (5) of [12] a similar decomposition to compute the density, according to the *Bateman–Horn Conjecture* (see Sect. 17.3), of the primes of the form $n^4 + 1$.

Proof We proceed by recursion on $L \geq 1$. When $L = 1$, our assumption is that $F(z) = 1 + e_1 z + O(z^2)$ for some integer e_1. Then $F(z)(1 - z)^{e_1} = 1 + O(z^2)$. This is rather obvious when $e_1 \geq 0$ and also holds true when $e_1 = -\ell < 0$. Indeed, we have in this case

$$(1 - z)^{e_1} = \left(1 + z + z^2 + \ldots\right)^\ell = 1 + \ell z + \frac{\ell(\ell + 1)}{2}z^2 + \ldots$$

and its product with $F(z)$ has a zero coefficient for z.

Let us assume that the claim has been proved for L. Our aim is to prove it for $L+1$. By applying our recursion hypothesis, we have $F(z) = \prod_{1\leq \ell \leq L}(1-z^\ell)^{-e_\ell}F_1(z)$ where $F_1(z) = 1 + O(z^{L+1})$. Because or our assumptions, we can write $F(z) =$

$1 + zQ(z) + bz^{L+1} + O(z^{L+2})$ where Q is a polynomial with integer coefficients and degree at most $L - 1$. One can also expand $\prod_{1 \leq \ell \leq L}(1 - z^\ell)^{-e_\ell}$ in power series and check that it has integer coefficients. Because of our hypothesis on F_1, these coefficients are equal to the Taylor coefficients of F, up to the one of z^{L+1} which may differ, i.e.

$$\prod_{1 \leq \ell \leq L}(1 - z^\ell)^{-e_\ell} = 1 + zQ(z) + cz^{L+1} + O(z^{L+2}).$$

Let us look at $G(z) = F(z)\prod_{1 \leq \ell \leq L}(1 - z^\ell)^{e_\ell}(1 - z^{L+1})^{b-c}$. We find that

$$
\begin{aligned}
G(z) &= \frac{1 + zQ(z) + bz^{L+1} + O(z^{L+2})}{1 + zQ(z) + cz^{L+1} + O(z^{L+2})}\left(1 + (b - c)z^{L+1} + O(z^{L+2})\right) \\
&= \left(1 + \frac{(c - b)z^{L+1}}{1 + zQ(z) + cz^{L+1}}\right)\left(1 + (b - c)z^{L+1}\right) + O(z^{L+2}) \\
&= \left(1 + (c - b)z^{L+1}\right)\left(1 + (b - c)z^{L+1}\right) + O(z^{L+2}) = 1 + O(z^{L+2}).
\end{aligned}
$$

Our claim is proved for $L + 1$, ending the proof of the existence of (e_ℓ).

We now turn our attention to the uniqueness claim. For the sake of contradiction, suppose there exists a different sequence of integers (f_ℓ) such that $F(z) = \prod_{\ell \leq L}(1 - z^\ell)^{-f_\ell}F_2(z)$. Let k be the smallest integer such that $f_\ell \neq e_\ell$. Put

$$H(z) = \prod_{\ell \leq k-1}(1 - z^\ell)^{-e_\ell}.$$

We then have, and that is the key observation,

$$F(z) = H(z)(1 - z^k)^{-e_k} + O(z^{k+1})$$

on the one hand, and similarly $F(z) = H(z)(1 - z^k)^{-f_k} + O(z^{k+1})$ on the other. As both expressions have different coefficients of z^k, we have reached a contradiction.   □

The coefficients $(e_\ell)_{\ell \leq L}$ may depend on L. Comparing them as L varies is readily done, thanks to their unicity. This leads to our next result.

Theorem 9.4

Let $F(z)$ be a function on a disc $|z| < r \leq 1$. Assume that $F(0) = 1$ and that, for any $L \geq 1$, F admits there a Taylor development of order $L \geq 1$ with integer coefficients and remainder $O(z^{L+1})$. Then there exist unique integers $(e_\ell)_{\ell \geq 1}$ such that (9.3) holds.

When such a situation arises, we say that $F(X) = \prod_{\ell \geq 1}(1 - z^\ell)^{-e_\ell}$ *formally*. Often, this formal writing can be transformed into an equality between two analytic functions; Theorem 9.5 is a prototype of such a result.

Proof The development $F(z) = 1 + \sum_{\ell \leq L} a_\ell z^\ell + O(z^{L+1})$ ensures us the existence, by Theorem 9.3, of a sequence of integers $(e_\ell^{(L)})_{\ell \leq L}$; Since $(e_\ell^{(L+1)})_{\ell \leq L}$ verifies also $F(z) = \prod_{1 \leq \ell \leq L} (1 - z^\ell)^{-e_\ell^{(L+1)}} (1 + O(z^{L+1}))$, the unicity proved in Theorem 9.3 ensures us that $e_\ell^{(L+1)} = e_\ell^{(L)}$ when $\ell \leq L$. In short, we have only one sequence $(e_\ell)_{\ell \geq 1}$ and its truncations. □

Here is the *Cyclotomic Identity*, also known as the *Necklace Identity*, that lies at the origin of Theorem 9.4.

Exercise 9-2. Let α be some real number. Prove that, formally, we have

$$1 - \alpha X = \prod_{n \geq 1} (1 - X^n)^{M(\alpha;n)} ,$$

where

$$M(\alpha; n) = \frac{1}{n} \sum_{d \mid n} \mu(n/d) \alpha^d. \tag{9.4}$$

Exercise 9-3. Prove that if conversely we fix $e_n = -24$, then $F(X) = \sum_{n \geq 0} \tau(n + 1) X^n$, where τ is the Ramanujan τ-function.

9.3 Explicit Witt Expansion of Polynomials

Theorem 9.4 ensures the existence of the exponents (e_n), but we need an explicit description and to move from formal equality to equality of functions. The next theorem treats the case when F is a polynomial.

Theorem 9.5

Let $F(t) = 1 + a_1 t + \ldots + a_\delta t^\delta \in \mathbb{Z}[t]$ be a polynomial of degree δ. Let $\alpha_1, \ldots, \alpha_\delta$ be the inverses of its roots. Put $s_F(k) = \alpha_1^k + \ldots + \alpha_\delta^k$. The $s_F(k)$ are integers and satisfy the Girard–Newton recursion

$$s_F(k) + a_1 s_F(k-1) + \ldots + a_{k-1} s_F(1) + k a_k = 0, \tag{9.5}$$

where we have defined $a_{\delta+1} = a_{\delta+2} = \ldots = 0$. Put

$$b_F(k) = \frac{1}{k} \sum_{d \mid k} \mu(k/d) s_F(d) . \tag{9.6}$$

Then $b_F(k) \in \mathbb{Z}$. Moreover, let $\beta \geq 2$ be larger than all the $|\alpha_i|$. We have

$$|b_F(k)| \leq 2 \deg F \cdot \beta^k / k . \tag{9.7}$$

Finally, when $|z| < 1/\beta$, the product $\prod_{j \geq 1}(1 - z^j)^{b_F(j)}$ converges absolutely in the sense of Godement and for any such z, we have

$$F(z) = \prod_{j \geq 1}(1 - z^j)^{b_F(j)} . \tag{9.8}$$

Proof The recursion relations (9.5) allow to compute easily the $s_F(k)$ and, moreover, they show that the $s_F(k)$ must be integers. By assumption we have

$$F(z) = 1 + a_1 z + \ldots + a_\delta z^\delta = \prod_{j=1}^{\delta}(1 - \alpha_j z) , \tag{9.9}$$

which by logarithmic differentiation yields, when $|z| < 1/\beta$,

$$-\frac{zF'(z)}{F(z)} = \sum_{j=1}^{\delta} \frac{\alpha_j z}{1 - \alpha_j z} = \sum_{j=1}^{\infty} s_F(j)z^j . \tag{9.10}$$

By Theorem 9.4 there are integers $c_F(j)$ such that, for any $L \geq 1$, the identity

$$F(z) = \prod_{j=1}^{L}(1 - z^j)^{c_F(j)} F_1(z) \tag{9.11}$$

holds true with $F_1(z) = 1 + O(z^{L+1})$. From (9.10) and (9.11) it follows that

$$\sum_{j \leq L} s_F(j)z^j + O(z^{L+1}) = \sum_{\ell \leq L} \ell c_F(\ell)\frac{z^\ell}{1 - z^\ell} - \frac{zF_1(z)'}{F_1(z)} .$$

Notice that $zF_1(z)'/F_1(z) = O(z^{L+1})$, hence

$$\sum_{j \leq L} s_F(j)z^j = \sum_{\ell \leq L} \ell c_F(\ell)\frac{z^\ell}{1 - z^\ell} + O(z^{L+1})$$

$$= \sum_{\ell \leq L} \ell c_F(\ell) \sum_{k \geq 1} z^{k\ell} + O(z^{L+1}) = \sum_{j \geq 1}\left(\sum_{\substack{\ell k = j, \\ \ell \leq L}} \ell c_F(\ell)\right)z^j + O(z^{L+1}) .$$

Therefore $s_F(j) = \sum_{\ell | j} \ell c_F(\ell)$ for $j \leq L$. But since L is arbitrary, this holds for every index j. By Möbius inversion of Sect. 6.1 we deduce that $\ell c_F(\ell) = \sum_{d | \ell} \mu(\ell/d)s_F(d)$, i.e. that $b_F(\ell) = c_F(\ell)$.

A bound for $b_F(\ell)$ is readily obtained: since $|s_F(j)| \leq \deg F \cdot \beta^j$, we get

$$|b_F(k)| \leq \frac{\deg F}{k} \sum_{1 \leq j \leq k} \beta^j \leq \frac{\deg F}{k} \beta \frac{\beta^k - 1}{\beta - 1}$$

$$\leq \frac{\deg F}{k} \frac{\beta^k}{1 - 1/\beta} \leq 2 \deg F \cdot \beta^k / k \ .$$

To prove that the product $H(z) = \prod_{j \leq 1}(1 - z^j)^{b_F(j)}$ is absolutely convergent in the sense of Godement, we remark that $(1 - z^j)^b$ is just the repetition $|b|$ times of $(1 - z^j)$, when $b \geq 0$, and of $(1 - z^j) = 1 - \dfrac{z^j}{1 - z^j}$ otherwise. In any case, it is of the form $1 - a_j$ with $|a_j| \leq 2|z|^j$. We next note that

$$\sum_{j \geq 1} |b_F(j)| |a_j| \leq 4 \deg F \sum_{j \geq 1} \beta^j |z|^j = 4 \deg F \frac{\beta|z|}{1 - \beta|z|} < \infty \ .$$

This establishes the convergence of our product and the existence of $H(z)$. To finally conclude that $H(z) = F(z)$, we notice that $F(z)/H(z) = 1 + O(z^{L+1})$ for every $L \geq 1$; this implies that the Taylor development at $z = 0$ of both F and H have equal coefficients for arbitrary large L, and since both F and H are equal to their power series, we conclude that $F(z) = H(z)$ when $\beta|z| < 1$, as required. This ends the proof. □

Consider the following example. The Artin constant A given by (9.2) is of the form $\prod_{p \geq 2} h_A(1/p)$ with

$$h_A(z) = \frac{1 - z - z^2}{1 - z} \ .$$

Let us first focus on $F(z) = 1 - z - z^2$. It is the reciprocal of $F^*(z) = z^2 - z - 1 = z^2 + a_1 z + a_2$. This has roots the golden ratio $\rho = (1 + \sqrt{5})/2$ and $-1/\rho = (1 - \sqrt{5})/2$. We have $s_F(n) = \rho^n + (-\rho)^{-n}$. Setting $k = 1$ in (9.5) we obtain $s_F(1) = -a_1 = 1$. Setting $k = 2$ we obtain $s_F(2) = -a_1 s_F(1) - 2a_2 = 3$. For $k \geq 3$ we have $s_F(k) = s_F(k - 1) + s_F(k - 2)$. We conclude that

$$1 - t - t^2 = \prod_{k=1}^{\infty} (1 - t^k)^{b_F(k)} \ ,$$

with $b_F(k)$ given by (9.6). We compute that the sequence $b_F(1), b_F(2), \ldots$ is $1, 1, 1, 1, 2, 2, 4, 5, 8, \ldots$ and infer that the Witt expansion of $h_A(z)$ is given by (9.8).

Exercise 9-4. Show that, for every prime p, the Fibonacci numbers F_p and F_{p-2} satisfy the congruence $3F_p - 2F_{p-2} \equiv 1[p]$.

Exercise 9-5.

1 ⋄ Show that, when α is an integer, the quantity $M(\alpha; n)$ defined in (9.4) is also an integer.

2 ⋄ Give a proof of Fermat's Little Theorem 5.1 by taking n to be a prime.

3 ⋄ Show that $M(-1; 1) = M(-1; 2) = 1$ and $M(-1; n) = 0$ when $n \geq 3$.

There are numerous easy upper estimates for the inverse of the modulus of all the roots of $F(t)$ in terms of its coefficients. Here is a simplistic one.

Exercise 9-6. Let $F(X) = 1 + a_1 X + \ldots + a_\delta X^\delta$ be a polynomial and let ρ be one of its roots. Show that either $|\rho| \geq 1$ or $1/|\rho| \leq |a_1| + |a_2| + \ldots + |a_\delta|$.

Exercise 9-7. Let $F(X) = 1 + 2X - 2X^2 + X^3$. The reciprocal of its roots are the roots of the polynomial $X^3 F(1/X)$. Find an approximation of $s_F(7)$ by using Pari/GP and the function `polroots`. Compare this approximation to the exact value obtained by using (9.5).

9.4 Euler Product Expansion in Terms of Zeta-Values

We define the *multiplicative partial zeta function* by

$$\zeta_P(s) = \zeta(s) \prod_{p \leq P} (1 - p^{-s}), \tag{9.12}$$

where $\zeta(s)$ denotes of course the Riemann zeta function.

Theorem 9.6

Let $F, G \in \mathbb{Z}[X]$ be two coprime polynomials satisfying $F(0) = G(0) = 1$ such that $(F(X) - G(X))/X^2 \in \mathbb{Z}[X]$. Let $\beta \geq 2$ be an upper bound for the maximum modulus of the inverses of the roots of F and of G. Let $P > \beta$ be a parameter. Then, for any parameter $J \geq 3$, we have

$$\prod_{p > P} \frac{F(1/p)}{G(1/p)} = \prod_{2 \leq j \leq J} \zeta_P(j)^{b_G(j) - b_F(j)} \times I,$$

where the integers $b_G(j)$ and $b_F(j)$ are defined in Theorem 9.5 and

$$\max(I, 1/I) \leq \exp\left(\frac{(\deg F + \deg G)(\beta/P)^J P}{(1 - \beta/P) J}\right).$$

The readers will find in [10] a more precise result of the same flavour. Both conditions of integrality of the coefficients of F and G and $F(0) = G(0) = 1$ can be quite binding. Theorem 17.2 shows that these conditions may be relaxed.

Proof The proof requires several steps. The very first one is a direct consequence of (9.8), which leads to the identity

$$\frac{F(z)}{G(z)} = \prod_{j=2}^{\infty} (1 - z^j)^{b_F(j)-b_G(j)} . \tag{9.13}$$

The absence of the $j = 1$ term is due to our assumption that $(F(X)-G(X))/X^2 \in \mathbb{Z}[X]$. Our second step is to control the rate of convergence of the product in (9.13). By Eq. (9.7), we know that $|b_F(j) - b_G(j)| \le (\deg F + \deg G)\beta^j/j$. Therefore, for any bound J, we have

$$\sum_{j \ge J} |z^j| |b_F(j) - b_G(j)| \le (\deg F + \deg G)\frac{|z\beta|^J}{(1 - |z\beta|)J} , \tag{9.14}$$

as soon as $|z| < 1/\beta$. We thus have

$$\frac{F(z)}{G(z)} = \prod_{2 \le j \le J} (1 - z^j)^{b_F(j)-b_G(j)} \times I_1 , \tag{9.15}$$

where $|\log I_1| \le (\deg F + \deg G)|t\beta|^J/[(1 - |z\beta|)J]$.

Now that we have the expansion (9.15) for each prime p, we may combine them. We immediately get

$$\prod_{p>P} \frac{F(1/p)}{G(1/p)} = \prod_{p>P} \prod_{2 \le j \le J} (1 - p^{-j})^{b_G(j)-b_F(j)} \times I_2 ,$$

where I_2 satisfies

$$|\log I_2| \le (\deg F + \deg G) \sum_{p \ge P} \frac{\beta^J}{1 - \beta/P} \frac{1}{p^J}$$

$$\le \frac{(\deg F + \deg G)\beta^J}{1 - \beta/P} \left(\frac{1}{P^J} + \int_P^{\infty} \frac{dt}{t^J} \right)$$

$$\le \frac{(\deg F + \deg G)(\beta/P)^J P}{1 - \beta/P} ,$$

since $P \ge 2$ and $J \ge 3$. We may rearrange the product over the primes p and get

$$\prod_{p>P} \frac{F(1/p)}{G(1/p)} = \prod_{2 \le j \le J} \zeta_P(j)^{b_G(j)-b_F(j)} \times I_2 ,$$

and this ends the proof. $\qquad\square$

Exercise 9-8. Recall that

$$\mathscr{C}_0 = \prod_{p \geq 2}\left(1 - \frac{3}{p(p+1)}\right). \tag{0.2}$$

Let $\{u_n\}_{n=1}^{\infty}$ be the sequence of integers determined by $u_1 = -1$, $u_2 = 7$ and $u_{n+2} = -u_{n+1} + 3u_n$ for $n \geq 1$.

1 $\diamond$ Show that $2\mathscr{C}_0 = \prod_{k \geq 2} \zeta_2(k)^{-v_k}$, where $kv_k = \sum_{d|k} \mu(k/d)(u_d - (-1)^d)$.

2 $\diamond$ By using the third question of Exer. 9-5, who show that, when $k \geq 2$, we have $kv_k = \sum_{d|k} \mu(k/d)u_d$.

3 $\diamond$ Use a Sage script to prove that

$\mathscr{C}_0 = 0.29261\ 98570\ 45154\ 91401\ 24089\ 97714\ 49464\ 17435\ 61519\ 76215\ldots$

Exercise 9-9. Recall that

$$\mathscr{C}_1 = \prod_{p \geq 2}\left(1 - \frac{1}{p(p+2)}\right). \tag{21.3}$$

Show that $8\mathscr{C}_1 = 7 \prod_{k \geq 2} \zeta_2(k)^{-v_k}$, where $kv_k = \sum_{d|k}(u_d - (-2)^d)\mu(k/d)$ and $\{u_n\}_{n=1}^{\infty}$ is the sequence of integers uniquely determined by $u_1 = -2, u_2 = 6$ and $u_{n+2} = -2u_{n+1} + u_n$. This leads to

$\mathscr{C}_1 = 0.75947\ 92316\ 30837\ 16720\ 48837\ 06032\ 50718\ 85299\ 46853\ 67848\ldots$

While investigating the density of members of some linear recurrence that have some divisibility property in [13, 14], P.J. Stephens encountered what is now known as the *Stephens constant*

$$S = \prod_{p}\left(1 - \frac{p}{p^3 - 1}\right). \tag{9.16}$$

See also [11] by P. Moree and P. Stevenhagen, or [7] by P. Kurlberg and C. Pomerance.

Exercise 9-10. Show that Stephens's constant can be evaluated with Theorem 9.6, the denominator being already in the right format. Prove that

$S = 0.57595\ 99688\ 92945\ 43964\ 31633\ 75492\ 49669\ 2506\ 13967\ 17649\ldots$

Exercise 9-11. Under the hypotheses of Theorem 9.6 show that the function

$$s \mapsto \prod_{p > P} \frac{F(1/p^s)}{G(1/p^s)} \tag{9.17}$$

defined at first for $\Re s \geq 1$ admits an analytic continuation to $\Re s > 0$.

The constant B of Exer. 9-1 was our motivating example, but our theory does not apply to this very case! This is not exactly true: a slight modification of it in fact *does*. It is provided by the next theorem.

Theorem 9.7

Let $F, G \in \mathbb{Z}[X]$ be two coprime polynomials satisfying $F(0) = G(0) = 1$ and such that $(F(X) - G(X))/X^3 \in \mathbb{Z}[X]$. Let $\beta \geq 2$ be an upper bound for the maximum modulus of the inverses of the roots of F and of G. Let $P > \beta^2$ be a parameter. Then, for any parameter $J \geq 3$, we have

$$\prod_{p>P} \frac{F(1/\sqrt{p})}{G(1/\sqrt{p})} = \prod_{3 \leq j \leq J} \zeta_P(j/2)^{b_G(j) - b_F(j)} \times I,$$

where the integers $b_G(j)$ and $b_F(j)$ are defined in Theorem 9.5 and

$$\max(I, 1/I) \leq \exp\left(\frac{(\deg F + \deg G)(\beta/\sqrt{P})^J \sqrt{P}}{(1 - \beta/\sqrt{P}) J}\right).$$

The proof follows precisely the previous one where we simply replace p by $\sqrt{p}$ everywhere. See also Theorem 10.1 for a version with $p^{1/3}$ instead of $p^{1/2}$.

9.5 Refining Exercise 9-1

Our aim is to compute the constant B of Exer. 9-1 with more accuracy. We assume that Pari/GP gives us the necessary accuracy for $\zeta(j/2)$. We employ Theorem 9.7 with

$$F(X) = 1 - X^2 + 4X^3, \quad \text{and} \quad G(X) = 1 - X^2. \tag{9.18}$$

The polynomial G is already in the proper format. Concerning F, we find that $s_F(1) = 0$, $s_F(2) = 2$, $s_F(3) = -12$ and $s_F(k) = s_F(k-2) - 4s_F(k-3)$ for $k \geq 3$. By Exer. 9-6, we may take $\beta = 5$. Whence, for any parameter $P > 25$, we have

$$B = \prod_{p \leq P} \left(1 + \frac{4}{\sqrt{p}(p-1)}\right) \prod_{3 \leq j \leq J} \zeta_P(j/2)^{-b_F(j)} \times I$$

where $|\log I| \leq (5(5/\sqrt{P})^J \sqrt{P})/((1 - 5/\sqrt{P})J)$. We use the following Pari/GP script to generate the values of $s_F(k)$ and of $b_F(k)$.

```
{innerGetsF(k) =
  /* returns [sF(k-2), sF(k-1), sF(k)] */
  if(k == 3, return([0, 2, -12]),
    my(aux = innerGetsF(k-1));
    return([aux[2], aux[3], aux[2]-4*aux[1]]));}

{GetsF(k) =
  if(k == 1, return(0),
    if(k == 2, return(2),
      return(innerGetsF(k)[3]),),);}

{GetbF(k) =
  my(res = 0);
  fordiv(k, d, res += moebius(k/d)*GetsF(d));
  return(res/k);}
```

The sequence $(s_F(k))_{k \geq 1}$ reads

$$0, 2, -12, 2, -20, 50, -28, 130, -228, 242, -748, 1154, -1716, 4146, -6332, \ldots$$

while the sequence $(b_F(k))_{k \geq 1}$ reads

$$0, 1, -4, 0, -4, 10, -4, 16, -24, 26, -68, 92, -132, 298, -420, \ldots$$

But since we want to control the error term precisely, it is better to use interval arithmetic and to switch to Sage:

```
R = RealIntervalField(1000)

def GetsF(k):
   if k == 1:
      return(R(0))
   elif k == 2:
      return(R(2))
   else:
      return innerGetsF(k)[2]

def innerGetsF(k):
   if k == 3:
      return(vector([R(0), R(2), R(-12)]))
   else:
     aux = innerGetsF(k-1)
     return(vector([aux[1], aux[2], aux[1] - 4*aux[0]]))
```

```
def GetbF(k):
    res = R(0)
    for d in divisors(k):
        res += moebius(k/d)*GetsF(d)
    return(res/k)

def ConstantB(bigP = 10000, bigJ = 10):
    res = R(1);
    for p in Primes():
        if p <= bigP:
            sqrtp = R(sqrt(p))
            res *= (1 + 4/sqrtp/(p-1))
            for j in range(3, bigJ):
                res *= (1 - 1/sqrtp^j)^(-GetbF(j))
        else:
            break
    for j in range(3, bigJ):
        res *= R(zeta(j/2))^(-GetbF(j))
    logerr =  R(5*(5/sqrt(bigP))^bigJ*sqrt(bigP)
    logerr = logerr/bigJ/(1-5/sqrt(bigP)))
    err = R(exp(logerr))
    out = [R(res/err).lower(), R(res*err).upper()]
    return([out[0], out[1], out[1]-out[0]])
```

The readers may store this script in a file named `GetConstantB.sage` and use, under Sage, the commands

```
load("GetConstantB.sage")
ConstantB(2000, 120)
```

This simplistic script produces the first *hundred ten* digits of B in three minutes!

$B = 28.78108\,84670\,86464\,52437\,86800\,17466\,17316\,89580\,11675\,79488\,55724$
$43201\,90516\,16396\,67019\,81888\,62787\,85739\,32143\,30001\,95674\,58325\ldots$

We take this opportunity to remind the readers that the suffix of the file given as argument to the `load`-function in Sage *is* important: it *should* be ".sage" as the way the file is parsed depends on this suffix. The reader should also consider using the `attach`-function that reloads the script automatically.

Exercise 9-12. Let $F(X) = 1 - X - X^3 + 2X^4$. Show that $s_F(1) = s_F(2) = 1$, $s_F(3) = 4$, $s_F(4) = -3$ and $s_F(k) = s_F(k-1) + s_F(k-3) - 2s_F(k-4)$. Adapt Theorem 9.7 for $1/p^{1/3}$ rather than $1/\sqrt{p}$ to prove that

$$\prod_{p \geq 2} \left(1 + \frac{1}{(p-1)(p^{1/3}-1)} \right) = 26.17426\ 30679\ 53288\ 62276\ 13918\ 04715\ldots$$

This value will be used in Chap. 27.

9.6 Euler Sums, with no log-Factor

What has been said for Euler products is, mutatis mutandis, also valid for sums of
the form

$$\sum_{p>P} h(1/p),\tag{9.19}$$

although the theory will rely on the identity (9.21) rather than on the Witt expansion.
If $h(z) = \sum_{j \geq 2} a_j z^j$, the Taylor series of $h(z)$ is valid for $|z| < \beta$, with $\beta > 0$, then
we have for $P > 1/\beta$,

$$\sum_{p>P} h(1/p) = \sum_{j=2}^{\infty} a_j \sum_{p>P} \frac{1}{p^j},\tag{9.20}$$

where the inner sum can be computed via the identity*

$$\sum_{p>P} \frac{1}{p^s} = \sum_{n=1}^{\infty} \frac{\mu(n)}{n} \log \zeta_P(ns),\tag{9.21}$$

valid for $s > 1$.

Exercise 9-13. Show that, when $\Re s > 1$, we have $\log \zeta_P(s) = \displaystyle\sum_{p>P}\sum_{k \geq 1} \frac{1}{kp^{ks}}$, the
series being absolutely convergent, and prove (9.21).

Consider the following example. Anticipating on the second Mertens Theorem,
namely, Theorem 12.3, we recall that

$$\sum_{p \leq x} \frac{1}{p} = \log \log x + b + o(1),$$

where b is called the Mertens constant (some authors call it the Meissel–Mertens
constant). As shown when proving Theorem 12.4, we have

$$b = \gamma + \sum_p \left(\log(1 - 1/p) + 1/p \right) = \gamma - \sum_p \sum_{k \geq 2} \frac{1}{kp^k}\,.$$

Using (9.21), we deduce that

---

* This identity is due to J.W.L. Glaisher in [5].

$$b = \gamma + \sum_{k=2}^{\infty} \frac{\mu(k)}{k} \log \zeta(k) .$$

Utilizing this expression and a table of zeta-values compiled by Legendre, Mertens showed that $b = 0.26149\,72128\ldots$, see also [8] by P. Lindqvist and J. Peetre. With a computer it is of course easy to go beyond Mertens:

$b = 0.26149\,2128\,47642\,78375\,54268\,38608\,69585\,90515\,66648\,26119\ldots$

The next exercise is inspired from the preprint [4] by P. Flajolet and I. Vardi that has been floating on the web since 1996.

Exercise 9-14. By using Exer. 7-2 and the series expansion of $\log(1 - x)$ show that $\gamma = 1 - \sum_{k\geq 2} \frac{1}{k}(\zeta(k) - 1)$.

We get the digits of γ_1 similarly.

Exercise 9-15.

1 ⋄ With $H_\ell = \sum_{u \leq \ell} 1/u$, show that $\log^2(1 - x) = \sum_{k\geq 2} 2H_{k-1}x^k/k$, valid for $|x| < 1$.

2 ⋄ By using Exer. 7-2, show that $\gamma_1 = \sum_{k\geq 2} \frac{\zeta'(k) + H_{k-1}(\zeta(k) - 1)}{k}$.

Exercise 9-16. (Based on [1]). Let $N_2(x)$ be the number of integers $n \leq x$ of the form $n = pq$, with p and q primes and $q \equiv \pm 1 \pmod{p}$. It can be shown that $N_2(x) = C\frac{x}{\log x} + O\left(\frac{x}{(\log x)^2}\right)$, with $C = \frac{1}{2} + \sum_{p\geq 3} \frac{2}{p(p - 1)}$.

1 ⋄ Establish the identity $\sum_{p\geq 2} \frac{1}{p(p - 1)} = \sum_{k=2}^{\infty} \log \zeta(k)\frac{(\varphi(k) - \mu(k))}{k}$.

2 ⋄ Use Pari/GP or Sage to check that

$$C = 1.04631\,33380\,99590\,25572\,87349\,19711\,8847\ldots.$$

9.7 Euler Sums, with log-Factor

The convolution method will often lead to constants of the type

$$\sum_{p>P} h(1/p) \log p \tag{9.22}$$

as in (10.4) or in Exer. 8-4. Let us start our stroll with an exercise.

Exercise 9-17.

1 ⋄ Find an accurate value for $\sum_{p \geq 2} \dfrac{\log p}{p^2 - 1}$ and for $\sum_{p \geq 2} \dfrac{\log p}{p^3 - 1}$.

2 ⋄ Show that $\displaystyle\sum_{p \geq 2} \frac{\log p}{p(p-1)} = \sum_{p \geq 2} \frac{\log p}{p^2 - 1} + \sum_{p \geq 2} \frac{\log p}{p^3 - 1} + \sum_{p \geq 2} \frac{\log p}{p^5 + p^4 - p^2 - p}$

and determine the first twenty digits of $\displaystyle\sum_{p \geq 2} \frac{\log p}{p(p-1)}$.

We take this opportunity to present a Pari/GP code that simplifies the computation of $-\zeta'(s)$ for *a real parameter s*:

```
{zetaprime(s) = -sumpos(n = 2, log(n)/n^s);}
```

Theorem 9.8

Let $h(z) = a_2 z^2 + a_3 z^3 + \ldots$ be a series such that the coefficients $(a_m)_{m \geq 2}$ (that may be arbitrary complex numbers) satisfy $|a_m| \leq C\beta^m$ for some parameters $C > 0$ and $\beta \geq 2$. Let $P > \beta$ and $J \geq 2$ be two further parameters. We have

$$\sum_{p > P} h(1/p) \log p = -\sum_{2 \leq j \leq J} b_j \frac{\zeta_P'(j)}{\zeta_P(j)} + E,$$

where the integers b_j are given by $b_j = \sum_{d \mid j, d \geq 2} \mu(j/d) a_d$ and

$$|E| \leq 12C \frac{|\beta/P|^J P}{1 - |\beta/P|}.$$

Note that $-\zeta_P'(j)/\zeta_P(j)$ is easily computed using

$$-\frac{\zeta_P'(j)}{\zeta_P(j)} = -\frac{\zeta'(j)}{\zeta(j)} - \sum_{p \leq P} \frac{\log p}{p^j - 1}. \tag{9.23}$$

Proof We follow the principle of the proof of Theorem 9.6. The first step is to express h as a *Lambert series* for t such that $|t| < 1/\beta$:

$$h(t) = \sum_{j \geq 2} b_j \frac{t^j}{1 - t^j} = \sum_{j \geq 2} b_j \sum_{d \geq 1} t^{jd} = \sum_{m \geq 1} \left(\sum_{\substack{j \mid m, \\ j \geq 2}} b_j \right) t^m.$$

This is valid provided that $\beta|t| < 1$ and gives us, via Möbius inversion described in Sect. 6.1, the value of the complex numbers b_j. Still following the previous argument and using the absolute convergence to move the summands as we deem necessary, we write

$$\sum_{p>P} h(1/p) \log p = \sum_{j \geq 2} b_j \sum_{p>P} \frac{\log p}{p^j - 1} = -\sum_{j \geq 2} b_j \frac{\zeta'_P(j)}{\zeta_P(j)} .$$

To truncate this expression at $j \leq J$, we first note that

$$|\zeta_P(j)| \geq 1 - \sum_{n \geq 2} \frac{1}{n^j} \geq 1 - \sum_{n \geq 2} \frac{1}{n^2} = 2 - \frac{\pi^2}{6} \geq 1/3 .$$

We proceed with

$$|\zeta'_P(j)| \leq \sum_{\substack{n>P, \\ p|n \Rightarrow p>P}} \frac{\log n}{n^j} .$$

When $j \log P \geq 1$, the function $n \mapsto (\log n)/n^j$ is non-increasing, hence

$$|\zeta'_P(j)| \leq \frac{\log P}{P^j} + \int_P^\infty \frac{(\log t)dt}{t^j} \leq \left(\frac{j}{j-1} \frac{\log P}{P} + \frac{1}{(j-1)^2} \right) \frac{1}{P^{j-1}} \leq \frac{2P}{P^j}$$

since $(\log P)/P \leq 1/e$. We now notice that

$$|b_j| \leq C \sum_{d \leq j} \beta^d \leq 2C\beta^j .$$

This finally gives us

$$\sum_{j \geq J} \frac{|b_j|}{P^j} \leq C \frac{|\beta/P|^J}{1 - |\beta/P|} .$$

This concludes the proof. $\square$

Exercise 9-18.

1 ⋄ Show that, for every complex number t such that $|t| < 1$, we have $\dfrac{1}{1 - t + t^2} = 1 + t + \sum_{k \geq 2} c_k t^k$, where (c_k) is defined by the recursion $c_k - c_{k-1} + c_{k-2} = 0$ and $c_0 = c_1 = 1$. Prove also that $|c_k| \leq 1$.

2 ⋄ Adapt the Sage script given in Sect. 9.5 to compute $\displaystyle\sum_{p \geq 2} \frac{\log p}{p^2 - p + 1}$.

This same constant is computed in the Postscript file available on the website [6] by X. Gourdon & P. Sebah. The constants we have computed arise if one wants to use (10.6) in practice. The readers may have a look at the next displayed equation there to see that in case of functions close to the divisor function, the convolution method also asks for sums of the type

$$\sum_{p>P} h(1/p)(\log p)^2 \qquad (9.24)$$

that one might want to compute with high numerical precision. We leave it to the interested readers to pursue this in detail.

References

[1] O.-M. Camburu et al. "Cyclotomic coefficients: gaps and jumps". In: *J. Number Theory* 163 (2016), pp. 211–237. https://doi.org/10.1016/j.jnt.2015.11.020 (cit. on p. 96).

[2] H. Cohen. "High precision computations of Hardy-Littlewood constants". In: *preprint* (1996), pp. 1–19. www.math.u-bordeaux1.fr/%5C~cohen/hardylw.dvi (cit. on p. 83).

[3] G. Dahlquist. "On the analytic continuation of Eulerian products". In: *Ark. Mat.* 1 (1952), pp. 533–554 (cit. on p. 83).

[4] P. Flajolet and I. Vardi. "Zeta Function expansions of Classical Constants". In: *preprint* (1996), pp. 1–10. http://algo.inria.fr/flajolet/Publications/landau.ps (cit. on p. 96).

[5] J.W.L. Glaisher. "On the sums of the inverse powers of the prime numbers." In: *Quart. J.* 25 (1891), pp. 347–362 (cit. on p. 95).

[6] X. Gourdon and P. Sebah. "Constants from number theory". In: http://numbers.computation.free.fr/Constants/constants.html (2010). http://numbers.computation.free.fr/Constants/Miscellaneous/constantsNumTheory.ps (cit. on p. 99).

[7] P. Kurlberg and C. Pomerance. "On a problem of Arnold: the average multiplicative order of a given integer". In: *Algebra Number Theory* 7.4 (2013), pp. 981–999. https://doi.org/10.2140/ant.2013.7.981 (cit. on p. 92).

[8] P. Lindqvist and J. Peetre. "On a number-theoretic sum considered by Meissel-a historical observation". In: *Nieuw Arch. Wisk. (4)* 15.3 (1997), pp. 175–179 (cit. on p. 96).

[9] M. Mazur and B.V. Petrenko. "Representations of analytic functions as infinite products and their application to numerical computations". In: *Ramanujan J.* 34.1 (2014), pp. 129–141. https://doi.org/10.1007/s11139-013-9546-3 (cit. on p. 83).

[10] P. Moree. "Approximation of singular series and automata". In: *Manuscripta Math.* 101.3 (2000). With an appendix by Gerhard Niklasch, pp. 385–399. https://doi.org/10.1007/s002290050222 (cit. on pp. 83, 90).

[11] P. Moree and P. Stevenhagen. "A two-variable Artin conjecture". In: *J. Number Theory* 85.2 (2000), pp. 291–304. https://doi.org/10.1006/jnth.2000.2547 (cit. on p. 92).

[12] D. Shanks. "On numbers of the form n^4 1". In: *Math. Comput.* 15 (1961), pp. 186–189 (cit. on p. 85).

[13] P.J. Stephens. "Prime divisors of second order linear recurrences. II". In: *J. Number Theory* 8.3 (1976), pp. 333–345. https://doi.org/10.1016/0022-314X(76)90011-1 (cit. on p. 91).

[14] P.J. Stephens. "Prime divisors of second-order linear recurrences. I". In: *J. Number Theory* 8.3 (1976), pp. 313–332. https://doi.org/10.1016/0022-314X(76)90010-X (cit. on p. 91).

[15] E.Witt. "Treue Darstellung Liescher Ringe". In: *J. Reine Angew. Math.* 177 (1937), pp. 152–160. https://doi.org/10.1515/crll.1937.177.152 (cit. on p. 85).

Chapter 10
Some Practice

We detail four examples in this chapter and provide a wide-ranging ready-made result. The way most of this chapter is written differs rather seriously from the prevailing one in the previous chapters: we essentially enunciate things and expect the readers to complete the proofs.

10.1 Four Examples

Example 1. Let us consider the function $f_2(n) = \mu^2(n)/\varphi(n)$. We readily find that

$$
\begin{aligned}
D(f_2, s) &= \prod_{p \geq 2} \left(1 + \frac{1}{(p-1)p^s} \right) \\
&= \prod_{p \geq 2} \left(1 - \frac{1}{(p-1)p^{s+1}} - \frac{1}{(p-1)p^{2s+1}} \right) \left(\frac{1}{1 - 1/p^{s+1}} \right) \\
&= C_2(s)\zeta(s+1),
\end{aligned}
$$

where $C_2(s)$ is holomorphic and bounded in absolute value above, and below away from 0 when $\Re s > -1/2$. We deduce from this expression that $D(f_2, s)$ is meromorphic when $\Re s > -1/2$ and admits a simple pole at $s = 0$.

Example 2. Let us consider the function $f_3(n) = 2^{\Omega(n)}$. We get

$$
\begin{aligned}
D(f_3, s) &= \prod_{p \geq 2} \left(1 - \frac{2}{p^s} \right)^{-1} = \prod_{p \geq 2} \left(1 + \frac{1}{p^{2s} - 2p^s} \right) \zeta^2(s) \\
&= C_3(s)\zeta^2(s),
\end{aligned}
$$

where $C_3(s)$ is holomorphic when $\Re s > 1/2$. This expression shows that $D(f_3, s)$ is meromorphic when $\Re s > 1/2$ and has a double pole at $s = 1$.

O. Ramaré, *Excursions in Multiplicative Number Theory*, Birkhäuser Advanced Texts Basler Lehrbücher, https://doi.org/10.1007/978-3-030-73169-4_10

Example 3. Let us consider the function

$$\mathcal{F}(s) = \prod_{p \geq 2} \left(1 - \frac{1}{p^s} + \frac{1}{p^{2s}}\right). \tag{10.1}$$

The Euler product is absolutely convergent in the sense of Godement when $\Re s > 1$, but is it expressible as a Dirichlet series? The problem is to find an arithmetical function f_4 such that $D(f_4, s) \overset{?}{=} \mathcal{F}(s)$. We propose three ways to do so. Firstly, the multiplicative function defined on prime powers by

$$f_4(p) = -1, \quad f_4(p^2) = 1, \quad \forall k \geq 2, f_4(p^k) = 0$$

solves our problem. Secondly, we notice that, for any indeterminate X, we have

$$(1 - X + X^2)\frac{1 - X^2}{1 - X} = \frac{1 - X^6}{1 - X^3},$$

whence we get

$$\mathcal{F}(s) = \frac{\zeta(3s)}{\zeta(6s)} \frac{\zeta(2s)}{\zeta(s)}.$$

This indeed answers our question, since

$$\zeta(3s) = D(\mathbb{1}_{X^3}, s), \ \zeta(6s)^{-1} = D(\mu \cdot \mathbb{1}_{X^6}, s), \ \zeta(2s) = D(\mathbb{1}_{X^2}, s)$$

and therefore $f_4 = \mathbb{1}_{X^3} \star (\mu \cdot \mathbb{1}_{X^6}) \star \mathbb{1}_{X^2} \star \mu$ solves our problem. A third way of writing this is possible by employing the unitary convolution, and indeed it is immediate that

$$f_4 = \mu \,\square\, (\mu^2 \cdot \mathbb{1}_{X^2}) \tag{10.2}$$

is a solution, hence *the* solution.

Example 4. Let us now consider the Dirichlet series

$$L(s) = \sum_{n \geq 1} \frac{(-1)^{n-1}}{n^s}. \tag{10.3}$$

This series converges absolutely when $\Re s > 1$, and is simply convergent as soon as $\Re s > 0$. We even have $L(s) = (2^{1-s} - 1)\zeta(s)$, which in particular implies that this series admits a holomorphic continuation to the whole complex plane.

10.2 Expansion as Dirichlet Series

We have come up with the functions $C_2(s)$ and $C_3(s)$, but we have not shown that they are indeed Dirichlet series. We make up for this now. Our aim is to write

$$C_i(s) = D(g_i, s) = \sum_{n \geq 1} \frac{g_i(n)}{n^s}, \tag{10.4}$$

when $i = 2$ or $i = 3$ and where the functions g_i are expected to be multiplicative. This is swiftly done and the exact values of $g_i(p^k)$ are obtained by identifying the coefficients in the local factor of the series. Indeed, we have

$$\begin{cases} g_2(p) = g_2(p^2) = -\dfrac{1}{p(p-1)} \\ g_2(p^k) = 0, \quad (k \geq 3) \end{cases} \qquad \begin{cases} g_3(p) = 0 \\ g_3(p^k) = 2^{k-2}, \quad (k \geq 2). \end{cases}$$

When $i \in \{1, 2, \}$, the series $D(|g_i|, s)$ converge there where we have shown that $D(g_i, s)$ exist, namely, respectively, when $\Re s > -1/2$ and when $\Re s > 1/2$.

10.3 Expansion as Convolution Product

We now hope the readers will feel comfortable with the expansion

$$f_2(n) = \sum_{\ell m = n} g_2(\ell)/m \,,$$

as well as with

$$f_3(n) = \sum_{\ell m = n} g_3(\ell) d(m) \,.$$

10.4 A General Lemma

The next lemma generalizes a result of Riesel & Vaughan from [6] and can be found in [4]. One of its main advantages is its completely explicit nature.

Lemma 10.1

Let g, h and k be three complex-valued functions. Let us assume that $g = h \star k$, $D(|h|, s)$ is absolutely convergent for $\Re(s) \geq -1/3$ and, finally, that there exist four constants A, B, C and D, such that

$$\sum_{n \leq t} k(n) = A \log^2 t + B \log t + C + O^*(Dt^{-1/3}) \text{ when } t > 0 \,.$$

Then, for any $t > 0$, we have

$$\sum_{n \leq t} g(n) = u \log^2 t + v \log t + w + O^*\left(Dt^{-1/3} D(|h|, -1/3)\right),$$

where $u = AH(0)$, $v = 2AH'(0) + BH(0)$ and $w = AH''(0) + BH'(0) + CH(0)$ with $H(s) = D(h, s)$. In addition, we have

$$\sum_{n \leq t} ng(n) = Ut \log t + Vt + W + O^*\left(2.5Dt^{2/3} D(|h|, -1/3)\right),$$

where

$$\begin{cases} U = 2AH(0), \quad V = -2AH(0) + 2AH'(0) + BH(0), \\ W = A(H''(0) - 2H'(0) + 2H(0)) + B(H'(0) - H(0)) + CH(0). \end{cases}$$

Proof. We write $\sum_{\ell \leq t} g(\ell) = \sum_m h(m) \sum_{n \leq t/m} k(n)$. All the subsequent regularity of our expressions comes from the fact that it is not necessary to restrict m by the condition $m \leq t$ in the $\sum_m h(m)$. The proof of the first claim readily follows.

In order to estimate $\sum_{\ell \leq t} \ell g(\ell)$ for $t > 0$, we simply write

$$\sum_{\ell \leq t} \ell g(\ell) = t \sum_{\ell \leq t} g(\ell) - \int_1^t \sum_{\ell \leq u} g(\ell) du,$$

and introduce therein the asymptotic expression of $\sum_{\ell \leq u} g(\ell)$ we have just obtained. $\square$

In order to apply the preceding lemma, we shall require the next one.

Lemma 10.2

For every $t > 0$, we have

$$\sum_{n \leq t} \frac{1}{n} = \log t + \gamma + O^*\left(0.9105 t^{-1/3}\right).$$

and

$$\sum_{n \leq t} \frac{d(n)}{n} = \frac{1}{2} \log^2 t + 2\gamma \log t + \gamma^2 - 2\gamma_1 + O^*\left(1.641 t^{-1/3}\right),$$

where $d(n)$ is the number of divisors of n and

$$\gamma_1 = \lim_{n \to \infty} \left(\sum_{m \leq n} \frac{\log m}{m} - \frac{\log^2 n}{2} \right).$$

Recall that $\gamma_1 = -0.07281\,58454\,83676\,72486\ldots$.

Proof. The proof of the second part of this lemma comes from [6, Corollary 2.2]. It relies on the hyperbola formula described in Chap. 11, see Exer. 11-5.

As for the first formula, we recall the classical estimate

$$\left| \sum_{n \leq t} \frac{1}{n} - \log t - \gamma \right| \leq \frac{7}{12t} \quad \text{when } t \geq 1. \tag{10.5}$$

We prove a better inequality (the constant $7/12$ is replaced by $6/11$) in [3, Lemma 2.1]. When $0 < t < 1$, we choose $a > 0$ such that $\log t + \gamma + a\,t^{-1/3} \geq 0$. This function decreases from 0 to $(a/3)^3$ and increases afterwards. This gives us the minimal value $a = 3 \exp(-\gamma/3 - 1) \leq 0.9105$. $\square$

In practice, the function g is multiplicative and satisfies $g(p) = b/p + o(1/p)$ with $b = 1$ or 2. In such a case, we select $\sum k(n)n^{-s} = \zeta(s+1)^b$, while h is the multiplicative function determined by $\sum h(n)n^{-s} = \sum g(n)n^{-s}\zeta(s+1)^{-b}$. Let us now give expressions for $H(0)$, $H'(0)$ and $H''(0)$ that will enable precise computations. First, we have

$$H(0) = \prod_p \left(1 + \sum_{m \geq 1} h(p^m)\right),$$

and we have seen how to compute an Euler product. The case of $H'(0)$ relies on a simple trick. Indeed computing the log-derivative of H show that

$$\frac{H'(0)}{H(0)} = \sum_p \frac{\sum_m mh(p^m)}{1 + \sum_m h(p^m)}(-\log p). \tag{10.6}$$

This process has been used in Exer. 8-4. We can proceed similarly for $H''(0)$:

$$\frac{H''(0)}{H(0)} = \left(\frac{H'(0)}{H(0)}\right)^2 + \sum_p \left\{\frac{\sum_m m^2 h(p^m)}{1 + \sum_m h(p^m)} - \left(\frac{\sum_m mh(p^m)}{1 + \sum_m h(p^m)}\right)^2\right\} \log^2 p.$$

We then approximate both series by restricting the variable p up to some P. The tail of both series (i.e. the part with $p > P$) may be bounded above in absolute value by extending the sum over every prime $p > P$ to one over every integer $n > P$. If one wants to be more precise, Sect. 9.7 offers in most cases a robust way to compute these sums.

Certifying an accurate value for $D(|h|, -1/3)$ may be difficult. Here is a modification of Theorem 9.7 that is suitable for this purpose.

Theorem 10.1

Let $F, G \in \mathbb{Z}[X]$ be two coprime polynomials satisfying $F(0) = G(0) = 1$ and such that $(F(X) - G(X))/X^4 \in \mathbb{Z}[X]$. Let $\beta \geq 2$ be an upper bound for the maximum modulus of the inverses of the roots of F and of G. Let $P > \beta^3$ be a parameter. For any parameter $J \geq 4$, we have

$$\prod_{p > P} \frac{F(1/p^{1/3})}{G(1/p^{1/3})} = \prod_{4 \leq j \leq J} \zeta_P(j/3)^{b_G(j) - b_F(j)} \times I,$$

where the integers $b_G(j)$ and $b_F(j)$ are defined in Theorem 9.5 and

$$\max(I, 1/I) \leq \exp\left(\frac{(\deg F + \deg G)(\beta/P^{1/3})^J P^{1/3}}{(1 - \beta/P^{1/3})J}\right).$$

The partial zeta-function $\zeta_P(s)$ is defined in (9.12).

10.5 A Final Example

Lemma 10.3

For every $x > 0$ and every integer $d \geq 1$, we have

$$\sum_{\substack{n \leq x \\ (n,d)=1}} \frac{\mu^2(n)}{\varphi(n)} = \frac{\varphi(d)}{d} \left\{ \log x + \gamma + \sum_{p \geq 2} \frac{\log p}{p(p-1)} + \sum_{p \mid d} \frac{\log p}{p} \right\} + O^* \left(7.38 \frac{k_1(d)}{x^{1/3}} \right),$$

where

$$k_1(d) = \prod_{p \mid d} (1 + p^{-2/3}) \left(1 + \frac{p^{1/3} + p^{2/3}}{p(p-1)} \right)^{-1}.$$

Remark 10.1

The function we are summing appears classically in sieve theory and is often denoted by $G_d(X)$. Its Dirichlet series reads

$$\sum_n \frac{\mu^2(n)}{\varphi(n) n^{s-1}} = \frac{\zeta(s)}{\zeta(2s)} \prod_{p \geq 2} \left(1 + \frac{1}{(p-1)(p^s + 1)} \right)$$

so that, anticipating on latter chapters, we see that the remainder term $O(x^{-1/2})$ is admissible (the method we present could give $O(x^{-1/2} \log^2 x)$), and that we cannot expect anything better than $O(x^{-3/4})$. See Exer. 22-6. Theorem 3.1 of [5] shows in particular that

$$\sum_{n \leq x} \frac{\mu^2(n)}{\varphi(n)} = \log x + \gamma + \sum_{p \geq 2} \frac{\log p}{p(p-1)} + O^* \left(\frac{11}{\sqrt{x \log x}} \right).$$

In [7] Eq. (2.11), Rosser and Schoenfeld gave the approximation

$$\gamma + \sum_{p \geq 2} \frac{\log p}{p(p-1)} = 1.332\,582\,275\,733\,220\ldots \tag{10.7}$$

The readers will find in Exer. 9-17 how to calculate these decimals on their own.

Proof. Define a multiplicative function h_d by

$$h_d(p) = \frac{1}{p(p-1)}, \quad h_d(p^2) = \frac{-1}{p(p-1)}, \quad h_d(p^m) = 0 \quad \text{if } m \geq 3,$$

for prime numbers p that do not divide d, and $h_d(p^m) = \mu(p^m)/p^m$ for every $m \geq 1$ when p divides d. We have

$$\sum_{n \geq 1} \frac{h_d(n)}{n^s} \zeta(s+1) = \sum_{\substack{n \geq 1 \\ (n,d)=1}} \frac{\mu^2(n)}{\varphi(n)n^s}.$$

Lemma 10.2 applies. We skip the details and only mention that

$$\prod_{p \geq 2} \left(1 + \frac{p^{1/3} + p^{2/3}}{p(p-1)} \right) \leq 8.1 . \tag{10.8}$$

Exercise 10-1. In the proof above, it is rather difficult to get a decent approximate value for (10.8). Use Theorem 10.1 to prove that this product equals 8.08355 85539 63356 08308 08933 52206 72340 67219 27332 46118 . . .

Exercise 10-2.

1 ◇ Show that $\displaystyle\sum_{n \leq x} \varphi(n) = (3/\pi^2)x^2 + O(x^{5/3})$ for $x \geq 2$.

2 ◇ Show that $\displaystyle\sum_{\substack{m,n \leq x \\ \gcd(m,n)=1}} 1 = -1 + 2 \sum_{n \leq x} \varphi(n)$ and then conclude that the sum on the left-hand side is equal to $(6/\pi^2)x^2 + O(x^{5/3})$.

Exercise 10-3. Prove that $\displaystyle\sum_{n \leq x} \sigma(n) = (\pi^2/12)x^2 + O(x^{5/3})$ for $x \geq 2$.

Exercise 10-4.

1 ◇ Let n be a square-free integer. Show that $\dfrac{1}{\varphi(n)} = \displaystyle\sum_{\ell \in E_n} \dfrac{1}{\ell}$, where E_n stands for the set of integers whose prime factors divide n.

2 ◇ Show that $\displaystyle\sum_{n \leq x} \frac{\mu^2(n)}{\varphi(n)} = \sum_{n \leq x} \mu^2(n) \sum_{\ell \in E_n} \frac{1}{\ell} \geq \sum_{\ell \leq x} \frac{1}{\ell}$.

3 ◇ From the above, deduce the classical inequality $\displaystyle\sum_{n \leq x} \frac{\mu^2(n)}{\varphi(n)} \geq \log x$.

Exercise 10-5.

1 ◇ Show that for any integer $n \geq 1$ we have $\dfrac{\pi^2}{6} \dfrac{\varphi(n)}{n} = \displaystyle\sum_{d \geq 1} \dfrac{\mu(d)c_d(n)}{\prod_{p|d}(p^2-1)}$ where $c_d(n)$ is the Ramanujan sum defined in Eq. (1.9).

2 ◇ Show that, when d is square-free, we have $\prod_{p|d}(p^2-1) = \sigma(d)\varphi(d)$.

Exercise 10-6.

1 ⋄ Show that the series $\sum\limits_{n \geq 1} \dfrac{\mu^2(n)}{\sigma(n)\varphi(n)}$ converges towards $\pi^2/6$.

2 ⋄ Use the ideas of the Exer. 10-4 to establish that, for every $x \geq 1$, we have
$$\sum_{n \leq x} \frac{\mu^2(n)}{\sigma(n)\varphi(n)} = \frac{\pi^2}{6} + O^*(1/x).$$

10.6 Handling Square-free Integers

The condition "n is square-free" can often be treated by using the following convolution identity:
$$\sum_{d^2 \mid n} \mu(d) = \begin{cases} 1 & \text{when } \mu^2(n) = 1, \\ 0 & \text{otherwise.} \end{cases} \tag{10.9}$$

Proof. Since both functions, the first one associating to n the value $\sum_{d^2 \mid n} \mu(d)$ and the second one being the characteristic function of the square-free integers, are multiplicative (see Exer. 10-7), it is enough to check the validity of the claimed identity on powers of primes, where it is obvious.   □

Exercise 10-7.

1 ⋄ Let f be a multiplicative function. Show that the map $n \mapsto \mathbb{1}_{\chi^2}(n)f(\sqrt{n})$ is equally multiplicative.

2 ⋄ Let f and g be two multiplicative functions. Show that the function $n \mapsto \sum\limits_{d^2 \mid n} f(d)g(n/d^2)$ is multiplicative.

The interest of identity (10.9) lies in that the variable d is much smaller than n. Here is a usage of it:
$$\sum_{n \leq x} \mu^2(n) = \sum_{n \leq x} \sum_{d^2 \mid n} \mu(d) = \sum_{d \leq \sqrt{x}} \mu(d) \sum_{\substack{n \leq x \\ d^2 \mid n}} 1$$
$$= \sum_{d \leq \sqrt{x}} \mu(d)\left(\frac{x}{d^2} + O^*(1)\right).$$

We continue by noticing that, since $|\mu(d)| \leq 1$, we have
$$\sum_{d \leq \sqrt{x}} \mu(d)O^*(1) = O^*(\sqrt{x})$$

and

$$\sum_{d \le \sqrt{x}} \frac{\mu(d)}{d^2} = \sum_{d \ge 1} \frac{\mu(d)}{d^2} + O^*\left(\sum_{d > \sqrt{x}} \frac{1}{d^2}\right).$$

A comparison to an integral tells us that the last error term is at most $1/(\sqrt{x} - 1)$. Finally we get

$$\sum_{n \le x} \mu^2(n) = \frac{6}{\pi^2} x + O(\sqrt{x}), \tag{10.10}$$

since

$$\sum_{d \ge 1} \mu(d)/d^2 = \prod_{p \ge 2} (1 - p^{-2}) = 1/\zeta(2) = 6/\pi^2. \tag{10.11}$$

It is easy enough to get an explicit error term in (10.10). It is however much more difficult to obtain a *small* constant therein. H. Cohen*, F. Dress and M. El Marraki addressed this question, after several others, in [1], and obtained that this error term, in absolute value, is $\le 0.02767\sqrt{x}$ when $x \ge 438\,653$. Even though this paper is based on research done more than 10 years earlier, no one has since been able to improve on this bound; the interested readers may consult [5]. In Pari/GP, the control statement `forsquarefree` loops over square-free integers (but look carefully at its syntax).

Exercise 10-8. Show that there exists a constant C such that, for any $x \ge 1$, we have $\displaystyle\sum_{n \le x} \frac{\mu^2(n)}{n} = \frac{6}{\pi^2} \log x + C + O((\log x)/\sqrt{x})$.

Exercise 10-9. Give an asymptotic for $\displaystyle\sum_{\substack{n \le x \\ (n,d)=1}} \frac{\mu^2(n)}{n}$ where d is some integer parameter.

Exercise 10-10. Let $x \ge 2$ and $h < \sqrt{x}$ be two real parameters. Let D be a positive parameter that we will choose at the end of the argument.

1 ◇ Show that the number N of square-free integers in the interval $(x - h, x]$ is given by $N = N_1 + N_2$ where

$$N_1 = \sum_{\substack{x-h<d^2\ell \le x, \\ d \le D}} \mu(d), \quad N_2 = \sum_{\substack{x-h<d^2\ell \le x, \\ d > D}} \mu(d).$$

2 ◇ Prove that $N_1 = \dfrac{6}{\pi^2} h + O^*(2hD^{-1} + D)$.

3 ◇ Prove that in $|N_2| \le x/D^2$.

---

* This is the same Henri Cohen who initiated the Pari/GP project, in Basic!

4 ◇ Select $D = x^{1/3}$ and show that, when $x \geq 7000$ the interval $(x - 4x^{1/3}, x]$ contains a square-free number.

The paper [2] by M. Filaseta is a good read on this topic.

References

[1]   H. Cohen, F. Dress, and M. El Marraki. "Explicit estimates for summatory functions linked to the Möbius μ-function". In: *Funct. Approx. Comment. Math.* 37.part 1 (2007), pp. 51–63. https://doi.org/10.7169/facm/1229618741 (cit. on p. 109).

[2]   M. Filaseta. "Short interval results for squarefree numbers". In: *J. Number Theory* 35.2 (1990), pp. 128–149. https://doi.org/10.1016/0022-314X(90)90108-4 (cit. on p. 109)

[3]   Akhilesh P and O. Ramaré. "Explicit averages of non-negative multiplicative functions: going beyond the main term". In: *Colloq. Math.* 147.2 (2017), pp. 275–313. https://doi.org/10.4064/cm6080-4-2016 (cit. on p. 104).

[4]   O. Ramaré. "On Snirel'man's constant". In: *Ann. Scu. Norm. Pisa* 21 (1995), pp. 645–706 (cit. on p. 103).

[5]   O. Ramaré. "Explicit average orders: news and problems". In: *Number theory week 2017.* Vol. 118. Banach Center Publ. Polish Acad. Sci. Inst. Math., Warsaw, 2019, pp. 153–176 (cit. on pp. 106, 109).

[6]   H.I. Riesel and R.C. Vaughan. "On sums of primes". In: *Arkiv för mathematik* 21 (1983), pp. 45–74 (cit. on pp. 103, 104).

[7]   J.B. Rosser and L. Schoenfeld. "Approximate formulas for some functions of prime numbers". In: *Illinois J. Math.* 6 (1962), pp. 64–94 (cit. on p. 106).

Chapter 11
Dirichlet's Hyperbola Principle

During a lecture given at the German Academy of Sciences, in 1849, Dirichlet described an efficient manner of computing the mean value of some arithmetical functions. His talk is reproduced in [3]. The procedure described here is now called *Dirichlet's Hyperbola Principle*.

11.1 First Error Term for the Divisor Function

Let us start by describing the state of the art before Dirichlet entered the scene.

Theorem 11.1

When $x \geq 1$, we have $\sum_{n \leq x} d(n) = x \log x + O^*(2x)$.

Proof Using the definition of $d(n)$ we write

$$\sum_{n \leq x} d(n) = \sum_{\ell \leq x} \sum_{m \leq x/\ell} 1 = \sum_{\ell \leq x} \left(\frac{x}{\ell} + O^*(1) \right) = x \log x + O^*(2x),$$

where we employed (7.4) to estimate $\sum_{\ell \leq x} 1/\ell$. □

The error term in the theorem above is not much smaller than the main term and, indeed, the main term we have found does not give a very precise estimate of the sum. For instance, when $x = 10^6$, we have $\sum_{n \leq 10^6} d(n) = 13\,970\,034$ whereas $10^6 \log(10^6) = 13\,815\,510.557\ldots$. The challenge is to get a better approximation in this problem.

© The Author(s), under exclusive license to Springer Nature Switzerland AG 2022
O. Ramaré, *Excursions in Multiplicative Number Theory*, Birkhäuser Advanced Texts Basler Lehrbücher, https://doi.org/10.1007/978-3-030-73169-4_11

11.2 The Dirichlet Hyperbola Principle

Here is the main result of this chapter. It is in essence a simple combinatorial identity, but it has so many applications that it deserves a full chapter.

Lemma 11.1

Let f and g be two arithmetical functions. Let L, M and N be positive real parameters such that $L = MN$. We have

$$\sum_{\ell \leq L} (f \star g)(\ell) = \sum_{m \leq M} f(m) \sum_{n \leq L/m} g(n) + \sum_{n \leq N} g(n) \sum_{m \leq L/n} f(m)$$

$$- \left(\sum_{m \leq M} f(m) \right) \left(\sum_{n \leq N} g(n) \right).$$

Fig. 11.1: We are initially summing over the domain $\{(m, n)/mn \leq L\}$. This domain is the union of the two partial domains $\{(m, n)/mn \leq L$ and $m \leq M\}$ and $\{(m, n)/mn \leq L$ and $n \leq N\}$. The points in the intersection $\{(m, n)/mn \leq L$ and $m \leq M$ and $n \leq N\}$ are counted twice, so we are to correct the formula by removing once the contribution of these points.

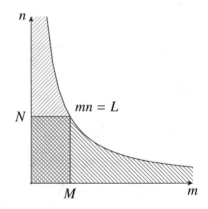

Proof We write

$$\sum_{\ell \leq L} (f \star g)(\ell) = \sum_{mn \leq L} f(m)g(n)$$

$$= \sum_{m \leq M} f(m) \sum_{n \leq L/m} g(n) + \sum_{M < m \leq L} \sum_{n \leq L/m} f(m)g(n)$$

$$= \sum_{m \leq M} f(m) \sum_{n \leq L/m} g(n) + \sum_{n \leq N} g(n) \sum_{M < m \leq L/n} f(m),$$

which concludes the proof. □

We can represent the method graphically, as illustrated in Fig. 11.1.

11.3 A Better Remainder in the Divisor Problem

> **Theorem 11.2**
>
> We have $\sum_{n \le x} d(n) = x \log x + (2\gamma - 1)x + O(\sqrt{x})$.

Now for $x = 10^6$ we check that $10^6 \log(10^6) + (2\gamma - 1)10^6 = 13\,969\,941.887\ldots$, which differs from the correct value just by $92.112\ldots$! Since $\sqrt{10^6} = 1000$, we conclude that the error term is probably of even smaller magnitude than what we prove here. Studying this error term is a central problem in analytic number theory. At the time of writing, M.N. Huxley has obtained in [8] the best bound, which is $O_\epsilon(x^{131/416+\epsilon})$, valid for any positive ϵ. We have $\frac{131}{416} = 0.3149\ldots$.

Proof Denote by $\sum_{n \le x} d(n)$ by $S(x)$. By the Dirichlet hyperbola formula we obtain

$$S(x) = \sum_{\ell m \le x} 1 = 2 \sum_{\ell \le \sqrt{x}} [x/\ell] - [\sqrt{x}]^2 ,$$

where $[y]$ stands for the integer part of real y. Using

$$[y] = y - \tfrac{1}{2} - \{y\} + \tfrac{1}{2} , \tag{11.1}$$

we get

$$S(x) = 2x \sum_{\ell \le \sqrt{x}} 1/\ell - [\sqrt{x}] - [\sqrt{x}]^2 - 2 \sum_{\ell \le \sqrt{x}} \left(\{x/\ell\} - \tfrac{1}{2} \right) . \tag{11.2}$$

Clearly the last sum is small, but we still have to work out the first three terms. Notice that on writing $[\sqrt{x}] = \sqrt{x} - \{\sqrt{x}\}$ we have

$$[\sqrt{x}] + [\sqrt{x}]^2 = \sqrt{x} + x - 2\sqrt{x}\{\sqrt{x}\} + \{\sqrt{x}\}^2 - \{\sqrt{x}\} .$$

Using the Euler–Maclaurin formula for the partial harmonic sum we have

$$\sum_{\ell \le \sqrt{x}} 1/\ell = \log[\sqrt{x}] + \gamma + \frac{1}{2[\sqrt{x}]} + O(1/\sqrt{x}) .$$

Hence, on multiplying by $2x$ we get

$$2x \sum_{\ell \le \sqrt{x}} \frac{1}{\ell} = 2x \log[\sqrt{x}] + 2\gamma x + \frac{x}{[\sqrt{x}]} + O(x/\sqrt{x})$$

$$= x \log x + 2\gamma x + 2x \log\big(1 - \{\sqrt{x}\}/\sqrt{x}\big) + \frac{\sqrt{x}}{1 - \{\sqrt{x}\}/\sqrt{x}} + O(1)$$

$$= x \log x + 2\gamma x - \sqrt{x}\{\sqrt{x}\} + \sqrt{x} + O(1) ,$$

which concludes the proof.   □

Here is a script to tabulate the error term in Theorem 11.2 and compare it with the much smaller $x^{1/4}$.

```
{Tabulate(lowerlimit, upperlimit, step = 1000) =
   my(res = 0, aux, maxi = 0, where = 1);
   for( n = 1, lowerlimit -1, res += numdiv(n));
   for( n = lowerlimit, upperlimit,
     res += numdiv(n);
     aux = (res -(n*log(n)+(2*Euler-1)*n))/n^(1/4);
     if( (n-lowerlimit)%step == 0,
         print("n = ", n, " remainder/n^(1/4) <= ", aux),);
     if(abs(aux) > maxi, maxi = aux; where = n,));
   print("-------------------------");
   print("For ", lowerlimit, " <= n <= ", upperlimit);
   print("|sum(m <= n)d(m)-(nlog n +(2Euler-1) n)|/n^(1/4)<=");
   print(maxi, " reached at ", where);
}
Tabulate(1000,1000000, 10000);
```

This error term is widely believed to be $\ll_\varepsilon x^{1/4+\varepsilon}$ for every positive ε. This should be the right order of magnitude, as K. Soundararajan, improving on a result of Hardy in [6], showed in [11] that it is often at least as large as $(x \log x)^{1/4}$ times a function going very slowly to infinity. The readers who want to know more on this issue should read about the Voronoi Summation Formula. Pari/GP proposes a faster way to achieve the above computation: replace lines 4 and 5 by the more mysterious code:

```
forfactored(N = lowerlimit, upperlimit,
   my(n = N[1]);
   res += vecprod( [k+1 | k<-N[2][,2]~]);
```

Exercise 11-1. Recover the result of Exer. 6-6, i.e. $\displaystyle\sum_{n \leq x} \frac{\mu(n)}{n} \log \frac{x}{n} \ll 1$, by studying the expression $\sum_{nm \leq x} \mu(n)d(m)$.

Exercise 11-2. (From [1, Lemma 3.1]). Define $\psi(x) = \{x\} - \frac{1}{2}$, where $\{x\}$ is the fractional part of x and $\psi_2(x) = \frac{1}{2}\psi(x)^2$.

1◇ Show that, when $0 \leq y < 1$, we have $\displaystyle\int_1^{1+y} \psi(t)dt + \frac{1}{8} = \psi_2(y)$.

2◇ Show that the function given by $\mathcal{R}(x) = \frac{1}{4} - 3\psi_2(\sqrt{x}) + 4x \displaystyle\int_{\sqrt{x}}^{\infty} \frac{\psi_2(t)}{t^3} dt$ satisfies $|\mathcal{R}(x)| \leq 1/2$.

3 ◇ Show that $\displaystyle\sum_{n\le x} d(n) = x(\log x + 2\gamma - 1) - 2\sum_{n\le\sqrt{x}} \psi\left(\frac{x}{n}\right) + \mathcal{R}(x).$

4 ◇ Using the above, show that $\displaystyle\sum_{n\le x} d(n) = x(\log x + 2\gamma - 1) + O^*(\sqrt{x} + \tfrac{1}{2}).$

Exercise 11-3. The exercise considers the *Lambert series* $L_0(z) = \displaystyle\sum_{n\ge1}\frac{z^n}{1 - z^n},$ where z is a complex parameter such that $|z| < 1$.

1 ◇ Show that we have $L_0(z) = \displaystyle\sum_{n\ge1} d(n)z^n.$

2 ◇ Let $Y \ge 1$ be a real parameter. Show that $L_0\!\left(e^{-1/Y}\right) = Y(\log Y + \gamma) + O(\sqrt{Y})$

3 ◇ Invoke Exer. 6-3 to prove that, when $Y \ge 1$, we have $\displaystyle\sum_{n\ge1}(1+\mu(n))\frac{e^{-n/Y}}{1 + e^{-n/Y}} =$
$Y\log 2 + O(\sqrt{Y}).$

Exercise 11-4. (From [10]). Recall that $d^*(n)$ denotes the number of unitary divisors of n. Show that

$$\sum_{n\le x} d^*(n) = \frac{6x}{\pi^2}\left(\log x + 2\gamma - 1 - \frac{2\zeta'(2)}{\zeta(2)}\right) + O(\sqrt{x}\log x).$$

This proof is reproduced by E. Cohen in [2]. By using Exer. 1-20 and a stronger remainder term for the average order of the divisor function, A.A. Gioia and A.M. Vaidya improved this error term to $O(\sqrt{x})$ in [4]. Stronger results are available.

Exercise 11-5. Prove that, when $x \ge 1$, we have

$$\sum_{n\le x}\frac{d(n)}{n} = \frac{1}{2}\log^2 x + 2\gamma\log x + \gamma^2 - 2\gamma_1 + O^*\left(\frac{3}{2\sqrt{x}} + \frac{1}{x}\right).$$

The readers may compare this result with Lemma 10.2.

Exercise 11-6. Let $d_3(n)$ be the number of triples (n_1, n_2, n_3) such that $n_1 n_2 n_3 = n$. Show that $d_3 = \mathbb{1} \star d$ and give an approximation for $\sum_{n\le x} d_3(n)$ with an explicit error term of size $x^{2/3}\log x$. Compare this with a direct computation up to $x = 10^6$.

Further Reading

There is an extensive literature on the divisor problem and on lattice points counting. We refer specifically to the two books [9] by E. Krätzel and [7] by M.N. Huxley. A large part of the developments relies on the *exponential sums* theory; the standard reference on this latter subject is the book [5] by S.W. Graham and G. Kolesnik.

References

[1] D. Berkane, O Bordellès, and O. Ramaré. "Explicit upper bounds for the remainder term in the divisor problem". In: *Math. of Comp.* 81.278 (2012), pp. 1025–1051 (cit. on p. 114).

[2] E. Cohen. "The number of unitary divisors of an integer". In: *Amer. Math. Monthly* 67 (1960), pp. 879–880. https://doi.org/10.2307/2309455 (cit. on p. 115).

[3] J.P.G.L. Dirichlet. "Über die Bestimmung des mittleren Werthe in der Zahlentheorie". In: *Gelesen in des Akademie der Wissenchaften am 9 August 1849* (1849), 15pp (cit. on p. 111).

[4] A.A. Gioia and A.M. Vaidya. "The number of squarefree divisors of an integer". In: *Duke Math. J.* 33 (1966), pp. 797–799. http://projecteuclid.org/euclid.dmj/1077376698 (cit. on p. 115).

[5] S.W. Graham and G. Kolesnik. *van der Corput's method of exponential sums*. Vol. 126. London Mathematical Society Lecture Note Series. Cambridge University Press, Cambridge, 1991, pp. vi+120. https://doi.org/10.1017/CBO9780511661976 (cit. on p. 115).

[6] G. H. Hardy. "On Dirichlet's Divisor Problem". In: *Proc. London Math. Soc. (2)* 15 (1916), pp. 1–25. https://doi.org/10.1112/plms/s2-15.1.1 (cit. on p. 114).

[7] M.N. Huxley. *Area, lattice points, and exponential sums*. Vol. 13. London Mathematical Society Monographs. New Series. Oxford Science Publications. The Clarendon Press, Oxford University Press, New York, 1996, pp. xii+494 (cit. on p. 115).

[8] M.N. Huxley. "Exponential sums and lattice points. III". In: *Proc. London Math. Soc. (3)* 87.3 (2003), pp. 591–609. https://doi.org/10.1112/S0024611503014485 (cit. on p. 113).

[9] E. Krätzel. *Lattice points*. Vol. 33. Mathematics and its Applications (East European Series). Kluwer Academic Publishers Group, Dordrecht, 1988, p. 320 (cit. on p. 115).

[10] F. Mertens. "Ueber einige asymptotische Gesetze der Zahlentheorie." In: *Journal für die reine und angewandte Mathematik* 77 (1874), pp. 289–338. http://eudml.org/doc/148241 (cit. on p. 115).

[11] K. Soundararajan. "Omega results for the divisor and circle problems". In: *Int. Math. Res. Not.* 36 (2003), pp. 1987–1998. https://doi.org/10.1155/S1073792803130309 (cit. on p. 114).

Chapter 12
The Mertens Estimates

We need estimates for the number of primes in the initial interval. Such estimates are long known. Efforts to make them explicit started in the thirties, see, for instance, the papers [2] by R. Breusch and [27] by J.B. Rosser. For the history of prime number theory, we refer to the excellent book [19] by W. Narkiewicz.

12.1 The von Mangoldt Λ-Function

We need a function introduced by H.C.F. von Mangoldt. This mathematician published the important memoir [16] on prime numbers in which he introduced this function with the notation $L(n)$. We stick to the modern notation $\Lambda(n)$. It is defined by

$$\Lambda(n) = \begin{cases} \log p & \text{when } n = p^a \text{ with } a \geq 1, \\ 0 & \text{otherwise.} \end{cases} \tag{12.1}$$

For instance, $\Lambda(2) = \Lambda(4) = \log 2$ and $\Lambda(15) = 0$. Let us specify explicitly that $\Lambda(1) = 0$. The appearance of this function is not at all mysterious when taking the viewpoint of Dirichlet series, since, following Riemann, the readers will immediately prove by taking the logarithmic derivative of the Euler identity (see Exer. 3-18) that

$$-\frac{\zeta'(s)}{\zeta(s)} = \sum_{n \geq 1} \frac{\Lambda(n)}{n^s} \tag{12.2}$$

valid for $\Re s > 1$. In case the readers are not comfortable with this proof, as it uses logarithms of complex numbers, they may first prove this identity when $s > 1$ is a real number and conclude by the unicity of the analytic continuation.

The Λ-function enables us to differentiate the powers of the primes from the other integers and attributes them a weight that varies slowly. We shall further see in Lemma 12.2 that the contribution of the higher prime powers is negligible when compared to the one of the primes themselves.

© The Author(s), under exclusive license to Springer Nature Switzerland AG 2022
O. Ramaré, *Excursions in Multiplicative Number Theory*, Birkhäuser Advanced Texts
Basler Lehrbücher, https://doi.org/10.1007/978-3-030-73169-4_12

The second interest comes from the identity

$$\forall n \geq 1, \quad \sum_{d \mid n} \Lambda(d) = \log n, \tag{12.3}$$

where the sum is over all the divisors $d \geq 1$ of n. Of course, the readers will recognize at this stage a convolution identity, namely, $\mathbb{1} \star \Lambda = \log$ which may be readily proved by noticing first that $-\zeta'(s) = \sum_{n \geq 1} (\log n)/n^s$ and secondly that

$$\left(-\frac{\zeta'(s)}{\zeta(s)} \right) \cdot \zeta(s) = -\zeta'(s) .$$

We present now a more pedestrian argument leading to the identity (12.3).

Proof When $n = 1$, the identity is obvious. When n is larger, we decompose it in prime factors $n = p_1^{a_1} p_2^{a_2} \cdots p_K^{a_K}$ where the p_i's are distinct primes and the a_i's are integers ≥ 1. We thus get

$$\log n = a_1 \log p_1 + a_2 \log p_2 + \ldots + a_K \log p_K .$$

The divisors d of n for which $\Lambda(d) \neq 0$ are the $p_1^{b_1}$'s with $1 \leq b_1 \leq a_1$ (there are a_1 such divisors), then the $p_2^{b_2}$'s with $1 \leq b_2 \leq a_2$ (there are a_2 such divisors), and so on. Since of course $a_1 \log p_1 = \sum_{1 \leq b_1 \leq a_1} \log p_1$, the readers will easily complete the proof. $\qquad\square$

12.2 From the Logarithm to the von Mangoldt Function

Let us start with a classical lemma.

Lemma 12.1

When $x \geq 1$, we have $\displaystyle\sum_{n \leq x} \log n = x \log x - x + O^*(\log(2x))$.

We prove in Lemma 20.5 a much more precise version of this asymptotics.

Proof Let N be the integer part of x. We proceed by comparing to an integral, i.e. we use the inequalities

$$\int_{n-1}^{n} \log t \, dt \leq \log n \leq \int_{n}^{n+1} \log t \, dt .$$

These are a simple consequence of the non-decreasing property of the logarithm. On adding these inequalities, we get

$$\int_{1}^{N} \log t \, dt \leq \sum_{2 \leq n \leq N} \log n \leq \int_{2}^{N+1} \log t \, dt \tag{12.4}$$

and some more work leads to the result, provided one remembers that an anti-derivative of $x \mapsto \log x$ is $x \mapsto x \log x - x$.

Figure 12.1 proposes a graphical interpretation of the method. The total area contained in the rectangles is the quantity we are interested in. We see graphically that it is bounded above by the integral of the function on the right-hand side of (12.4) and bounded below by the integral of this same function shifted by one unit towards the right, and this is the left-hand side of (12.4).

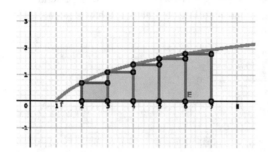

Fig. 12.1: A graphical interpretation

We now write

$$\sum_{n \leq x} \log n = \sum_{n \leq x} \sum_{d \mid n} \Lambda(d) = \sum_{d \leq x} \Lambda(d)[x/d] .$$

The idea of all that follows is based on the equality

$$\sum_{d \leq x} \Lambda(d)[x/d] = x \log x - x + O^*(\log(2x)) . \tag{12.5}$$

12.2.1 An Upper Bound à la Chebyshev

Theorem 12.1

For $x \geq 1$ we have $\displaystyle\sum_{x/2 < d \leq x} \Lambda(d) \leq 7x/10.$

Proof A direct application of (12.5) gives

$$\sum_{d \leq x} \Lambda(d)\big([x/d] - 2[x/(2d)]\big) = x \log 2 + O^*(2 \log(2x)) .$$

We observe that $[x] - 2[x/2] \geq 0$ for every real number x: indeed, it is a periodical function of period 2 which takes integer values; furthermore this quantity vanishes when x is in $[0, 1)$ and takes the value 1 when x is in $[1, 2)$. Therefore, Eq. (12.5) implies that

$$\sum_{x/2 < d \le x} \Lambda(d)\big([x/d] - 2[x/(2d)]\big) \le x \log 2 + 2 \log(2x) \,.$$

When d lies between $x/2$ and x, we have $[x/d] - 2[x/(2d)] = 1$ which leads to

$$\sum_{x/2 < d \le x} \Lambda(d) \le x \log 2 + 2 \log(2x) \,.$$

This inequality proves the theorem when $x \ge 1150$. To extend it to every real number $x \ge 1$, we first note that it suffices to check it when x is an integer. We then use the next script.

```
{Lambda(d) = my(dec = factor(d), P = dec[,1]);
    if(#P != 1, return(0), return(log(P[1])));}

{check( upperlimit ) = my(res = 0.0, lambdaxbytwo, lambdax);
    for( x = 2, upperlimit,
    if(x%2==1, lambdaxbytwo = 0, lambdaxbytwo = Lambda(x/2));
    lambdax = Lambda(x);
    res = res - lambdaxbytwo + lambdax;
    if( res <= 7*x/10,,
        print("The inequality is violated at x = ", x));}
```

□

Exercise 12-1. Show that $[2x] + [2y] \ge [x] + [y] + [x + y]$ for any real numbers x and y.

Exercise 12-2. Show that the three quantities $[y] - [y/2] - \frac{3}{2}[y/3]$, $[y] - [y/2] - [y/3] - \frac{5}{6}[y/5]$ and $[y] - [y/2] - [y/3] - [y/5] + [y/30]$ are non-negative. Show further that $[y] - [y/2] - \frac{3}{2}[y/3] \ge 1$ when $0 < y < 3$. Deduce from this fact an upper bound for $\displaystyle\sum_{x/3 < n \le x} \Lambda(d)$.

More usage of integer parts inequalities to infer bounds on $\sum_d \Lambda(d)$ can be found in the paper [4] by N. Costa Pereira as well as in Chap. 14.

Exercise 12-3. (From [12], Lemma 1) Show that when $a_1, \ldots, a_k$ are positive integers such that $\sum_i a_i^{-1} \le 1$, the inequality

$$[x] \ge \sum_{1 \le i \le k} \left[\frac{x}{a_i}\right]$$

holds true for any positive real number x and that we have equality if and only if $\sum_i a_i^{-1} = 1$ and $\mathrm{lcm}(a_i, 1 \le i \le k)|[x]$.

The readers will find in Exers. 1 to 20 of Part VIII of the book [22] by G. Pólya and G. Szegő more properties of the integer part function.

Exercise 12-4. (Beatty's Theorem) Given any irrational number $x > 1$, we define $y = x/(x - 1)$. Define the S. Beatty sequences by $u_n = [nx]$ and $v_n = [ny]$

1 ◇ Show that $u_{n+1} - u_n - [x]$ takes only the values 0 and 1. The generated sequence is known to be a *Sturmian word*.

2 ◇ Show that if a positive integer j can be written as $j = u_n$ for some n, then it cannot be written as $j = v_m$.

3 ◇ Show that every positive integer j can be written either as u_n or as v_m.

By splitting the interval $[1, x]$ in $(x/2, x]$ union $(x/4, x/2]$ union and so on, we deduce from Theorem 12.1 the following classical corollary.

Corollary 12.1 (Chebyshev's bound)

We have $\sum_{d \le x} \Lambda(d) \le 7x/5$ for every real number $x \ge 1$.

The name "Chebyshev's bound" comes from P. Chebyshev, who was the first to establish bounds of a similar nature in [3] by a method close to the one we present, though more refined and leading to better constants. Let us mention that J.B. Rosser has shown in 1941 that the maximum of the function $\sum_{d \le x} \Lambda(d)/x$ is reached at $x = 113$ and is somewhat less than 1.04.

Exercise 12-5. Deduce from Exers. 6-5 and 6-6 and the Möbius inversion formula that there exists a constant C such that, for any real number $x \ge 1$, we have

$$\sum_{n \le x} \Lambda(n) \le Cx.$$

See Chap. 14 for more on this issue and for yet another proof of the above.

A Chebyshev's bound automatically yields an estimate for the number of primes below a given bound.

Corollary 12.2

For every $x > 1$, the number of primes below x is at most $3x/(2 \log x)$.

We recall the traditional notations $\pi(x) = \sum_{p \le x} 1$ and $\psi(x) = \sum_{d \le x} \Lambda(d)$.

Proof For $x \ge 5$, we get from the previous corollary the bound $\sum_{p \le x} \log p \le 7x/5$. We remove the weight $\log p$ by summation by parts, i.e. we write

$$\frac{1}{\log p} = \frac{1}{\log x} + \int_p^x \frac{dt}{t \log^2 t},$$

which gives us

$$\pi(x) = \sum_{p \le x} \frac{\log p}{\log p} \le \frac{7x}{5 \log x} + \int_2^x \frac{7dt}{5 \log^2 t}.$$

For the last integral, we start by noticing that a (classical!) integration by parts gives us, for any $k \ge 0$, that

$$J_k = \int_2^x \frac{dt}{\log^k t} \le \frac{x}{\log^k x} + \int_2^x \frac{k\,dt}{\log^{k+1} t}$$

from which we deduce that $(\log x - 3)(\log^2 x)J_3 \le x$ and $J_2 \le x/\log^2 x + J_3$. This gives the claimed estimate when $x \ge \exp(5)$. To check the result for the remaining values of x, we need only worry about prime values of x when $x \ge e$, as $3x/(2\log x)$ increases monotonically when $x \ge e$. This is swiftly done. In the range $(1, e]$, we have $\pi(x) \le 1$ and $3x/(2\log x) \ge 4$, which is enough to end the proof.   □

Exercise 12-6. Let $p_1 < p_2 < \ldots$ be the sequence of prime numbers with $p_1 = 2$ We want to prove that $p_n \ge \frac{2}{3}n\log n$.

1 ⋄ Establish this inequality when $n \le 100$ (we have $p_{100} = 541$).

2 ⋄ Show that the function $x \mapsto \dfrac{3x}{2\log x}$ is increasing when $x \ge e$.

3 ⋄ Show that, when $x = \frac{2}{3}n\log n$, we have $\frac{3x}{2\log x} \le n$ as soon as $n \ge 5$. Conclude

Exercise 12-7. Prove that the series $\sum_p 1/(p\log p)$ is convergent.

The Λ-function gives a non-zero weight not only to the primes but also to their powers. Most of the time the contribution of the squares, cubes and higher powers can be neglected as the following lemma makes clear.

Lemma 12.2

For $x \ge 1$ we have $\displaystyle\sum_{\substack{d \le x \\ d \text{ is not a prime}}} \Lambda(d) \le 3\sqrt{x}$.

Proof Indeed, the counted integers d can be written as p^a with $a \ge 2$ and $p \le \sqrt{x}$. The primes appear in p^2, p^3, and so on up to p^a, where $a \le (\log x)/\log p$. The previous corollary concludes the proof.   □

The next exercise is inspired by a remark made at the end of [28] by A. Selberg. We take it from [26, Chap. 2].

Exercise 12-8. Let $x \ge 1$ be a real parameter and f be a function. Suppose that there exist three constants $M(f)$, $M^+(f)$ and $\delta > 0$ (that can all depend on x just like f), such that for all $y \le x$ we have

$$\sum_{d \le y} f(d) = M(f)y + O(y^{1-\delta}) \quad \text{and} \quad \sum_{d \le x^{\frac{1}{1+\delta}}} |f(d)| \ll (M^+(f) + \delta^{-1})x^{\frac{1}{1+\delta}}.$$

Assume in addition that $|M(f)| \ll M^+(f)$. We define $c_1(\lambda) = \lambda \displaystyle\int_\lambda^\infty \{t\}\frac{dt}{t^2}$.

1 ⋄ Show that $c_1(1) = 1 - \gamma$. Prove that c_1 is differentiable except at integer points, where we have $|c_1(\lambda) - 1/2| \le 1/(4\lambda)$.

2 ⋄ Show that $\displaystyle\sum_{\ell \le \lambda}\left(\frac{1}{\ell} - \frac{1}{\lambda}\right) = \int_1^\lambda [t]\frac{dt}{t^2} = \log\lambda - \int_1^\lambda \{t\}\frac{dt}{t^2}.$

3 ⋄ Show that for $\lambda > 0$, we have

$$\sum_{d \le x/\lambda} f(d)\{x/d\} = c_1(\lambda)M(f)x/\lambda + O\big((M^+(f) + \delta^{-1})x^{\frac{1}{1+\delta}}\big).$$

4 ⋄ Define $c_2(\lambda) = \lambda\displaystyle\int_\lambda^\infty \{t\}\log t\,\frac{dt}{t^2}.$ Show that for $\lambda \ge 1$ we have

$$\sum_{d \le x/\lambda} f(d)\{x/d\}\log\frac{x}{d} = c_2(\lambda)M(f)\frac{x}{\lambda} + O\big((M^+(f) + \delta^{-1})x^{\frac{1}{1+\delta}}\log x\big).$$

5 ⋄ Show that for $\lambda \ge 1$ we have $\displaystyle\sum_{d \le x/\lambda}\mu^2(d)\left\{\frac{x}{d}\right\} = \frac{6}{\pi^2}c_1(\lambda)\frac{x}{\lambda} + O(x^{2/3}).$

6 ⋄ Show that we have $\displaystyle\sum_{d \le x}\mu^2(d)\left\{\frac{x}{d}\right\}\log\frac{x}{d} = \frac{6}{\pi^2}c_2(1)x + O(x^{2/3}\log x).$

7 ⋄ Prove that $\displaystyle\sum_{d \le x}\frac{\mu(d)}{d}\left(\log\frac{x}{d} + \gamma\right) = 1 + O^*\left(\frac{7}{12}\frac{6}{\pi^2}\right) + O(x^{-1/2}).$

8 ⋄ Prove that

$$\sum_{n \le x}\Lambda(n) = \left(1 + O^*\left(\frac{7}{12}\frac{6}{\pi^2} + \frac{6}{\pi^2}c_2(1) + \frac{6}{\pi^2}\gamma(1-\gamma)\right)\right)x + O(x^{2/3}\log x).$$

where $c_2(1) = \int_1^\infty \{t\}\log t\,dt/t^2 = 0.49\ldots$

9 ⋄ Deduce from the above that $\displaystyle\sum_{n \le x}\Lambda(n) \ge 0.19x$ when x is large enough.

The readers will find in [5] by de la Vallee Poussin, in [17] by A. Mercier and in [20] and [21] by F. Pillichshammer variants around average estimates of $\{x/d\}$.

12.2.2 The Three Theorems of Mertens

Now we present three theorems in the spirit of the results of Mertens in his 1874-memoir.

Theorem 12.2

For $x \ge 2$ the following holds $\displaystyle\sum_{d \le x}\frac{\Lambda(d)}{d} = \log x - \frac{2}{3} + O^*(\frac{1}{2})$. Furthermore, we have

$$\sum_{p \le x}\frac{\log p}{p} = \log x - \frac{16}{15} + O^*(\frac{9}{10}).$$

One can show that $\sum_{d \leq x} \Lambda(d)/d$ is equal to $\log x - \gamma + o(1)$. As a consequence of Theorem 12.2 we obtain the elegant bound

$$\sum_{p \leq x} \frac{\log p}{p} \leq \sum_{d \leq x} \frac{\Lambda(d)}{d} \leq \log x . \qquad (12.6)$$

Proof We start with Eq. (12.5) and notice that $x - 1 \leq [x] \leq x$. Hence

$$7/10 + \log(2x)/x \geq \sum_{d \leq x} \Lambda(d)/d - \log x + 1 \geq -\log(2x)/x .$$

Thus, for $x \geq 120$ we get the upper bound $\log x - 4/15$ and the lower one $\log x - 21/20$, which is better that the result stated. For small x we proceed by the direct calculation.

```
{check( upperlimit ) =
  my( res = 1, aux, ispatnbyfour, ispatn, coef =20, where);
  for(n = 3, upperlimit,
    if(n%4!=0, ispatnbyfour=0, ispatnbyfour = isprime(n/4));
    ispatn = isprime(n);
    res = res - ispatnbyfour + ispatn;
    aux = (n+1)/log(n+1);
    if(res/aux < coef, coef = res/aux; where = n,));
  print("We have sum_(x/4< p <= x)1/(x/log x) >= ", coef);
  print("for ", 3 ," < x <= ", upperlimit, ".");
  print("It is reached just before x = ", where+1);
}
```

Next, we have

$$-\frac{7}{6} - \sum_{k \geq 2, p \geq 2} \frac{\log p}{p^k} \leq \sum_{p \leq x} \frac{\log p}{p} - \log x \leq -\frac{1}{6} .$$

After summing over k and a bit of numerical computation, we find that the sum over k and p above is at most 0.8. $\qquad \square$

Here is the second theorem.

Theorem 12.3 (Mertens's second Theorem)

For $x \geq 2$ we have $\sum_{p \leq x} \frac{1}{p} = \log \log x + b + O^*(4/\log x)$, where $b = 0.2614972128476427\ldots$ is the Mertens–Meissel Constant.

The argument below ensures the existence of the constant b, but says nothing about its value. However, during the proof of Theorem 12.4, we will show that

$$b = \gamma - \sum_{k \geq 2} \sum_{p \geq 2} \frac{1}{kp^k} . \qquad (12.7)$$

We have explained in Sect. 9.6 how to compute this constant with high accuracy, the value of γ being obtained through Ex. 9-14.

Proof We use summation by parts on the second estimate of Theorem 12.2. to obtain the estimate

$$\sum_{p \leq x} \frac{1}{p} = \sum_{p \leq x} \frac{\log p}{p} \left(\int_p^x \frac{dt}{t \log^2 t} + \frac{1}{\log x} \right)$$

$$= \int_2^x \sum_{p \leq t} \frac{\log p}{p} \frac{dt}{t \log^2 t} + \frac{1}{\log x} \sum_{p \leq x} \frac{\log p}{p} .$$

This leads to

$$\sum_{p \leq x} \frac{1}{p} = \log \log x - \log \log 2 + 1 + \int_2^\infty \left(\sum_{p \leq t} \frac{\log p}{p} - \log t \right) \frac{dt}{t \log^2 t}$$

$$- \int_x^\infty \left(\sum_{p \leq t} \frac{\log p}{p} - \log t \right) \frac{dt}{t \log^2 t} + O^* \left(\frac{2}{\log x} \right) .$$

The second integral is bounded above in absolute value by $2/\log x$, as we again use

$$\left| \sum_{p \leq t} \frac{\log p}{p} - \log t \right| \leq 2 .$$

The proof ends here. □

Exercise 12-9. Decide whether the series $\sum_{n \geq 1} 1/(nd(n))$ is convergent or not What about $\sum_{n \geq 1} 1/(nd(n)^2)$?

Exercise 12-10. Show that $\omega(n)$, the number of prime factors of n, is very small on average, and more precisely that $\dfrac{1}{x} \sum_{n \leq x} \omega(n) = \log \log x + O(1)$.

The previous exercise shows that the average order of $\omega(n)$ is $\log \log n$. Hardy and Ramanujan proved in [13] that much more is true and that for nearly every integer $\omega(n)$ is close to $\log \log n$. In 1934, P. Turán in [29] gave a very simple proof of this fact, later generalized by J. Kubilius in [14] and [15]. We present this proof in the next exercise.

Exercise 12-11. When f is a function on the primes $\leq N$, we define

$$M(f) = \sum_{p \leq N} f(p)/p , \quad D(f) = \sum_{p \leq N} |f(p)|^2/p .$$

1 ⋄ Show that

$$\sum_{p \leq N} |f(p)| \leq \left(D(f) \sum_{p \leq N} p \right)^{1/2} , \quad M(f) \leq \left(D(f) \sum_{p \leq N} \frac{1}{p} \right)^{1/2} ,$$

and conclude that $\sum_{p \leq N} |f(p)| M(f) = o(ND(f))$.

2 ⋄ Let f be a real-valued and non-negative function on the primes. Show that

$$\sum_{n \leq N} \left| \sum_{p|n} f(p) - M(f) \right|^2 \leq (1 + o(1))(ND(f)) . \tag{12.8}$$

This inequality, though maybe with a different constant than $1 + o(1)$, is called the *Turán-Kubilius inequality*.

3 ⋄ Let f be a real-valued function on the primes. Let $f^+ = max(0, f)$ and $f^- = -min(0, f)$. On using $f = f^+ - f^-$ and the inequality $|a + b|^2 \leq 2(a^2 + b^2)$, establish (12.8), but with the constant $1 + o(1)$ replaced by $2 + o(1)$. Proceed similarly and establish the same inequality when f is complex-valued, again with the constant $2 + o(1)$.

A different approach to this result is presented in Exer. 28-6. We have computed the *normal order* of the function $n \mapsto \omega(n)$ in the previous exercise. Much more is known and one can compute the k-th moment as well as is done in [6] by H. Delange. The next exercise brushes on this subject, by following an approach due to A. Granville & K. Soundararajan in [11].

Exercise 12-12. Let us define, for every prime p, the function f_p on integers by

$$f_p(n) = \begin{cases} 1 - \frac{1}{p} & \text{when } p|n, \\ -\frac{1}{p} & \text{when } p \nmid n. \end{cases}$$

1 ⋄ Let $d \leq x^{4/5}$ be square-free and let δ be a divisor of d. Prove that

$$\sum_{\substack{n \leq x, \\ (n,d)=\delta}} 1 = \frac{\varphi(d)x}{d\varphi(\delta)} + O\left(2^{\omega(d)-\omega(\delta)}\right) .$$

2 ⋄ Evaluate $\sum_{n \leq x} (f_{p_1} f_{p_2} f_{p_3} f_{p_4})(n)$ when the p_i's are all distinct and none is more than $x^{1/5}$.

3 ⋄ Evaluate $\sum_{n \leq x} f_{p_1}(n)^2 f_{p_2}(n) f_{p_3}(n)$ when the p_i's are all distinct and none is more than $x^{1/5}$.

4 ⋄ Evaluate $\sum_{n \leq x} f_{p_1}(n)^2 f_{p_2}(n)^2$ when the p_i's are distinct and $\leq x^{1/5}$.

5 ⋄ Evaluate $\sum_{n \leq x} f_{p_1}(n)^3 f_{p_2}(n)$ when the p_i's are distinct and $\leq x^{1/5}$.

6 ⋄ Evaluate $\sum_{n \leq x} f_{p_1}(n)^4$ when $p_i \leq x^{1/5}$.

7 ⋄ Evaluate $\sum_{n \leq x} \left| \sum_{p \leq x^{1/5}} f_p(n) \right|^4$ and finally determine $\sum_{n \leq x} \left| \omega(n) - \sum_{p \leq x} \frac{1}{p} \right|^4$.

The interested readers should continue this journey by studying the *Erdős–Kac Theorem* from [10].

Theorem 12.4 (Mertens's third Theorem)

For $x \geq 2$, we have $\prod_{p \leq x}\left(1 - \frac{1}{p}\right) \sim \frac{e^{-\gamma}}{\log x}$.

Proof In Theorem 12.3, we have proved that, when $x \geq 2$, we have

$$\sum_{p \leq x} \frac{1}{p} = \log \log x + b + O^*(4/\log x),$$

but we do not know yet the final expression for b. From this, we immediately deduce that

$$-\sum_{p \leq x} \log\left(1 - \frac{1}{p}\right) = \log \log x + c + O^*\left(\sum_{p > x}\left(\log\left(1 - \frac{1}{p}\right) + \frac{1}{p}\right)\right) + O^*\left(\frac{4}{\log x}\right),$$

where the constant c is given by

$$c = b + \sum_{p \geq 2}\sum_{k \geq 2} \frac{1}{kp^k} .$$

We easily check that the function $x \mapsto (-x - \log(1 - x))/x^2$ is non-decreasing over $(0, 1/2]$ (it has a Taylor expansion having only non-negative coefficients), hence

$$-\sum_{p > x}\left(\log\left(1 - \frac{1}{p}\right) + \frac{1}{p}\right) \leq \frac{4}{5}\sum_{p > x}\frac{1}{p^2} \leq \frac{8}{5x} .$$

Therefore

$$\prod_{p \leq x}\left(1 - \frac{1}{p}\right)\log x = e^c \exp O^*\left(\frac{4}{\log x} + \frac{8}{5x}\right) = e^c\left(1 + O\left(\frac{1}{\log x}\right)\right),$$

as required. Let us now proceed to identify the constant c. We first notice that

$$\sum_{n \leq x} \frac{\Lambda(n)}{n \log n} = \log \log x + c + O\left(\frac{1}{\log(2x)}\right) .$$

Indeed, the summation restricted to the prime values of n amounts to $\sum_{n \leq x} 1/p$ while the remaining one is

$$\sum_{k \geq 2}\sum_{p^k \leq x} \frac{1}{kp^k} = \sum_{k \geq 2}\sum_{p \leq x} \frac{1}{kp^k} - \sum_{k \geq 2}\sum_{p^k > x} \frac{1}{kp^k} .$$

But we find that, for any fixed p, we have

$$\sum_{\substack{k \geq 2, \\ p^k > x}} \frac{1}{p^k} \leq \frac{1}{p^{k_0(p)}} \frac{p}{p-1} \, ,$$

where $k_0(p)$ is the first integer at least 2 for which $p^{k_0(p)} > x$. Hence

$$\sum_{p} \sum_{\substack{k \geq 2, \\ p^k > x}} \frac{1}{p^k} \leq \sum_{p \geq \sqrt{x}} \frac{2}{p^2} \ll x^{-1/2} \, . \tag{12.9}$$

We are ready to finish the argument. Let δ be a positive real parameter. We have

$$\delta \int_1^\infty \sum_{n \leq x} \frac{\Lambda(n)}{n \log n} \frac{dx}{x^{1+\delta}} = \log \zeta(1 + \delta) = -\log \delta + O(\delta)$$

A simple exchange of summation and integral establishes that

$$\delta \int_1^\infty \sum_{1 \leq n \leq \log x} \frac{1}{n} \frac{dx}{x^{1+\delta}} = \log(1 - e^{-\delta}) = -\log \delta + O(\delta) \, .$$

This gives us

$$\delta \int_1^\infty \left(\sum_{n \leq x} \frac{\Lambda(n)}{n \log n} - \sum_{1 \leq n \leq \log x} \frac{1}{n} \right) \frac{dx}{x^{1+\delta}} = O(\delta) \, .$$

The left-hand side is however of size $c - \gamma + O(\delta)$, from which we deduce, on letting δ go to zero, that $c = \gamma$. This concludes the proof. □

Exercise 12-13. Prove that $\displaystyle\prod_{p \leq x}\left(1 + \frac{1}{p}\right) \sim \frac{6e^\gamma}{\pi^2} \log x$.

Exercise 12-14 (From [18]). For any positive integer k, we denote by $f(k)$ be the sum of the reciprocals of the primes dividing k and by $S(k)$ the sum of the reciprocals of the primes not exceeding k.

1 ◇ Show that for $k \geq 8$ we have $f(k) \leq \log \log(5 \log k) + 0.262 + \dfrac{21/5}{\log(5 \log k)}$.

2 ◇ Show that, when $k \geq 10^{16}$, we have $S(k) > f(k) + 1$.

3 ◇ Show that, when $k \leq 10^{16}$, then k has less than 14 prime factors and $f(k) \leq 1.65$. Show that this implies that $S(k) > f(k) + 1$ when $k \geq 10^7$.

4 ◇ Show that integers k below 10^7 have at most 8 prime factors. Therefore $f(k) \leq 1.46$ for those integers. Show that this implies that $S(k) > f(k) + 1$ when $k \geq 3 \cdot 10^5$.

5 ◇ Show that 840 is the largest integer for which $S(k) - f(k) < 1$.

Exercise 12-15 (Lemma 11 of [25]). Let us define, for any integer h, the two functions $S(h) = \displaystyle\sum_{2<p|h} \frac{\log p}{p-2}$ and $\Pi(h) = \displaystyle\prod_{2<p|h} \frac{p-2}{p-1}$.

1 ◇ Show by appealing to Theorem 12.2 that $\displaystyle\sum_{2<p\le x} \frac{\log p}{p-2} < \log x$.

2 ◇ Let p be a prime strictly larger than any of the primes that divide h. Show that $S(hp)\Pi(hp) > S(h)\Pi(h)$ when $h \ge 4$ and also when $h = 3$. Conclude that

$$\prod_{2<p|h} \frac{p-2}{p-1} \sum_{2<p|h} \frac{\log p}{p-2} \le 2e^{-\gamma} \prod_{p\ge 3}\left(1 - \frac{1}{(p-1)^2}\right).$$

We stop here our journey in Merten's garden. Two distinct continuations will be presented in Sect. 15.3 and in Exer. 24-7.

12.3 A Bertrand Postulate Type Result

In 1845 at the beginning of Sect. V of his paper [1], J. Bertrand conjectured that there is a prime number in the interval $[n, 2n - 3]$ for every integer $n \ge 4$. This conjecture was proved to be correct by P. Chebyshev in 1850. The results we presented before are slightly weaker than those of Chebyshev, but they still allow us to establish a result of the same kind, namely,

> **Theorem 12.5**
>
> For $x \ge 2$ we have $\displaystyle\sum_{x/4<p\le x} 1 \ge 2x/(25\log x)$.

In particular, there is a prime number in the interval $(x/4, x]$ when $x \ge 2$. The simple modification of our argument proposed by S. Ramanujan in [24] leads to a proof of Bertrand's postulate.

Proof Applying Theorem 12.2 for x and $x/4$ we obtain

$$\sum_{x/4<d\le x} \Lambda(d)/d \ge \log 4 - 1 . \tag{12.10}$$

Consequently, $\sum_{x/4<d\le x} \Lambda(d) \ge x(\log 4 - 1)/4 \ge x/11$ for $x \ge 16$. We extend this inequality to $x \ge 2$ by computation, but first we should pass from $\Lambda(d)$ to the sum over primes using Lemma 12.2, which gives

$$\sum_{x/4<p\le x} \log p \ge x/11 - 3\sqrt{x} \ge 2x/25 \tag{12.11}$$

for $x \ge 76\,000$, which implies the theorem in this case. The case $x < 76\,000$ can be easily dealt with by calculation. Since $x/\log x$ is a non-decreasing function when $x \ge e$, it is enough to check the claimed inequality for integer values of x, though we have to be careful about the lower bound! The worst case may be $2y/(25 \log y)$, where y is close to $x + 1$, i.e. $y = x + 1 - o(1)$.

```
{check( upperlimit ) =
  my( res = 1, aux, ispatnbyfour, ispatn, coef =20, where);
  for(n = 3, upperlimit,
    if(n%4!=0, ispatnbyfour=0, ispatnbyfour = isprime(n/4));
    ispatn = isprime(n);
    res = res - ispatnbyfour + ispatn;
    aux = (n+1)/log(n+1);
    if(res/aux < coef, coef = res/aux; where = n,));
  print("We have sum_(x/4< p <= x)1/(x/log x) >= ", coef);
  print("for ", 3 ," < x <= ", upperlimit, ".");
  print("It is reached just before x = ", where+1);
}
```

By running this script, we get that the minimum of our quantity when $x \le 100\,000$ is $0.58\ldots$ obtained just before $x = 29$. When x is between 2 and 3, we have $C(x) = 1$, while the minimum of $1/(x/\log x)$ is, by convexity, the minimum of $1/(2/\log 2)$ and of $1/(3/\log 3)$, i.e. $0.34\ldots$ $\square$

P. Dusart showed in [7] that for $x \ge 598$ we have

$$1 + \frac{0.992}{\log x} < \frac{\log x}{x} \sum_{p \le x} 1 < 1 + \frac{1.2762}{\log x} . \tag{12.12}$$

Exercise 12-16. Show that $\psi(x) = \vartheta(x) + \vartheta(x^{1/2}) + \vartheta(x^{1/3}) + \vartheta(x^{1/4}) + \ldots$, where $\vartheta(y) = \sum_{p \le y} \log p$.

Exercise 12-17. Prove that we can also define $\psi(n)$ as being the logarithm of the lcm of all the integers between 1 and n.

Further Reading

We only brushed probabilistic number theory. The readers will find much more material in [23] by A.G. Postnikov and in the large treaty [8, 9] by P.D.T.A.Elliott.

References

[1]   J. Bertrand. "Mémoire sur le nombre de valeurs que peut prendre une fonction quand on permute les lettres qu'elle renferme". In: *Journal de l'École Royale Polytechnique* 18. Cahier 30 (1845), pp. 123–140 (cit. on p. 131).

[2]   R. Breusch. "Zur Verallgemeinerung des Bertrandschen Postulates, daßzwischen x und 2 x stets Primzahlen liegen". In: *Math. Z.* 34.1 (1932), pp. 505–526. https://doi.org/10.1007/ BF01180606 (cit. on p. 119).

[3]   P.L. Chebyshev. "Mémoire sur les nombres premiers". In: *Journal de mathématiques pures et appliquées, Sér. 1* 17 (1852), pp. 366–390 (cit. on p. 123).

[4]   N. Costa Pereira. "Elementary estimates for the Chebyshev function $\psi(x)$ and for the Möbius function $M(x)$". In: *Acta Arith.* 52.4 (1989), pp. 307–337. https://doi.org/10.4064/aa-52-4-307-337 (cit. on p. 122).

[5]   C.-J.G.N.B. de la Vallée-Poussin. "Sur la valeur de certaines constantes arithmétiques". In: *Brux. S. sc.* 22 A (1898), pp. 84–90 (cit. on p. 125).

[6]   H. Delange. "Sur le nombre des diviseurs premiers de n". In: *C. R. Acad. Sci. Paris* 237 (1953), pp. 542–544 (cit. on p. 128).

[7]   P. Dusart. "Inègalitès explicites pour $\psi(X)$, $\theta(X)$, $\pi(X)$ et les nombres premiers". In: *C. R. Math. Acad. Sci., Soc. R. Can.* 21.2 (1999), pp. 53–59 (cit. on p. 132).

[8]   P.D.T.A. Elliott. *Probabilistic number theory. I.* Vol. 239. Grundlehren der Mathematischen Wissenschaften. Mean-value theorems. Springer-Verlag, New York-Berlin, 1979, xxii+359+xxxiii pp. (2 plates) (cit. on p. 132).

[9]   P.D.T.A. Elliott. *Probabilistic number theory. II: Central limit theorems.* English. Vol. 240. Springer, Berlin, 1980 (cit. on p. 132).

[10]   P. Erdös and M. Kac. "The Gaussian law of errors in the theory of additive number theoretic functions". In: *Amer. J. Math.* 62 (1940), pp. 738–742 (cit. on p. 129).

[11]   A. Granville and K. Soundararajan. "Sieving and the Erdős-Kac theorem". In: *Equidistribution in number theory, an introduction.* Vol. 237. NATO Sci. Ser. II Math. Phys. Chem. Springer, Dordrecht, 2007, pp. 15–27. https://doi.org/10.1007/978-1-4020-5404-4_2 (cit. on p. 128).

[12]   D. Hanson. "On the product of the primes". In: *Canad. Math. Bull.* 15 (1972), pp. 33–37. https://doi.org/10.4153/CMB-1972-007-7 (cit. on p. 122).

[13]   G.H. Hardy and S.A. Ramanujan. "The normal number of prime factors of a number n". English. In: *Quart. J.* 48 (1917), pp. 76–92 (cit. on p. 127).

[14]   J. Kubilius. "Probabilistic methods in the theory of numbers". In: *Uspehi Mat. Nauk (N.S.)* 11.2(68) (1956), pp. 31–66 (cit. on p. 127).

[15]   J. Kubilius. *Probabilistic methods in the theory of numbers.* Translations of Mathematical Monographs, Vol. 11. American Mathematical Society, Providence, R.I., 1964, pp. xviii+182 (cit. on p. 127).

[16]   H. von Mangoldt. "Extract from a paper entitled: Zu Riemann's Abhandlung "Ueber die Anzahl der Primzahlen unter einer gegebenen Grösse". (On Riemann's study "On the number of primes less than a given bound".). (Auszug aus einer Arbeit unter dem Titel: Zu Riemann's Abhandlung "Ueber die Anzahl der Primzahlen unter einer gegebenen Grösse".)" In: *Berl. Ber. 1894.* (1894), pp. 883–896 (cit. on p. 119).

[17]   Armel Mercier. "Comportement asymptotique de $\sum_{n \le x} n^a \{f(x/n)\}$". In: *Ann. Sci. Math. Québec* 9.2 (1985), pp. 199–202 (cit. on p. 125).

[18]   P. Moree. "Bertrand's postulate for primes in arithmetical progressions". In: *Comput. Math. Appl.* 26.5 (1993), pp. 35–43. https://doi.org/10.1016/0898-1221(93)90071-3 (cit. on p. 130).

[19]   W. Narkiewicz. *The development of prime number theory.* Springer Monographs in Mathematics. From Euclid to Hardy and Littlewood. Springer-Verlag, Berlin, 2000, pp. xii+448. https://doi.org/10.1007/978-3-662-13157-2 (cit. on p. 119).

[20]   F. Pillichshammer. "Euler's constant and averages of fractional parts". In: *Amer. Math. Monthly* 117.1 (2010), pp. 78–83. https://doi.org/10.4169/000298910X475014 (cit. on p. 125).

[21]   F. Pillichshammer. "A generalisation of a result of de la Vallée Poussin". In: *Elem. Math.* 67.1 (2012), pp. 26–38. https://doi.org/10.4171/EM/190 (cit. on p. 125).

[22]   G. Pólya and G. Szegő. *Problems and theorems in analysis. II.* Classics in Mathematics. Theory of functions, zeros, polynomials, number theory, geometry, Translated

from the German by C. E. Billigheimer, Reprint of the 1976 English translation. Springer-Verlag, Berlin, 1998, pp. xii+392. https://doi.org/10.1007/978-3-642-61905-2_7 (cit. on p. 122).

[23]   A. G. Postnikov. *Introduction to analytic number theory. Transl. from the Russian by G. A. Kandall. Ed. by Ben Silver. Appendix by P. D. T. A. Elliott.* English. Vol. 68. Providence, RI: American Mathematical Society, 1988, pp. vi + 320 (cit. on p. 132).

[24]   S. Ramanujan. "A proof of Bertrand's postulate". In: *J. Indian Math. Soc.* XI (1919), pp. 181–182 (cit. on p. 131).

[25]   O. Ramaré. "Approximate Formulae for $L(1, \chi)$, II". In: *Acta Arith.* 112 (2004), pp. 141–149 (cit. on p. 131).

[26]   O. Ramaré. *Un parcours explicite en théorie multiplicative.* vii+100 pp. Éditions universitaires europénnes, 2010 (cit. on p. 124).

[27]   J.B. Rosser. "The n-th prime is greater than $n \log n$". In: *Proc. Lond. Math. Soc., II. Ser.* 45 (1938), pp. 21–44 (cit. on p. 119).

[28]   A. Selberg. "On elementary problems in prime number-theory and their limitations". In: *C.R. Onziéme Congrés Math. Scandinaves, Trondheim, Johan Grundt Tanums Forlag* (1949), pp. 13–22 (cit. on p. 124).

[29]   P. Turán. "On a theorem of Hardy and Ramanujan". English. In: *J. Lond. Math. Soc.* 9 (1934), pp. 274–276. https://doi.org/10.1112/jlms/s1-9.4.274 (cit. on p. 129).

Part III
The Levin–Faĭnleĭb Walk

Chapter 13
The Levin–Faĭnleĭb Theorem and Analogues

13.1 A First Upper Bound

Our first theorem is efficient when the value of the non-negative multiplicative function g we are summing is about constant on prime numbers. This result exists in several brands, the most precise of which is due to Halberstam & Richert in [8]. The one we prove is a slight modification of the version found in Tenenbaum's book [15].

The initial idea comes from the celebrated Levin–Faĭnleĭb Theorem from [9]. The paper [11] explores and improves on the explicit aspect of the next two results.

Theorem 13.1

Let $D \geq 2$ be some fixed parameter. Assume g is a non-negative multiplicative function such that

$$\forall Q \in [1, D], \qquad \sum_{\substack{p \geq 2, v \geq 1 \\ p^v \leq Q}} g(p^v) \log(p^v) \leq KQ + K'$$

for two constants $K, K' \geq 0$ such that $D > \exp K'$. Then

$$\sum_{d \leq D} g(d) \leq (K+1) \frac{D}{\log D - K'} \sum_{d \leq D} \frac{g(d)}{d} .$$

See Exer. 13-11 for a lower bound.

Proof Let us set $G(D) = \sum_{d \leq D} g(d)$. On using $\log(D/d) \leq D/d$, we find that

$$G(D) \log D = \sum_{d \leq D} g(d) \log \frac{D}{d} + \sum_{d \leq D} g(d) \log d$$

$$\leq D \sum_{d \leq D} \frac{g(d)}{d} + \sum_{\substack{p \geq 2, v \geq 1 \\ p^v \leq D}} g(p^v) \log(p^v) \sum_{\substack{\ell \leq D/p^v \\ (\ell, p) = 1}} g(\ell) ,$$

O. Ramaré, *Excursions in Multiplicative Number Theory*, Birkhäuser Advanced Texts Basler Lehrbücher, https://doi.org/10.1007/978-3-030-73169-4_13

where the second summand is obtained by writing

$$\log d = \sum_{p^\nu \| d} \log(p^\nu) \,.$$

We continue by again exchanging the summation signs:

$$\sum_{\substack{p \geq 2, \nu \geq 1 \\ p^\nu \leq D}} g(p^\nu) \log(p^\nu) \sum_{\substack{\ell \leq D/p^\nu \\ (\ell, p)=1}} g(\ell) = \sum_{\ell \leq D} g(\ell) \sum_{\substack{p \geq 2, \nu \geq 1 \\ p^\nu \leq D/\ell \\ (p, \ell)=1}} g(p^\nu) \log(p^\nu)$$

$$\leq \sum_{\ell \leq D} g(\ell) \left(K \frac{D}{\ell} + K' \right) \,.$$

The proof is now easily concluded.   □

Exercise 13-1. Prove a better version of Theorem 13.1 by using the improved inequality $\log \frac{D}{d} \leq \frac{D}{d} - 1$.

Theorem 13.2

Let $D \geq 2$ be a fixed parameter. Assume g is a non-negative multiplicative function. Then

$$\sum_{d \leq D} g(d)/d \leq \exp\left(\sum_{\substack{\nu \geq 1, \\ p^\nu \leq D}} g(p^\nu)/p^\nu \right),$$

where the right-hand side sum is over all prime powers $p^\nu \leq D$.

Proof Indeed, the development of the Euler product

$$\prod_{p \leq D} \left(\sum_{\nu \geq 0, p^\nu \leq D} g(p^\nu)/p^\nu \right)$$

contains at least all the terms of the sum $\sum_{d \leq D} g(d)/d$. We thus get an upper bound in this manner which we simplify by using that, for fixed p, we have

$$\log\left(\sum_{\substack{\nu \geq 0, \\ p^\nu \leq D}} g(p^\nu)/p^\nu \right) \leq \sum_{\substack{\nu \geq 1, \\ p^\nu \leq D}} g(p^\nu)/p^\nu \,. \tag{$\boxtimes$}$$

Exercise 13-2. Assume g is a multiplicative non-negative function that satisfies $0 \leq g(p) \leq \lambda < 1$ for every prime p, where λ is some fixed parameter. Prove that

$$\sum_{d \leq D} \frac{\mu^2(d) g(d)}{d} \geq (1 - \lambda) \prod_{p \leq D} \left(1 + \frac{g(p)}{p} \right).$$

Exercise 13-3.

1 ⋄ Show that we have, for any $x \geq 1$, $\log x \leq \sum_{n \leq x} n^{-1} \leq \prod_{p \leq x} (1 - 1/p)^{-1}$.

2 ⋄ Show that, when $y \in (0, 1]$, the inequality $-\log(1 - y) \leq y + y^2/2$ holds true

3 ⋄ Show that $\sum_{p \geq 2} p^{-2} \leq 1/2$.

4 ⋄ Deduce from above that, for any $x > 1$, we have $\sum_{p \leq x} p^{-1} \geq \log \log x - 1/4$.

Exercise 13-4. Let g be a (non especially non-negative) multiplicative function Assume that for any $Q \in [1, D]$ we have

$$\sum_{\substack{p \geq 2, v \geq 1 \\ p^v \leq Q}} \max(|g|, |\mathbb{1} \star g|)(p^v) \log(p^v) \leq KQ + K',$$

where $K, K' \geq 0$ are two constants such that $D > \exp K'$.

1 ⋄ Show that $\left| \sum_{n \leq x} \frac{g(n)}{n} \right| \leq \frac{1}{x} \sum_{n \leq x} (|g| + |\mathbb{1} \star g|)(n)$.

2 ⋄ Show that $\left| \sum_{d \leq D} g(d) \right| \leq (2K + 2) \frac{D}{\log D - K'} \sum_{d \leq D} \frac{|g|(d) + |\mathbb{1} \star g|(d)}{d}$.

3 ⋄ Prove that $\left| \sum_{d \leq x} \frac{\mu(d)}{2^{\omega(d)} d} \right| = O(1)$.

4 ⋄ Fix a parameter $\xi \in (0, 1]$. Show that $\sum_{n \leq x} \frac{\mu(n) \xi^{\omega(n)}}{n} \ll (\log x)^{\max(1-\xi, \xi)-1}$.

5 ⋄ Fix a parameter $\xi \in (0, 1]$. Show that $\sum_{n \leq x} \mu(n) \xi^{\omega(n)} \ll x (\log x)^{\max(1-\xi, \xi)-1}$.

6 ⋄ Show that the two upper bounds above are uniform in $\xi \in [a, 1]$ for any constant $a > 0$.

Exercise 13-5. (From S. Selberg in [14]). Consider the sum $S(x) = \sum_{n \leq x} \frac{\mu(n)}{n 2^{\omega(n)}}$.

1 ⋄ Prove the identity $xS(x) = \sum_{n \leq x} \frac{1}{2^{\omega(n)}} + \sum_{n \leq x} \frac{\mu(n)}{2^{\omega(n)}} \left\{ \frac{x}{n} \right\}$.

2 ⋄ Show that $S(x) \geq 0$.

3 ⋄ Similarly, prove that $\sum_{n \leq x} \frac{\mu(n)}{n \lambda^{\omega(n)}} \geq 0$ for every real number $\lambda \geq 2$.

4 ⋄ Anticipating on Theorem 13.3, prove that there is a constant $c_0 > 0$ such that $|S(x)| \leq c_0/\sqrt{\log x}$.

13.2 An Asymptotic Formula

Our third theorem applies to non-negative multiplicative functions g that satisfy $g(p) \simeq \kappa/p$ for every primes p and some $\kappa > 0$. The required hypotheses are much stronger, but the pay-off is that we get an asymptotic formula. The following two facts are responsible for complicating the argument:

- We keep track of the dependence on all the parameters to get a fully explicit result.
- We allow g not to vanish on powers of primes rather than restricting the sum to square-free integers. The readers will see in Sects. 13.3 and 18.2 the usefulness of such an extension.

Most of the next result can be found in [6] by Halberstam and Richert, and a somewhat simplified proof in the book [7] by the same two authors.

Theorem 13.3

Let g be a non-negative multiplicative function. Let κ, L and A be three non-negative real parameters such that

$$\begin{cases} \sum_{\substack{p \geq 2, v \geq 1 \\ p^v \leq Q}} g(p^v) \log(p^v) = \kappa \log Q + O^*(L) & (Q \geq 1), \\ \sum_{p \geq 2} \sum_{v, k \geq 1} g(p^k) g(p^v) \log(p^v) \leq A. \end{cases}$$

Then, when $D \geq \exp(2(L + A))$, we have

$$\sum_{d \leq D} g(d) = C \left(\log D\right)^{\kappa} \left(1 + O^*(B/\log D)\right),$$

with
$$\begin{cases} C = \frac{1}{\Gamma(\kappa+1)} \prod_{p \geq 2} \left\{ \left(1 + \sum_{v \geq 1} g(p^v)\right)\left(1 - \frac{1}{p}\right)^{\kappa} \right\}, \\ B = 2(L + A)(1 + 2(\kappa + 1)e^{\kappa+1}). \end{cases}$$

In most applications, the dependence on L may be important. The dependence on A can almost always be handled in a trivial manner. We remark that Rawsthorne in [13] and Greaves & Huxley in [4] provide a lower bound for the final sum by assuming only lower bounds on the primes.

Proof The starting point is similar to the one we used in the proof of Theorem 13.1 (and in fact predates it):

$$G(D) \log D = \sum_{d \leq D} g(d) \log \frac{D}{d} + \sum_{d \leq D} g(d) \log d$$

$$= \sum_{d \leq D} g(d) \log \frac{D}{d} + \sum_{\substack{p \geq 2, v \geq 1 \\ p^v \leq D}} g(p^v) \log(p^v) \sum_{\substack{\ell \leq D/p^v \\ (\ell, p)=1}} g(\ell).$$

We now define

$$G_p(X) = \sum_{\substack{\ell \le X \\ (\ell,p)=1}} g(\ell) \quad \text{and} \quad T(D) = \sum_{d \le D} g(d) \log \frac{D}{d} = \int_1^D G(t) \frac{dt}{t} \,,$$

so that we may rewrite the above in the form

$$G(D) \log D = T(D) + \sum_{\substack{p \ge 2, v \ge 1 \\ p^v \le D}} g(p^v) \log(p^v) G_p(D/p^v) \,.$$

Moreover, we check easily that $G_p(X) = G(X) - \sum_{k \ge 1} G_p(X/p^k)$ which, together with our hypotheses, leads to

$$G(D) \log D = T(D) + \sum_{\substack{p \ge 2, v \ge 1 \\ p^v \le D}} g(p^v) \log(p^v) G(D/p^v) + O^*(AG(D))$$

$$= T(D) + \sum_{d \le D} g(d) \sum_{\substack{p \ge 2, v \ge 1 \\ p^v \le D/d}} g(p^v) \log(p^v) + O^*(AG(D))$$

$$= T(D)(\kappa + 1) + O^*((L + A)G(D)) \,.$$

Therefore we have

$$(\kappa + 1)T(D) = G(D) \log D\,(1 + r(D)) \quad \text{with} \quad r(D) = O^*\left(\frac{L + A}{\log D}\right).$$

The idea is now to consider this relation as a differential equation. We set

$$\exp E(D) = \frac{(\kappa + 1)T(D)}{(\log D)^{\kappa+1}} = \frac{G(D)}{(\log D)^\kappa}(1 + r(D))$$

and obtain, for $D \ge D_0 = \exp(2(L + A))$,

$$E'(D) = \frac{T'(D)}{T(D)} - \frac{(\kappa + 1)}{D \log D} = \frac{r(D)(\kappa + 1)}{(1 - r(D))D \log D} = O^*\left(\frac{2(L + A)}{D(\log D)^2}\right), \quad (13.1)$$

since $|r(D)| \le 1/2$ as soon as $D \ge D_0$. We infer from the above that for $D \ge D_0$:

$$E(\infty) - E(D) = \int_D^\infty E'(t)dt = O^*\left(\frac{2(L + A)}{\log D}\right).$$

We should stop at this equation: it holds because E is *absolutely continuous* over every finite interval $\subset [1, \infty)$, where this notion has been detailed in Sect. 7.2. Summarizing, we have reached

$$\frac{G(D)}{(\log D)^\kappa} = \frac{\exp E(D)}{1 + r(D)} = \frac{e^{E(\infty)}}{1 + r(D)}\left(1 + O^*\left(\frac{2(L + A)}{\log D}(\kappa + 1)e^{\kappa+1}\right)\right).$$

Since the inequality $1/(1 + x) \le 1 + 2x$ holds true whenever $0 \le x \le 1/2$, this estimate simplifies to

$$\frac{G(D)}{(\log D)^\kappa} = e^{E(\infty)}\left(1 + O^*\left(\frac{2(L + A)}{\log D}(1 + 2(\kappa + 1)e^{\kappa+1})\right)\right),$$

provided that $D \ge D_0$. The proof is complete at this stage, save for the value of the constant $e^{E(\infty)} = C$.

Identifying C:

Let us first note that, when s is any positive real number, we have

$$D(g, s) = \sum_{d \ge 1}\frac{g(d)}{d^s} = s\int_1^\infty G(D)\frac{dD}{D^{s+1}}$$

$$= sC\int_1^\infty (\log D)^\kappa \frac{dD}{D^{s+1}} + O\left(sC\int_1^\infty (\log D)^{\kappa-1}\frac{dD}{D^{s+1}}\right)$$

$$= C\left(s^{-\kappa}\Gamma(\kappa + 1) + O(s^{1-\kappa}\Gamma(\kappa))\right).$$

It follows that

$$C = \lim_{s \to 0^+} D(g, s)s^\kappa \Gamma(\kappa + 1)^{-1} = \lim_{s \to 0^+} D(g, s)\zeta(s + 1)^{-\kappa}\Gamma(\kappa + 1)^{-1}.$$

It is then rather straightforward to show that the product

$$\prod_{p \ge 2}\left\{\left(\sum_{\nu \ge 0} g(p^\nu)\right)\left(1 - \frac{1}{p}\right)^\kappa\right\}$$

is convergent and equals $C\Gamma(\kappa + 1)$ as claimed.   □

13.3 Proof of Theorem $\mathscr{B}$

This section is devoted to a detailed proof of Theorem $\mathscr{B}$. We set $g(n) = \sqrt{d(n)}/n$. We readily compute that

$$\sum_{p^\nu \le Q}\frac{\sqrt{d(p^\nu)}}{p^\nu}\log(p^\nu) = \sum_{p^\nu \le Q}\frac{\sqrt{2}}{p^\nu}\log p + \sum_{\substack{p^\nu \le Q, \\ \nu \ge 2}}\frac{\nu\sqrt{\nu + 1} - \sqrt{2}}{p^\nu}\log p.$$

The first sum amounts to $\sqrt{2}(\log x + O^*(7/6))$ when $x \ge 2$ by Theorem 12.2. As for the second one, we check that, when $|z| \le 1$, we have

$$\sum_{v\geq 2} v(v+1)z^{v-2} = \sum_{v\geq 2} v(v-1)z^{v-2} + \frac{2}{z}\left(\sum_{v\geq 1} vz^{v-1} - 1\right)$$

$$= \frac{2}{(1-z)^3} + \frac{4-2z}{(1-z)^2} = \frac{2(z^2-3z+3)}{(1-z)^3}$$

so that we may write

$$\sum_{\substack{p^v\leq Q, \\ v\geq 2}} \frac{v\sqrt{v+1}-\sqrt{2}}{p^v}\log p \leq \sum_{p\leq\sqrt{Q}} \frac{6\log p}{p^2(1-1/p)^3} \leq 12\log 2 + \int_2^\infty \frac{12\log t\, dt}{(t-1)^2} \leq 25 .$$

Hence we select $\kappa = \sqrt{2}$ and $L = 27$. We next see that

$$\sum_{\substack{p\geq 2 \\ v,k\geq 1}} \frac{\sqrt{d(p^v)d(p^k)}}{p^{v+k}}\log(p^v) = \sum_{\substack{p\geq 2, \\ h\geq 2}} \frac{\log p}{p^h} \sum_{v+k=h} v\sqrt{(v+1)(k+1)} .$$

We immediately check that $\sqrt{(v+1)(k+1)} \leq (h+2)/2$ with $v+k=h$, and so

$$\sum_{v+k=h} v\sqrt{(v+1)(k+1)} \leq h(h+1)(h+2)/2 .$$

Further

$$\sum_{h\geq 2} h(h+1)(h+2)|z|^{h-2} \leq 4\sum_{h\geq 2}(h-1)h(h+1)|z|^{h-2} = \frac{24}{(1-|z|)^4} .$$

We thus get

$$\sum_{\substack{p\geq 2 \\ v,k\geq 1}} \frac{\sqrt{d(p^v)d(p^k)}}{p^{v+k}}\log(p^v) \leq \sum_{p\leq\sqrt{Q}} \frac{24\log p}{p^2(1-1/p)^4}$$

$$\leq 96\log 2 + \int_2^\infty \frac{96\log t\, dt}{(t-1)^2} \leq 150 .$$

Therefore we may select $A = 150$. When $x \geq \exp(20\,000)$ it follows by a direct usage of Theorem 13.3 that

$$\sum_{n\leq x} \frac{\sqrt{d(n)}}{n} = \mathscr{C}_1(\log x)^{\sqrt{2}}(1 + O^*(20000/\log x)) ,$$

where $\mathscr{C}_1$ is given by (0.3). This completes the proof of Theorem $\mathscr{B}$.

Please notice that the bound $\exp(20000)$ is extremely large and that the constant $\mathscr{C}_1$ may not be dealt with by Theorem 9.6. Computing the first hundred digits of $\mathscr{C}_1$ is an interesting challenge, though the initial problem looks simple enough.

Exercise 13-6. Show that, when D goes to infinity, we have $\sum\limits_{d \leq D} 9^{\omega(d)}/\varphi(d) \sim$ $C(\log D)^9$, where C is a positive constant that is to be given explicitly.

Exercise 13-7.

$1 \diamond$ Give an asymptotic for $\sum\limits_{n \leq N} \sum\limits_{d|n} \mu^2(n/d)3^{-\omega(d)}/n$.

$2 \diamond$ Give an asymptotic for $\sum\limits_{n \leq N} \sum\limits_{d|n} \mu^2(n/d)3^{-\omega(d)}/(n+1)$.

Exercise 13-8.

$1 \diamond$ Show that $\dfrac{n}{\varphi(n)} = \sum\limits_{d|n} \dfrac{\mu^2(d)}{\varphi(d)}$.

$2 \diamond$ Show that $\sum\limits_{n \leq N} \dfrac{n}{\varphi(n)} = \dfrac{\zeta(2)\zeta(3)}{\zeta(6)} N + O(\log(2N))$.

$3 \diamond$ Show that there exists a constant C such that

$$\sum_{n \leq N} \frac{1}{\varphi(n)} = \frac{\zeta(2)\zeta(3)}{\zeta(6)} \log N + C + O(\log(2N)/N).$$

$4 \diamond$ Show that $\sum\limits_{d \geq 1} \dfrac{\mu^2(d) \log d}{d\varphi(d)} = \sum\limits_{p \geq 2} \dfrac{\log p}{p^2 - p + 1} \prod\limits_{p \geq 2}\left(1 + \dfrac{1}{p(p-1)}\right)$ and deduce that the constant C from the previous question is given by

$$C = \frac{\zeta(2)\zeta(3)}{\zeta(6)}\left(\gamma - \sum_{p \geq 2} \frac{\log p}{p^2 - p + 1}\right).$$

Exer. 9-18 explains how to evaluate the constant C with high accuracy.

$5 \diamond$ Show that the remainder term in the asymptotic evaluation of the average of $1/\varphi(n)$ cannot be better than $O(1/N)$.

Exer. 22-6 provides a sequel to the previous exercise.

Exercise 13-9. Give an asymptotic for $\sum\limits_{d \leq D} 1/(1 + \varphi(d))$.

Exercise 13-10.

$1 \diamond$ Show that $\dfrac{\varphi(n)}{n} = \sum\limits_{d|n} \dfrac{\mu(d)}{d}$.

2 ◇ Show that, when $N \geq 1$, we have $\displaystyle\sum_{n \leq N} \frac{\varphi(n)}{n} = CN + O(\log(2N))$ for some explicit constant C.

3 ◇ Show that, when $N \geq 1$, we have $\displaystyle\sum_{n \leq N} \frac{\varphi(n)}{n^2} = C \log N + C_2 + O\left(\frac{\log(2N)}{N}\right)$ for some explicit constant C_2.

13.4 A Stronger Theorem

Our fourth theorem applies to non-negative multiplicative functions g that satisfy $g(p) \simeq \kappa$ for every prime p and some $\kappa > 0$. It is a simple consequence of Theorem 13.3 by following the path initiated by E. Wirsing in [16]. We borrow this result from [12, Theorem 21.2].

Theorem 13.4

Let f be a non-negative multiplicative function and κ be non-negative real parameter such that

$$
\begin{cases}
\displaystyle\sum_{\substack{p \geq 2, \nu \geq 1 \\ p^\nu \leq Q}} f(p^\nu) \log(p^\nu) = \kappa Q + O(Q/\log(2Q)) & (Q \geq 1), \\[4ex]
\displaystyle\sum_{p \geq 2} \sum_{\substack{\nu, k \geq 1, \\ p^{\nu+k} \leq Q}} f(p^k) f(p^\nu) \log(p^\nu) \ll \sqrt{Q} .
\end{cases}
$$

Then $\displaystyle\sum_{d \leq D} f(d) = \kappa C \cdot D \left(\log D\right)^{\kappa-1} (1 + o(1))$, where C is as in Theorem 13.3.

Though this theorem has a satisfactory conclusion, its hypotheses may be hard to meet at this level: we have not yet proved the Prime Number Theorem! This is the topic of Chap. 23. Theorem 13.4 will then prove its power. See, for instance, Exer. 23-9.

Proof We proceed as in Theorem 13.3. Write

$$
S(D) = \sum_{d \leq D} f(d) .
$$

By using Theorem 13.1 followed by an application of Theorem 13.3, we readily obtain the following a priori bound

$$
S(D) \ll D(\log(2D))^{\kappa-1} . \tag{13.2}
$$

Consider now $S^*(D) = \sum_{d \leq D} f(d) \log d$. Proceeding as in the proof of Theorem 13.3, we get

$$S^*(D) = \sum_{\substack{p \geq 2, v \geq 1 \\ p^v \leq D}} f(p^v) \log(p^v) \sum_{\substack{\ell \leq D/p^v \\ (\ell, p) = 1}} f(\ell)$$

$$= \sum_{\ell \leq D} f(\ell) \sum_{\substack{p \geq 2, v \geq 1 \\ p^v \leq D/\ell, \\ (p, \ell) = 1}} f(p^v) \log(p^v) ,$$

so that $S^*(D)$ equals

$$\sum_{\ell \leq D} f(\ell) \sum_{\substack{p \geq 2, v \geq 1 \\ p^v \leq D/\ell}} f(p^v) \log(p^v) - \sum_{\ell \leq D} f(\ell) \sum_{\substack{p \geq 2, v, k \geq 1 \\ p^{v+k} \leq D/\ell, \\ (p, \ell) = 1}} f(p^v) f(p^k) \log(p^v) .$$

We use our hypothesis on this expression and conclude that

$$S^*(D) = \kappa D \sum_{\ell \leq D} f(\ell)/\ell + O\!\left(Q \sum_{\ell \leq D} \frac{f(\ell)}{\ell \log(2Q/\ell)}\right) + O\!\left(\sqrt{Q} \sum_{\ell \leq D} \frac{f(\ell)}{\sqrt{\ell}}\right) .$$

Both error terms are shown to be $O(Q \log(2Q)^{\kappa-1})$ by appealing to (13.2), whereas the main term is evaluated via Theorem 13.3. Now the claimed asymptotic is obtain by integration by parts, i.e. by using

$$S(D) = 1 + \int_2^D S^*(t) \frac{dt}{t \log^2 t} + \frac{S^*(D)}{\log D} . \qquad \square$$

Exercise 13-11. (A lower estimate). Let g be a non-negative multiplicative function such that there exist three positive parameters C, κ and $Q_0 \geq 3$ such that
- For every prime power p^k, we have $g(p^k) \leq C$ for some fixed constant C.
- For $Q \geq Q_0$, we have $\sum_{p^k \leq Q} g(p^k) \log(p^k) \geq \kappa Q$.

1 ◇ Show that, as soon as $D \geq D_0 \geq Q_0$, we have

$$\sum_{d \leq D} g(d) \log d \geq \left(\kappa - \frac{C \log D_0}{2\sqrt{D_0}}\right) D \sum_{m \leq \sqrt{D}} g(m)/m .$$

2 ◇ Using Theorem 13.1 prove that $\sum_{d \leq D'} g(d) \ll D'(\log(2D'))^{K'}$ for any $D' \geq 1$ and some constant K'.

3 ◇ Deduce from the above two questions that, when D is large enough, we have

$$\sum_{d \leq D} g(d) \gg \frac{D}{\log D} \sum_{m \leq \sqrt{D}} g(m)/m.$$

The readers will find a more precise version of this last result in the work [3] of Elliott and J. Kish, around Lemma 20 and 21. A further usage of the Levin–Faĭnleĭb mechanism can be found in the paper [10] by Montgomery and Vaughan, see the comments following Lemma 26.5 for more details.

Before closing this chapter, we mention here that averages of multiplicative functions have been the topic of the theorem of Ikehara, which in its generalized form is due to Delange in [1], of the theorems of Wirsing in [16] and Delange in [2], and of the innovative work [5] of Halász.

Further Reading

The theory of averages of multiplicative functions has known huge advances and modifications in recent years. We refer to the freely available electronic documents by Granville and Soundararajan on the *pretentiousness* concept that is derived from the aforementioned work of Halász.

References

[1]   H. Delange. "Généralisation du théorème de Ikehara". In: *Ann. Sci. Ecole Norm. Sup. (3)* 71 (1954). www.numdam.org/item?id=ASENS_1954_3_71_3_213_0, pp. 213–242 (cit. on p. 145).

[2]   H. Delange. "Un théorème sur les fonctions arithmétiques multiplicatives et ses applications". In: *Ann. Sci. école Norm. Sup.* (3) 78 (1961). www.numdam.org/item?id=ASENS_1961_3_78_1_1_0, pp. 1–29 (cit. on p. 145).

[3]   P.D.T.A. Elliott and J. Kish. "Harmonic analysis on the positive rationals II: Multiplicative functions and Maass forms". In: *J. Math. Sci. Univ. Tokyo* 23.3 (2016), pp. 615–658 (cit. on p. 144).

[4]   G.R.H. Greaves and M.N. Huxley. "One sided sifting density hypotheses in Selberg's sieve". In: *Turku symposium on number theory in memory of Kustaa Inkeri*. Ed. by M. Jutila. Turku, Finland: Berlin: de Gruyter, May 1999, pp. 105–114 (cit. on p. 138).

[5]   G. Halász. "Über die Mittelwerte multiplikativer zahlentheorischer Funktionen". In: *Acta Math. Acad. Sci. Hungar.* 19 (1968), pp. 365–403 (cit. on p. 145).

[6]   H. Halberstam and H.-E. Richert. "Mean value theorems for a class of arithmetic functions". In: *Acta Arith.* 43 (1971), pp. 243–256 (cit. on p. 138).

[7]   H. Halberstam and H.-E. Richert. "Sieve methods". In: *Academic Press (London)* (1974), 364pp (cit. on p. 138).

[8]   H. Halberstam and H.-E. Richert. "On a result of R. R. Hall". In: *J. Number Theory* 11 (1979), pp. 76–89 (cit. on p. 135).

[9]   B.V. Levin and A.S. Faĭnleĭb. "Application of some integral equations to problems of number theory". In: *Russian Math. Surveys* 22 (1967), pp. 119–204 (cit. on p. 135).

[10] H.L. Montgomery and R.C. Vaughan. "Exponential sums with multiplicative coefficients". In: *Invent. Math.* 43.1 (1977), pp. 69–82. https://doi.org/10.1007/BF01390204 (cit. on p. 145).

[11] P. Moree. "Chebyshev's bias for composite numbers with restricted prime divisors". In: *Math. Comp.* 73.245 (2004), pp. 425–449 (cit. on p. 135).

[12] O. Ramaré. *Arithmetical aspects of the large sieve inequality*. Vol. 1. Harish-Chandra Research Institute Lecture Notes. With the collaboration of D.S. Ramana. New Delhi: Hindustan Book Agency, 2009, pp. x+201 (cit. on p. 143).

[13] D.A. Rawsthorne. "Selberg's sieve estimate with a one-sided hypothesis". In: *Acta Arith.* 49 (1982), pp. 281–289 (cit. on p. 138).

[14]   S. Selberg. *Über die Summe.* $\sum_{n \leq x} \frac{\mu(n)}{nd(n)}$. German. 12. Skand. Mat.-Kongr., Lund 1953, 264–272 (1954). (Cit. on p. 137).

[15]   G. Tenenbaum. *Introduction à la théorie analytique et probabiliste des nombres.* Second. Vol. 1. Cours Spécialisés. Paris: Société Mathématique de France, 1995 (cit. on p. 135).

[16]   E.A. Wirsing. "Das asymptotische Verhalten von Summen über multiplikative Funktionen". In: *Math. Ann.* 143 (1961), pp. 75–102 (cit. on pp. 143, 145).

Chapter 14
Variations on a Theme of Chebyshev

In [2], Chebyshev* proved among other things the Bertrand Postulate from [1], namely that there exists a prime in any interval $[n, 2n - 3]$, when $n > 3$ is an integer. In 1932, the globe-trotting mathematician P. Erdős (who was only 19 at that time) produced an ingenious proof of this result. Thanks to it, Erdős was noticed and invited to continue his studies in Berlin. We now present some variations around this proof.

The key observation was the following lemma.

> **Lemma 14.1**
>
> For any integer $m \geq 1$ the product $\prod\limits_{m+1 < p \leq 2m+1} p$ divides the binomial coefficient $\binom{2m+1}{m}$.

Proof Each prime number p in the interval $(m + 1, 2m + 1]$ divides $(2m + 1)!$, but does not divide $m!(m + 1)!$. It follows that such a prime divides

$$\binom{2m + 1}{m + 1} = \frac{(2m + 1)!}{m!(m + 1)!},$$

concluding the proof. $\qquad\square$

Exercise 14-1. Let n be a positive integer. Determine the gcd of $\binom{2n}{1}$, $\binom{2n}{2}$, ..., $\binom{2n}{2n-1}$.

The following auxiliary lemma is rather technical.

---

* This influential scientist is also renowned for ... having his name spelled in print in the largest number of different ways!

© The Author(s), under exclusive license to Springer Nature Switzerland AG 2022   151
O. Ramaré, *Excursions in Multiplicative Number Theory*, Birkhäuser Advanced Texts
Basler Lehrbücher, https://doi.org/10.1007/978-3-030-73169-4_14

Lemma 14.2

If m is a positive integer, then $\binom{2m+1}{m+1} \leq 4^m$.

The proof we present is elementary; the reader may get an explicit asymptotic expression by appealing to the Complex Stirling Formula cf. Lemma 20.4.

Proof By using the binomial expansion, we get

$$(1 + 1)^{2m+1} = \binom{2m+1}{0} + \binom{2m+1}{1} + \ldots + \binom{2m+1}{2m} + \binom{2m+1}{2m}$$

$$\geq \binom{2m+1}{m} + \binom{2m+1}{m+1} = 2\binom{2m+1}{m+1},$$

which concludes the proof.   □

Exercise 14-2. Show that, when $x \geq 1$, the product $\prod_{x/2 < p \leq x} p$ is smaller than $(7/3)^x$.

Exercise 14-3. (Legendre's Formula). Show that the power of the prime p that divides $m!$ is given by the next formula

$$v_p(m!) = \sum_{a \geq 1} [m/p^a] .$$

We deduce from our preparatory lemmata the following result:

Lemma 14.3

For any integer $n \geq 1$, the product $\prod_{p \leq n} p$ is less than 4^n.

D. Hanson in [6] found an original elementary manipulation of multinomial coefficients that reduces the 4^n above to 3^n.

Proof We proceed by induction on $\ell \geq 1$ and first show the following property:

$$\forall k \leq 2\ell, \quad \prod_{p \leq k} p \leq 4^k.$$

This is true for $\ell = 1$. As the product over the empty set is 1, we have $\prod_{p \leq 1} p = 1$ and $\prod_{p \leq 2} p = 2 \leq 4^1$. Now suppose that the property is verified for ℓ. Let us show it for $\ell + 1$. It is enough to prove that $\prod_{p \leq 2\ell+1} p \leq 4^{2\ell+1}$ and $\prod_{p \leq 2\ell+2} p \leq 4^{2\ell+2}$. As $2\ell + 2$ is not prime, the second sum can be reduced to the first one. By the recurrence hypotheses we have $\prod_{p \leq \ell+1} p \leq 4^{\ell+1}$, and on combining Lemma 14.1 and Lemma 14.2 we get $\prod_{\ell+1 < p \leq 2\ell+1} p \leq \binom{2\ell+1}{\ell+1} \leq 4^\ell$. Hence, we have

$$\prod_{p\le 2\ell+1} p = \prod_{p\le \ell+1} p \prod_{\ell+1<p\le 2\ell+1} p \le 4^{\ell+1}4^{\ell} = 4^{2\ell+1}.$$

The above inequality verifies the induction step and concludes the proof.   □

Lemma 14.4

For any real number $x > 1$ we have $\sum_{p\le x} \log p \le x \log 4$.

Proof Let x be a real number and let us denote by N its integer part. By Lemma 14.3 we have

$$\prod_{p\le x} p = \prod_{p\le N} p \le 4^N \le 4^x .$$

On taking logarithms of both the left and the right-hand side, we obtain the desired inequality.   □

Lemma 14.5

For any real number $x \ge 2$ we have $\int_2^x \frac{dt}{(\log t)^2} \le \frac{12x}{(\log x)^2}.$

Proof For $x \le x_0 = 10$ we have

$$\int_2^x \frac{dt}{(\log t)^2} \le \int_2^x \frac{dt}{(\log 2)^2} \le \left(\frac{\log x_0}{\log 2}\right)^2 \frac{x}{(\log x)^2} \le \frac{12x}{(\log x)^2} .$$

For $x \ge x_0$ consider the function

$$f : [x_0, \infty) \to \mathbb{R},$$
$$x \mapsto \frac{12x}{(\log x)^2} - \int_2^x \frac{dt}{(\log t)^2} .$$

Its derivative satisfies

$$f'(x) = \frac{12}{(\log x)^2} - \frac{24}{(\log x)^3} - \frac{1}{(\log x)^2} \ge \left(11 - \frac{24}{\log 10}\right)\frac{1}{(\log x)^2} \ge 0 .$$

Thus the function f is increasing and as we have already seen that it is positive at x_0, the lemma follows.   □

A similar inequality has been established in the different manner during the proof of Cor. 12.2.

Exercise 14-4. Show that, when $x \ge 2$, we have $\int_2^x \frac{dt}{(\log t)^2} \le \frac{12x}{5(\log x)^2}.$

Here is a result similar to Cor. 12.2 that we can get through the present approach.

Theorem 14.1

For any real number $x \geq 1$ we have $\sum_{p \leq x} 1 \leq 9x/\log x$.

Proof (*of Theorem* 14.1) We proceed by summation by parts. For any prime $p \leq x$, one has

$$\frac{1}{\log p} = \frac{1}{\log x} + \int_p^x \frac{dt}{t(\log t)^2} .$$

On using Lemma 14.4 we obtain

$$\sum_{p \leq x} 1 = \sum_{p \leq x} \log p \left(\frac{1}{\log x} + \int_p^x \frac{dt}{t(\log t)^2} \right)$$

$$= \frac{\sum_{p \leq x} \log p}{\log x} + \int_2^x \sum_{p \leq t} \log p \frac{dt}{t(\log t)^2} \leq \frac{x \log 4}{\log x} + \int_2^x \frac{dt}{(\log t)^2} \log 4 .$$

We next apply Lemma 14.5 to get that, for $x \geq 2$,

$$\sum_{p \leq x} 1 \leq \frac{x \log 4}{\log x} + \frac{12x \log 4}{(\log x)^2} \leq \frac{x \log 4}{\log x} \left(1 + \frac{12}{\log x} \right) .$$

Thus, if $x \geq 11$ we have $\sum_{p \leq x} 1 \leq 9x/\log x$. For the remaining x, the following holds:

- For $2 \leq x < 3$ we have $\sum_{p \leq x} 1 = 1 \leq x/\log x$;
- For $3 \leq x < 5$ we have $\sum_{p \leq x} 1 = 2 \leq x/\log x$;
- For $5 \leq x < 7$ we have $\sum_{p \leq x} 1 = 3 \leq x/\log x$;
- For $7 \leq x < 11$ we have $\sum_{p \leq x} 1 = 4 \leq (3x)/(2 \log x)$, though the stronger $\sum_{p \leq x} 1 \leq^? x/\log x$ is not satisfied in this interval.

The theorem is proven.   □

Exercise 14-5. Show that, when $x > 1$, we have $\sum_{p \leq x} 1 \leq 5x/\log x$.

The Erdős approach has been pursued by M. El Bachraoui in [4] and A. Loo in [8]. The next exercise gives a flavour of their work.

Exercise 14-6.

1 ◇ Prove that, when n is an even integer, the product $\prod_{\frac{n}{2} < p \leq \frac{3n}{4}} p \prod_{n < p \leq \frac{3n}{2}} p$ divides the binomial coefficient $\binom{3n/2}{n}$.

2 ◇ Show by induction that, when n is even, we have $\binom{3n/2}{n} \leq (6.75)^{n/2}$ and that, when in addition n is larger than 154, we have $\binom{3n}{2n} \geq 6.5^n$.

3 ◇ Prove that the prime decomposition of $\binom{3n}{2n}$ contains all the primes from the union of $[2n + 1, 3n]$, $(n, 3n/2]$ and $(\sqrt{3n}, 3n/4]$ with an exponent equal to 1. The primes below $\sqrt{3n}$ divide $\binom{3n}{2n}$ to a power that we denote by $\beta(p)$.

4 ◇ By combining Exer. 12-4 together with Exer. 14-3, show that the power of any prime p in $\binom{3n}{2n}$ is at most $(\log 3n)/\log p$. Deduce that $\prod_{p \le \sqrt{3n}} p^{\beta(p)} \le (3n)^{\pi(\sqrt{3n})}$,

where $\pi(m)$ counts the number of primes $p \le m$.

5 ◇ Combine the above to show that, when n is even, the interval $(n, 3n/2]$ contains a prime.

6 ◇ Adapt the above argument to cover the case when n is odd.

J. Nagura in [11] continued the approach of Chebyshev, which is perfectly described and extended by H. Diamond and P. Erdős in [3]. Regarding small intervals containing prime numbers, see [7] by Habiba Kadiri and Allysa Lumley.

The next exercise presents a very simple and elegant way (due to M. Nair in [13]), to obtain a lower bound for $\psi(x)$.

Exercise 14-7. Let us define, for any positive integer n, $d_n = \text{lcm}_{1 \le m \le n} m!$. It is shown in Exer. 12-17 that $d_n = \exp(\psi(n))$. Consider now

$$I_n = \int_0^1 x^n(1 - x)^n dx .$$

1 ◇ Show that $d_{2n+1} I_n$ is a positive integer.

2 ◇ Show that $I_n \le 1/4^n$.

3 ◇ Deduce from the above that $\psi(2n + 1) \ge 2n \log 2$.

In [9, Chap. 10], Montgomery recounts the history of this approach and proves several results, either improving on the above or exhibiting some limiting factors. The readers will find in [12] yet another proof of Nair that yields an even sharper lower estimate.

On anticipating on the next chapter, we finally mention that P. Erdős in [5] extended this work to prove that some arithmetic progressions indeed contain primes. This approach has been inspected from an explicit viewpoint in [10],

References

[1] J. Bertrand."Mémoire sur le nombre de valeurs que peut prendre une fonction quand on permute les lettres qu'elle renferme". In: *Journal de l'École Royale Polytechnique* 18. Cahier 30 (1845), pp. 123–140 (cit. on p. 147).

[2] P.L. Chebyshev. "Mémoire sur les nombres premiers". In: *Journal de mathématiques pures et appliquées, Sér. 1* 17 (1852), pp. 366–390 (cit. on p. 147).

[3]   H.G. Diamond and P. Erdős. "On sharp elementary prime number estimates". In: *Enseign. Math.* 26.3-4 (1980), pp. 313–321 (cit. on p. 150).

[4]   M. El Bachraoui. "Primes in the interval $[2n, 3n]$". In: *Int. J. Contemp. Math. Sci.* 1.13-16 (2006), pp. 617–621 (cit. on p. 150).

[5]   P. Erdős. "Über die Primzahlen gewisser arithmetischer Reihen". German. In: *Math. Z.* 39 (1935), pp. 473–491 (cit. on p. 151).

[6]   D. Hanson. "On the product of the primes". In: *Canad. Math. Bull.* 15 (1972), pp. 33–37. https://doi.org/10.4153/CMB-1972-007-7 (cit. on p. 148).

[7]   H. Kadiri and A. Lumley. "Short effective intervals containing primes". In: *Integers* 14 (2014). arXiv:1407:7902, Paper No. A61, 18 (cit. on p. 150).

[8]   A. Loo. "On the primes in the interval $[3n, 4n]$". In: *Int. J. Contemp. Math. Sci.* 6.37-40 (2011), pp. 1871–1882 (cit. on p. 150).

[9]   H.L. Montgomery. *Ten lectures on the interface between analytic number theory and harmonic analysis*. Vol. 84. CBMS Regional Conference Series in Mathematics. Published for the Conference Board of the Mathematical Sciences, Washington, DC, 1994, pp. xiv+220 (cit. on p. 151).

[10] P. Moree. "Bertrand's postulate for primes in arithmetical progressions". In: *Comput. Math. Appl.* 26.5 (1993), pp. 35–43. https://doi.org/10.1016/0898-1221(93)90071-3 (cit. on p. 151).

[11] J. Nagura. "On the interval containing at least one prime number". In: *Proc. Japan Acad.* 28 (1952), pp. 177–181 (cit. on p. 150).

[12] M.K.N. Nair. "A new method in elementary prime number theory". In: *J. London Math. Soc. (2)* 25.3 (1982), pp. 385–391. https://doi.org/10.1112/jlms/s2-25.3.385 (cit. on p. 151).

[13] M.K.N. Nair. "On Chebyshev-type inequalities for primes". In: *Amer. Math. Monthly* 89.2 (1982), pp. 126–129. https://doi.org/10.2307/2320934 (cit. on p. 150).

Chapter 15
Primes in Arithmetical Progressions

We first gently steer the readers through the general notions and then inspect them more closely in two special cases.

15.1 The Multiplicative Group of $\mathbb{Z}/q\mathbb{Z}$

We introduced in Chap. 5 the multiplicative group $(\mathbb{Z}/q\mathbb{Z})^\times$ when q is a prime number. This notion extends to the case when q is not necessarily prime and we quickly discuss this extension now.

The quotient set $\mathbb{Z}/q\mathbb{Z}$ is canonically equipped with a *ring* structure as mentioned in Sect. 5.1. We recall that, given any two integers m and n, they are said to be *congruent* modulo q when $q|m - n$. This relation separates the integers in q subsets since every integer can be uniquely written as $kq + a$ with $a \in \{0, \ldots, q - 1\}$. The equivalence class, or, equivalently, the *congruence class* of n is denoted by $n + q\mathbb{Z}$ or by $\overline{n}$, where this time the modulus q is hidden. Note that, when $m \in n + q\mathbb{Z}$, we have $\overline{m} = \overline{n}$; hence the representation $\overline{n}$ is not unique, and whenever we give a definition involving this representation, we are to check that it is independent on the choice of n. The addition over $\mathbb{Z}/q\mathbb{Z}$ is given by

$$\overline{a} + \overline{b} = \overline{a + b}$$

and this makes sense because it does not depend of the choice either of a or b. The product is defined similarly:

$$\overline{a} \cdot \overline{b} = \overline{ab} \ .$$

Here again, this definition is coherent because it does not depend on the choice of either a or of b. This ends our retelling of the classical theory.

When a is prime to q, every other representative a' of $\overline{a}$ is also prime to q. The coprimality of a to q thus extends as a property of the class. Let $(\mathbb{Z}/q\mathbb{Z})^\times$ be the set of the classes modulo q that are prime to q. It turns out that, when we multiply any two elements of this set, the result still belongs to this set. Moreover it is not difficult to

© The Author(s), under exclusive license to Springer Nature Switzerland AG 2022
O. Ramaré, *Excursions in Multiplicative Number Theory*, Birkhäuser Advanced Texts
Basler Lehrbücher, https://doi.org/10.1007/978-3-030-73169-4_15

see that $((\mathbb{Z}/q\mathbb{Z})^\times, \cdot)$ is a finite abelian group whose neutral element is $\overline{1}$. This means that the multiplication is associative, commutative (that's what *abelian* means), that multiplication by $\overline{1}$ leaves the elements unchanged and that every element has an inverse. The existence of this inverse is commonly proved by invoking the Bézout relation, but the argument below avoids it.

Proof We prove that, if $\overline{a}\overline{b} = \overline{a}\overline{b}'$ for a, b and b' all prime to q, then $\overline{b} = \overline{b}'$. This is obvious since the initial equation means that $ab - ab' \in \overline{0}$, i.e. that $q|a(b - b')$. Since q is prime to a, it has to divide $b - b'$, meaning that $\overline{b} = \overline{b}'$. Therefore the multiplication by $\overline{a}$ is injective over $(\mathbb{Z}/q\mathbb{Z})^\times$; as this set is finite, our multiplication map is thus one-to-one and in particular, there exists $\overline{b}$ such that $\overline{a}\overline{b} = \overline{1}$, concluding the proof. □

Exercise 15-1. We show in this exercise that the cardinality of $(\mathbb{Z}/q\mathbb{Z})^\times$ is $\varphi(q)$ We define $f(q) = \#\{a \in \{1, \ldots, q\}, \gcd(a, q) = 1\}$, which is the cardinality of $(\mathbb{Z}/q\mathbb{Z})^\times$.

1 ◇ Show that, when $d|q$, we have $\#\{a \le q, \gcd(a, q) = d\} = f(q/d)$.

2 ◇ Show that $\displaystyle\sum_{d|q} f(q/d) = q$.

3 ◇ Conclude.

The Lagrange Theorem tells us that

$$\forall a \in \mathbb{Z} / (a, q) = 1, \quad a^{\varphi(q)} \equiv 1 \bmod q.$$

We present a proof of this result in the following exercise.

Exercise 15-2. (Lagrange Theorem). For any $\overline{x} \in (\mathbb{Z}/q\mathbb{Z})^\times$, we define the set $H(x) = \{\overline{ax}, \overline{a}^2\overline{x}, \ldots, \overline{a}^{\varphi(q)}\overline{x}\}$, where $\overline{a}$ is some fixed invertible class modulo q.

1 ◇ Show that $H(1)$ contains $\overline{1}$. Deduce that $H(1)$ is a subgroup.

2 ◇ For any $\overline{x} \in (\mathbb{Z}/q\mathbb{Z})^\times$, show that $H(x)$ has the same cardinality as $H(1)$. Show further that either $H(x) = H(y)$, or they do not intersect.

3 ◇ Show that $\overline{x} \in H(x)$ and that $|H(1)|$ divides $\varphi(q)$.

4 ◇ Conclude.

The result of the next exercise is useful in many situations.

Exercise 15-3. Let q and q' be two positive integers, and let a be an integer coprime to q. Show that the class of a modulo q contains an integer b that is prime to qq' in two different ways (assuming, without loss of generality, that q and q' are coprime):

1 ◇ Consider the element $a + (qq' + 1 - a)q^{\varphi(q')}$.

2 ◇ Use the Chinese Remainder Theorem from Sect. 18.1 to show that there exists an integer k such that $a + kq \equiv 1 \bmod q'$.

Many mathematicians conjectured and tried to prove that, given any modulus q, the prime numbers are *equidistributed* among the $\varphi(q)$ residue classes of $(\mathbb{Z}/q\mathbb{Z})^\times$. In order to prove this equidistribution, Dirichlet introduced in [5] the *Dirichlet characters* modulo q: these are the functions $\chi : \mathbb{Z}/q\mathbb{Z} \mapsto \mathbb{C}$ that are such that

$$\chi(x) = 0 \quad \text{if and only if} \quad x \notin (\mathbb{Z}/q\mathbb{Z})^\times \quad \text{and} \quad \forall(x, y), \ \chi(xy) = \chi(x)\chi(y) \ .$$

When χ is a Dirichlet character modulo q, we have

$$\chi(1) = 1 \quad \text{and} \quad \forall a/\gcd(a, q) = 1, \quad \chi(a)^{\varphi(q)} = 1 \ ,$$

which implies that the non-zero values of χ are roots of unity. As a matter of fact, the name *Dirichlet character* is used to denote three distinct kind of functions:

- As above, it is a completely multiplicative function defined on $\mathbb{Z}/q\mathbb{Z}$ that vanishes outside $(\mathbb{Z}/q\mathbb{Z})^\times$.
- Given a Dirichlet character χ, as initially defined, we may consider the function $\tilde{\chi}$ over $\mathbb{Z}$ given by $\tilde{\chi}(n) = \chi(n \bmod q) = \chi(n + q\mathbb{Z})$.
- The third concept will not be used in this book but is of interest to the readers if they look at other sources: the restriction to $(\mathbb{Z}/q\mathbb{Z})^\times$ of a Dirichlet character as initially defined is also called by this name!

In short, to understand what is under scrutiny, the readers should keep in mind where the variable lives, whether $\mathbb{Z}/q\mathbb{Z}$, $\mathbb{Z}$ or $(\mathbb{Z}/q\mathbb{Z})^\times$. Note that the Legendre symbol introduced Sect. 5.4 *is* a Dirichlet character. The theory of characters then expands in two different directions: towards the theory of representations, and towards the Pontryagin duality. Notice also that Dirichlet's Theorem preceded the proof of the Prime Number Theorem by nearly 60 years.

For every q, the function given by

$$\chi_0(n) = \begin{cases} 1 & \text{when } n \text{ is prime to } q, \\ 0 & \text{otherwise,} \end{cases}$$

is a character, called *the principal character modulo q*. Please note that this character depends on q though the traditional notation χ_0 does not indicate this.

Exercise 15-4. Prove that the set of all the Dirichlet characters modulo q is a finite abelian group where the product of χ_1 and χ_2 is simply given by $(\chi_1\chi_2)(n) = \chi_1(n)\chi_2(n)$. The neutral element is the principal character χ_0.

Dirichlet characters are accessible in Pari/GP, but we have not studied enough the structure of $(\mathbb{Z}/q\mathbb{Z})^\times$ and its characters to be able to explain the description used. Let us only say that these characters are identified uniquely by an integer m between 1 and q and coprime to q. Here is how to get the list of values (as a `vector`) of the character indexed by m:

```
myq = 3;
G = znstar(myq, 1);
{VectorValChar(q, G, m) =
    vector(q, n, if(gcd(n, q) > 1, 0,
            exp(2*I*Pi*chareval(G, znconreychar(G, m), n))))}
VectorValChar(myq, G, 2)
```

The above script specifically gives the values of the character numbered 2 (i.e. $m = 2$) modulo 3. The readers will check that the character numbered 1 is the principal character.

In Sage, the set of Dirichlet characters is built via the call `DirichletGroup(q)`. The characters are numbered from 0 to $\varphi(q) - 1$ and several functions are accessible that we do not enumerate here. As an example, the script

```
DG = DirichletGroup(35)
mychi = DG[6]
[mychi(n) for n in range(1,35)]
C = ComplexIntervalField(200)
[C(mychi(n)) for n in range(1,35)]
```

first gives an exact answer, the values being polynomials in the symbol `zeta12` that represents a twelvth root of unity, and then the requested numerical approximation.

Exercise 15-5. An integer is said to be a *Carmichael number* if n is not prime but, for every integer a, we have $a^n \equiv a \bmod n$.

1 ◊ Prove the Theorem of Korselt: n is a Carmichael number if and only if (1) n is square-free and (2) for every prime factor p of n, the integer $p - 1$ divides $n - 1$.

2 ◊ Show that 561 is a Carmichael number (it is the smallest such number).

A. Korselt published this criterion in 1899, without really identifying the modulus. R.D. Carmichael established a criterion close to the one of Korselt in 1910/1912 and started giving the list 561, 1 105, 1 729, etc. Etc? Yes, since, effectively, there are infinitely many Carmichael numbers though this is far from being obvious! This was finally proved by Alford, Granville & Pomerance in [1]. As a matter of fact, Alford was writing a program to extend the list of known Carmichael numbers when he found that his algorithm was producing *really many* examples. He decided to go and knock on the doors of his colleagues, Granville and Pomerance, to show them his program. The trio then succeeded in turning this practical success into a theoretical one. Here are the first Carmichael numbers:

$$561 = 3 \cdot 11 \cdot 17, \quad 1729 = 7 \cdot 13 \cdot 19, \quad 2821 = 7 \cdot 13 \cdot 31,$$
$$1105 = 5 \cdot 13 \cdot 17, \quad 2465 = 5 \cdot 17 \cdot 29, \quad 6601 = 7 \cdot 23 \cdot 41.$$

15.2 Dirichlet Characters Modulo 3 and 4

Aside from the principal character, there is only one Dirichlet character modulo 4. It is defined by

$$\forall n \equiv 1 \bmod 2 , \quad \chi_4(n) = (-1)^{(n-1)/2} . \tag{15.1}$$

When n is even, we set $\chi_4(n) = 0$. Here is a property that we use very often:

$$\sum_{M < n \leq M+4} \chi_4(n) = 0 . \tag{15.2}$$

The script above initialized with $G = \texttt{znstar(4, 1)}$ refers to this character as the character numbered 3 (i.e. $m = 3$). Similarly, aside from the principal character, there is only one Dirichlet character modulo 3; it is given by

$$\chi_3(n) = \begin{cases} 0 & \text{if } n \equiv 0 \bmod 3, \\ 1 & \text{if } n \equiv 1 \bmod 3, \\ -1 & \text{if } n \equiv 2 \bmod 3. \end{cases} \tag{15.3}$$

The script above initialized with $G = \texttt{znstar(3, 1)}$ refers to this our character as the one numbered 2. We again have

$$\sum_{M < n \leq M+3} \chi_3(n) = 0 . \tag{15.4}$$

Both (15.2) and (15.4) are to be compared with Theorem 5.3.

Exercise 15-6. Write a Pari/GP code starting by `'chi3(n)='` and containing at most 23 block letters to compute χ_3.

Exercise 15-7. Show that the only non-principal characters modulo 8 are $\chi'_4(n) = (-1)^{(n-1)/2}$ and the two characters χ_8 and $\chi'_4\chi_8$ given by

$$\forall n \equiv 1 \bmod 2 , \quad \chi_8(n) = (-1)^{(n^2-1)/8} , \quad \chi'_4\chi_8(n) = (-1)^{(n-1)/2+(n^2-1)/8} .$$

Of course, we assign 0 as the value of these two characters on even integers.

We then consider the two Dirichlet series given by

$$L(s, \chi_3) = \sum_{n \geq 1} \frac{\chi_3(n)}{n^s} , \quad L(s, \chi_4) = \sum_{n \geq 1} \frac{\chi_4(n)}{n^s} . \tag{15.5}$$

Both sums converge at $s = 1$ by (15.4) and (15.2). We describe in Sect. 16 how to compute approximate values of $L(s, \chi)$ by using Hurwitz zeta functions, but since we introduced in Chap. 3 the `Dokchitser` package of T. Dokchitser, let us tell the readers what are the magic words required to compute both $L(s, \chi_3)$ and $L(s, \chi_4)$ through it:

```
myL3 = Dokchitser(conductor=3, gammaV=[1],
                  weight=1, eps=1, prec=300)
myL3.init_coeffs('chi3(k)',
                 pari_precode='chi3(n)=[0,1,-1][1+n%3]')

myL4 = Dokchitser(conductor=4, gammaV=[1],
                  weight=1, eps=1, prec=300)
myL4.init_coeffs('chi4(k)',
                 pari_precode='chi4(n)=[0,1,0,-1][1+n%4]')

print(myL3(1.1+I*2))
```

Please note that the command `myL3.init_coeffs(...)` has to be run every time you change the precision. The value taken by $L(s, \chi)$ at $s = 1$ is of particular significance: showing that $L(1, \chi_3)$ and $L(1, \chi_4)$ do not vanish is crucial to our analysis. Here is a lemma that enables us to compute these two values.

Lemma 15.1

Let $\alpha \in (0, 1)$ be some real number. We have

$$\lim_{K \to \infty} \sum_{1 \le k \le K} \frac{e(k\alpha)}{k} = -\log |2 \sin \pi\alpha| + i(\tfrac{1}{2} - \alpha)\pi,$$

where we use the classical notation

$$e(\alpha) = \exp(2i\pi\alpha) . \tag{15.6}$$

Proof This series converges towards $-\log(1 - e(\alpha))$ and our lemma belongs to a calculus course. Here is however an argument that uses the following explicit analytic continuation of the logarithm:

$$\log(x + iy) = \log \sqrt{x^2 + y^2} + 2i \arctan \frac{y}{x + \sqrt{x^2 + y^2}} , \quad x + iy \in \mathbb{C} \setminus \mathbb{R}^- .$$

On setting $1 - e(\alpha) = x + iy$ and taking $\beta = \pi\alpha$, we get $x = 1 - \cos \beta = 2 \sin^2(\beta/2)$, $x^2 + y^2 = 4 \sin^2(\beta/2)$ and $y = -\sin \beta = -2 \sin(\beta/2) \cos(\beta/2)$. It follows that

$$\log(1 - e(\alpha)) = \log |2 \sin(\beta/2)| + 2i \arctan \frac{-\cos(\beta/2)}{\sin(\beta/2) + 1} .$$

Define $t = \tan(\beta/4)$. Then using $\cos(\beta/2) = (1-t^2)/(1+t^2)$ and $\sin(\beta/2) = 2t/(1+t^2)$ we find that

$$\arctan \frac{-\cos(\beta/2)}{\sin(\beta/2) + 1} = \arctan \frac{\sin(\beta/4) - \cos(\beta/4)}{\sin(\beta/4) + \cos(\beta/4)} = \arctan \frac{\sin((\beta - \pi)/4)}{\cos((\beta - \pi)/4)} .$$

Finally, we obtain $\log(1 - e(\alpha)) = \log |2 \sin(\beta/2)| + i(\alpha - \tfrac{1}{2})\pi$.

Theorem 15.1

We have $L(1, \chi_3) = \dfrac{\pi}{3\sqrt{3}}$, $\quad L(1, \chi_4) = \dfrac{\pi}{4}$.

Proof We have $\chi_3(n) = (e(n/3) - e(2n/3))/(i\sqrt{3})$, which can be verified for any n via

$$e(1/3) = \frac{-1 + i\sqrt{3}}{2}, \quad e(2/3) = \frac{-1 - i\sqrt{3}}{2}.$$

Similarly, we see that $\chi_4(n) = (e(n/4) - e(3n/4))/(2i)$. Lemma 15.1 gives us

$$\sum_{k \geq 1} \frac{e(k/3)}{k} = -\log\left|2 \sin \frac{\pi}{3}\right| + i\frac{\pi}{6}, \quad \sum_{k \geq 1} \frac{e(2k/3)}{k} = -\log\left|2 \sin \frac{2\pi}{3}\right| - i\frac{\pi}{6}.$$

The conclusion is immediate. We point out that the result is easy to check numerically; indeed, a summation by parts directly gives us

$$\sum_{k \geq 1} \frac{e(k/3)}{k} = \sum_{1 \leq k \leq K} \frac{e(k/3)}{k} + \int_{K+1}^{\infty} \sum_{K < k \leq t} e(k/3) \frac{dt}{t^2}$$

$$= \sum_{1 \leq k \leq K} \frac{e(k/3)}{k} + O^*(1/(K+1)).$$

Taking, say, $K = 100$ and calculating with Pari/GP, we obtain a rather good approximation of the stated result.   □

Exercise 15-8. Let $\chi = \chi_4'\chi_8$ be the Dirichlet character defined in Exer. 15-7 Prove that $L(1, \chi) = \pi/(2\sqrt{2})$.

15.3 The Theorems of Mertens Modulo 3 and 4

We start by noticing that the Dirichlet characters are completely multiplicative functions, therefore the Euler product representation holds:

$$L(s, \chi) = \prod_{p \geq 2}\left(1 - \frac{\chi(p)}{p^s}\right)^{-1} \quad (\Re s > 1).$$

Exercise 15-9.

1 ⋄ Show that for $\chi = \chi_3$ or χ_4 we have $-\dfrac{L'}{L}(s, \chi) = \displaystyle\sum_{n \geq 1} \frac{\chi(n)\Lambda(n)}{n^s}$, where s is such that $\Re s > 1$.

2 ⋄ Show that for $\chi = \chi_3$ or χ_4 we have $\log L(s, \chi) = \displaystyle\sum_{n \geq 2} \frac{\chi(n)\Lambda(n)}{n^s \log n}$, where $\Re s > 1$.

Here is a discovery due to Dirichlet:

Lemma 15.2

For every n coprime to 3 we have

$$\mathbb{1}_{n\equiv 1[3]} = \tfrac{1}{2}\left(\mathbb{1}_{(n,3)=1}(n) + \chi_3(n)\right), \quad \mathbb{1}_{n\equiv 2[3]} = \tfrac{1}{2}\left(\mathbb{1}_{(n,3)=1}(n) - \chi_3(n)\right).$$

Similarly, for every n we have

$$\mathbb{1}_{n\equiv 1[4]} = \tfrac{1}{2}\left(\mathbb{1}_{(n,2)=1}(n) + \chi_4(n)\right), \quad \mathbb{1}_{n\equiv 3[4]} = \tfrac{1}{2}\left(\mathbb{1}_{(n,2)=1}(n) - \chi_4(n)\right).$$

By the lemma above, we may detect the membership of an integer to an arithmetic progression through a linear combination of multiplicative functions. Let us now proceed with a proof due to F. Mertens in [8].

Lemma 15.3

For $\chi = \chi_3$ or χ_4 and for $x \geq 1$, we have

$$L(1,\chi) = \sum_{n\leq x}\frac{\chi(n)}{n} + O(1/x), \; -L'(1,\chi) = \sum_{n\leq x}\frac{\chi(n)\log n}{n} + O\!\left(\frac{\log(2x)}{x}\right).$$

Proof Use summation by parts for both formulas.   □

Lemma 15.4

For $\chi = \chi_3$ or χ_4 and for $x \geq 1$, we have

$$-\frac{L'(1,\chi)}{L(1,\chi)} = \sum_{n\leq x}\frac{\Lambda(n)\chi(n)}{n} + O(1).$$

Note that the lemma can be expressed in the simpler form $\sum_{n\leq x}\Lambda(n)\chi(n)/n = O(1)$! However, we expect the sum on the right-hand side to approximate $-L'(1,\chi)/L(1,\chi)$ and that the $O(1)$ can be converted to $o(1)$. This is indeed the case, but we do not prove it in this book.

Proof We begin with Lemma 15.3:

$$-L'(1,\chi) = \sum_{n\leq x}\frac{\chi(n)\log n}{n} + O(\log(2x)/x)$$

and use $\log = \mathbb{1} \star \Lambda$. Therefore

$$-L'(1,\chi) = \sum_{\ell\leq x}\frac{\chi(\ell)\Lambda(\ell)}{\ell}\sum_{m\leq x/\ell}\frac{\chi(m)}{m} + O(\log(2x)/x).$$

The previous lemma brings us to

$$-L'(1,\chi) = L(1,\chi) \sum_{\ell \le x} \frac{\chi(\ell)\Lambda(\ell)}{\ell} + O\left(\sum_{\ell \le x} \Lambda(\ell)/x\right) + O(\log(2x)/x).$$

The Chebyshev Theorem (i.e. Cor. 12.1) together with the non-vanishing of $L(1,\chi)$ allow us to finish the proof. $\quad\square$

Exercise 15-10. Exploit Exer. 6-5 to show that, for $\chi = \chi_3$ or χ_4 and for $x \ge 1$, we have

$$\frac{1}{L(1,\chi)} = \sum_{n \le x} \frac{\mu(n)\chi(n)}{n} + O(1).$$

In [6], Eqs. (9) and (10) therein, Gel'fond added another tool to the Mertens toolbox. We present it in the next exercise.

Exercise 15-11.

$1 \diamond$ Show that $\displaystyle\sum_{d|n} \mu(d) \log \frac{x}{d} = \begin{cases} \log x & \text{when } n = 1, \\ \Lambda(n) & \text{when } n > 1. \end{cases}$

$2 \diamond$ Show that

$$\sum_{n \le x} \frac{\Lambda(n)\chi(n)}{n} = -\log x + L(1,\chi) \sum_{d \le x} \frac{\mu(d)\chi(d)}{d} \log \frac{x}{d} + O(1).$$

Theorem 15.2

We have

$$\sum_{\substack{n \le x, \\ n \equiv 1[3]}} \frac{\Lambda(n)}{n} = \tfrac{1}{2} \log x + O(1), \qquad \sum_{\substack{n \le x, \\ n \equiv 2[3]}} \frac{\Lambda(n)}{n} = \tfrac{1}{2} \log x + O(1).$$

The same holds for primes modulo 4

$$\sum_{\substack{n \le x, \\ n \equiv 1[4]}} \frac{\Lambda(n)}{n} = \tfrac{1}{2} \log x + O(1), \qquad \sum_{\substack{n \le x, \\ n \equiv 3[4]}} \frac{\Lambda(n)}{n} = \tfrac{1}{2} \log x + O(1).$$

This proves that the primes are evenly distributed among the two possible (i.e. where there may be infinitely many primes) classes modulo 3, and similarly modulo 4.

Proof Using Lemma 15.2 we write

$$\sum_{\substack{n \le x, \\ n \equiv 1[3]}} \frac{\Lambda(n)}{n} = \frac{1}{2} \sum_{\substack{n \le x, \\ \gcd(n,3)=1}} \frac{\Lambda(n)}{n} + \frac{1}{2} \sum_{\substack{n \le x, \\ \gcd(n,3)=1}} \frac{\Lambda(n)\chi_3(n)}{n}.$$

We apply Lemma 12.2 for the first sum (the contribution of powers of 3 being $O(1)$) and Lemma 15.4 for the second one. The statement follows. $\quad\square$

Exercise 15-12.

1 ⋄ Show that there exist two constants $c_1, c_2 > 0$, such that for $x \geq 7$, we have
$$c_2 x \geq \sum_{\substack{n \leq x, \\ n \equiv 1[3]}} \Lambda(n) \geq c_1 x.$$

Here is a Pari/GP script that proves that $\displaystyle \sum_{\substack{p \leq x, \\ p \equiv 1[3]}} 1 \leq \frac{1.15\, x}{2 \log x}$ when $x \leq 10^7$.

```
mymax = 0; res = 0;
forprimestep(p = 7, 10000000, 3,
  res ++; mymax = max(mymax, res/ (p/2/log(p))));
mymax
```

The sequence of primes of the form $3n + 1$ is referenced A003627 in [7], where some more information is available. The readers may also look at the sequences A005098, A002145 and A002476.

Exercise 15-13. In this exercise, the readers are led to prove a first and third Mertens' theorem (i.e. Theorem 12.2 and 12.4) for primes congruent to 1 modulo 4.

1 ⋄ Show that there exists a constant $b_{1,4}$ such that
$$\sum_{\substack{p \leq x, \\ p \equiv 1[4]}} \frac{1}{p} = \frac{1}{2} \log \log x + b_{1,4} + O(1/\log(2x))$$

for every $x \geq 1$.

2 ⋄ Deduce that there exists a constant $c_{1,4}$ such that for every $x \geq 1$, we have
$$-\sum_{\substack{p \leq x, \\ p \equiv 1[4]}} \log\left(1 - \frac{1}{p}\right) = \frac{1}{2} \log \log x + c_{1,4} + O(1/\log(2x)).$$

3 ⋄ Conclude that we have $\displaystyle \prod_{\substack{p \leq x, \\ p \equiv 1[4]}} \left(1 - \frac{1}{p}\right) = \frac{e^{-c_{1,4}}}{\sqrt{\log x}}(1 + O(1/\log x))$, which is valid for $x \geq 2$.

4 ⋄ Show that there exists a constant $c'_{1,4}$ such that
$$\prod_{\substack{p \leq x, \\ p \equiv 1[4]}} \left(1 + \frac{1}{p}\right) = e^{c'_{1,4}} \sqrt{\log x}(1 + O(1/\log x))$$

for every $x \geq 2$.

The same applies for primes congruent to 3 mod 4. Concerning the general case, the interested readers may take a look at the paper [2] by O. Bordellès. We complete this part with a product description for $L(1, \chi_4)$, where χ_4 is the Dirichlet character defined in (15.2).

Exercise 15-14. For any real parameter x larger than 2, we set

$$S(x) = \prod_{p \leq x} \left(1 - \frac{\chi_4(p)}{p}\right)^{-1}$$

with the aim of showing that $L(1, \chi_4) = \lim_{x \to \infty} S(x)$.

1 ⋄ By using the previous exercise and the remark following it show that $S(x)$ tends to some limit ℓ when x goes to infinity.

2 ⋄ Show that $S(x) - \sum_{n \leq x} \frac{\chi(n)\Lambda(n)}{n \log n} \ll x^{-1/2}$ and conclude that $S(x)$ also tends to ℓ.

3 ⋄ Show that, for any $\delta > 0$, we have $\delta \int_1^\infty S(x) \frac{dx}{x^{1+\delta}} = \log L(1+\delta, \chi)$. Conclude

Notice that the conclusion of this exercise is that

$$L(1, \chi_4) = \prod_{p \geq 2} \left(1 - \frac{\chi_4(p)}{p}\right)^{-1}, \tag{15.7}$$

where the product is convergent only in the sense that the partial products converge to a limit.

In the next two exercises, we show by elementary means that some arithmetic progressions do contain primes. These proofs yield quantitative bounds of a much lesser quality and cannot be generalized to every progression as shown by R. Murty in an early work (as this memoir is not publicly available, we refer to [9]). These proofs are however interesting in the connections they establish.

Exercise 15-15. Show that there are infinitely many primes of the form $4m + 3$.

Exercise 15-16. Show that there are infinitely many primes of the form $6m + 5$.

We restricted our attention to characters modulo 3 and 4 to express the argument more clearly. Here is a generalization that is easily within reach of the readers. It can be seen as a continuation of Exer. 5-10.

Exercise 15-17. Let b be a prime number. We define $\chi(n) = (\frac{n}{b})$. The function $L(s, \chi)$ has been defined in (5.9).

1 ⋄ Show that, for $x \geq 1$, we have $-\frac{L'(1, \chi)}{L(1, \chi)} = \sum_{n \leq x} \frac{\Lambda(n)\chi(n)}{n} + O(1)$.

2 ⋄ Show that

$$\sum_{\substack{n \leq x, \\ (\frac{n}{b})=1}} \frac{\Lambda(n)}{n} = \tfrac{1}{2} \log x + O(1), \qquad \sum_{\substack{n \leq x, \\ (\frac{n}{b})=-1}} \frac{\Lambda(n)}{n} = \tfrac{1}{2} \log x + O(1).$$

Conclude that about half of the primes are quadratic residues modulo b.

Epilogue

What about a Prime Number Theorem for Arithmetic Progression, that is to say an analogue of our Theorem 23.1 but with, say, primes congruent to 1 modulo 4? What we have proven here is only a logarithmic version of it. This is weaker as, for instance, Theorem 23.1 implies Theorem 12.2 via a summation by parts, but the reverse is false. A full version of the Prime Number Theorem for Arithmetic Progression can however be achieved and the readers may prove one by themselves. Historically, this is due to de la Vallée-Poussin in [4], who was barely twenty years old! The readers should first obtain an analogue of Theorem 23.3 for $L(s, \chi_4)$, and then aim at establishing that

$$\sum_{n \leq x} \chi_4(n)\mu(n) \ll_A x/(\log x)^A.$$

The road is open for the willing, but *there are* difficulties!

Further Reading

The readers who want to study prime in arithmetic progressions should foremost read the monograph [3] by H. Davenport and revised by H.L. Montgomery. The road is then open to a large body of work that lies outside the scope of the present book.

References

[1] W.R. Alford, A. J. Granville, and C. Pomerance. "There are infinitely many Carmichael numbers". In: *Ann. of Math. (2)* 139.3 (1994), pp. 703–722. https://doi.org/10.2307/2118576 (cit. on p. 156).

[2] O. Bordellès. "An explicit Mertens' type inequality for arithmetic progressions". In: *JIPAM. J. Inequal. Pure Appl. Math.* 6.3 (2005), Article 67, 10 (cit. on p. 162).

[3] H. Davenport. *Multiplicative number theory*. Third. Vol. 74. Graduate Texts in Mathematics. Revised and with a preface by Hugh L. Montgomery. Springer-Verlag, New York, 2000, pp. xiv+177 (cit. on p. 164).

[4] Ch. de la Vallée Poussin. "Recherches analytiques sur la théorie des nombres premiers. Deuxième partie. Les fonctions de Dirichlet et les nombres premiers de la forme linéaire

$Mx + N$". In: *Ann. Soc. Scient. Bruxelles, deuxième partie* 20 (1896), pp. 281–362 (cit. on p. 164).

[5]   J.P.G.L. Dirichlet. "Beweis des Satzes, das jede unbegrenzte arithmetische Progression, deren erstes Glied und Differenz ganze Zahlen ohne gemeinschaftlichen Factor sind, unendlich viele Primzahlen enthält". In: *Abhandlungen der Königlichen Preussischen Akademie der Wissenschaften zu Berlin* (1837) (cit. on p. 155).

[6]   A.O. Gel'fond. "On the arithmetic equivalent of analyticity of the Dirichlet L-series on the line Re $s = 1$". In: *Izv. Akad. Nauk SSSR. Ser. Mat.* 20 (1956), pp. 145–166 (cit. on p. 161).

[7]   OEIS Foundation Inc. *The On-Line Encyclopedia of Integer Sequence.* oeis.org/. 2019 (cit. on p. 162).

[8]   F. Mertens. "On Dirichlet's proof of the theorem that every infinite arithmetic progression of integers, whose difference is prime to its members, contains infinitely many prime numbers. (Ueber Dirichlet's Beweis des Satzes, dass jede unbegrenzte ganzzahlige arithmetische Progression, deren Differenz zu ihren Gliedern teilerfremd ist, unendlich viele Primzahlen enthält.)" In: *Wien. Ber.* 106 (1897), pp. 254–286 (cit. on p. 160).

[9]   M.R.P.M. Murty and N. Thain. "Prime numbers in certain arithmetic progressions". In: *Funct. Approx. Comment. Math.* 35 (2006), pp. 249–259. https://doi.org/10.7169/facm/1229442627 (cit. on p. 163).

Chapter 16
Computing a Famous Constant

Let $B(x)$ denote the number of integers $n \leq x$ that can be written as a sum of two integer squares. In early 1913 a then unknown clerk by the name of S. Ramanujan made the following claim in his first letter to the very famous mathematician Hardy.

Claim The number of integers between A and x which are either squares or sums of two squares is

$$K \int_A^x \frac{dt}{\sqrt{\log t}} + \theta(x),$$

where $K = 0.764\ldots$ and $\theta(x)$ is very small compared with the previous integral. K and $\theta(x)$ have been exactly found, though complicated...

In the ensuing correspondence it became clear that Ramanujan thought that $\theta(x) = O_\epsilon(x^{1/2+\epsilon})$ with $\epsilon > 0$ arbitrary, and that

$$K = \frac{1}{\sqrt{2}} \prod_{p \equiv 3[4]} (1 - p^{-2})^{-1/2}. \tag{16.1}$$

Thus, in effect, Ramanujan claimed that

$$B(x) = K \int_2^x \frac{dt}{\sqrt{\log t}} + O_\epsilon(x^{1/2+\varepsilon}), \tag{16.2}$$

with $\varepsilon > 0$ arbitrary and K as in (16.1). Note that it follows by partial integration from (16.2) that

$$B(x) \sim K \frac{x}{\sqrt{\log x}}. \tag{16.3}$$

Unknown to Ramanujan, E. Landau in 1908 had established this and the readers may try to use Theorem 13.3 to prove it. We recall that n is a sum of two squares if and only if the exponents of the primes congruent to 3 modulo 4 that divide n are even (see Exer. 5-12). On combining Theorem 13.4 together with Theorem 15.2, the readers should be able to prove the weaker estimate

$$\sum_{n \le x} \frac{b(n)}{n} \sim 2K \sqrt{\log x},$$

where $b(n)$ is the characteristic function of sums of two squares. The constant K is now called the *Landau–Ramanujan constant*. It is sometimes given in the alternative form

$$K = \frac{\pi}{4} \prod_{p \equiv 1[4]} (1 - p^{-2})^{1/2},$$

which is an easy consequence of (16.1) and the value $L(1, \chi_4) = \pi/4$ given in Theorem 15.1, χ_4 having been defined in (15.1).

Note that, for $\Re(s) > 1/2$,

$$\prod_{p \equiv 3[4]} (1 - p^{-2s})^{-2} = \frac{\zeta(2s)(1 - 2^{-2s})}{L(2s, \chi_4)} \prod_{p \equiv 3[4]} (1 - p^{-4s})^{-1}. \tag{16.4}$$

How to compute $L(2s, \chi_4)$? We can do so by using the Hurwitz zeta function, defined for any positive real number x by:

$$\zeta(s, x) = \sum_{m \ge 0} \frac{1}{(x + m)^s}. \tag{16.5}$$

Once we have this function at our disposal, here is how to obtain $L(2s, \chi_4)$:

$$L(2s, \chi_4) = \sum_{k \ge 0} \frac{1}{(1 + 4k)^{2s}} - \sum_{k \ge 0} \frac{1}{(3 + 4k)^{2s}} = \frac{1}{4^{2s}} \left(\zeta(2s, 1/4) - \zeta(2s, 3/4) \right).$$

In Pari/GP, the necessary function is `zetahurwitz`. Here is the corresponding Sage script:

```
C = ComplexIntervalField(500)

def GetL(s):
    return(C((hurwitz_zeta(s, 1/4)-hurwitz_zeta(s, 3/4))/4^s))

GetL(C(2.1+I))
```

By recursion we infer from (16.1) and (16.4) the following formula,

$$K = \frac{1}{\sqrt{2}} \prod_{n=1}^{\infty} \left((1 - 2^{-2^n}) \frac{\zeta(2^n)}{L(2^n, \chi_4)} \right)^{1/2^{n+1}},$$

which was already known to Ramanujan, see [1, pp. 60-66] edited by B.C. Berndt, and in [10, p. 78] by D. Shanks. Using this expression one computes that $K = 0.76422365358922066299\ldots$, confirming the decimals given by Ramanujan.

Indeed, as of 2019 there are more than 125 000 decimals of K known, cf. sequence A064533 of the OEIS [3].

The curious readers will wonder whether Ramanujan's claim (16.2) is actually true. It is not difficult to use Landau's method* to show, like J.-P. Serre in [7], that for every $m \geq 2$ there exists constants $c_1, \ldots, c_{m-1}$ such that

$$B(x) = K \frac{x}{\sqrt{\log x}} \left(1 + \frac{c_1}{\log x} + \ldots + \frac{c_{m-1}}{\log^{m-1} x} + O\left(\frac{1}{\log^m x} \right) \right). \tag{16.6}$$

It follows by partial integration from (16.2) that

$$B(x) = K \frac{x}{\sqrt{\log x}} \left(1 + \frac{1}{2 \log x} + O\left(\frac{1}{\log^2 x} \right) \right), \tag{16.7}$$

and thus Ramanujan's claim implies that $c_1 = 1/2$.

An expression for c_1 can be given involving some easy constants and the infinite sum

$$S = \sum_{p \equiv 3[4]} \frac{\log p}{p^2 - 1}.$$

Applying (16.4) m times, we obtain

$$S = \sum_{p \equiv 3[4]} \frac{\log p}{p^{2m+1} - 1} + \frac{1}{2} \sum_{n=1}^{m} \left\{ \frac{L'(2^m, \chi)}{L(2^m, \chi)} - \frac{\zeta'(2^m)}{\zeta(2^m)} - \frac{\log 2}{2^{2m} - 1} \right\}.$$

Using this Shanks in [10] computed S with high precision and gave a rigorous estimate for c_1, showing that $c_1 \neq 1/2$, thus disproving the claim of Ramanujan. We encourage the readers to do so by using something similar to (9.23).

16.1 The Hexagonal Versus the Square Lattice

The number $a^2 + b^2$ is the square of the distance of the point (a, b) to the origin. Thus the function $B(x)$ counts the squares of the *different* distances in the so-called *square lattice*, having the points (a, b) with both a and b integers, as its lattice points. Using analytically similar, however algebraically much more involved, techniques one can count the number of different squares of distances $\leq x$ in *any* two-dimensional lattice. A special role is played here by the *hexagonal lattice* Σ of covolume 1. It is given by

$$\Sigma = \sqrt{\frac{2}{\sqrt{3}}} \left(\mathbb{Z} \begin{pmatrix} 1 \\ 0 \end{pmatrix} \oplus \mathbb{Z} \begin{pmatrix} 1/2 \\ \sqrt{3}/2 \end{pmatrix} \right).$$

---

* Though this method is nowadays usually refered to as the *Selberg–Delange Method*, it is Landau who initiated the usage of Mellin inversion together with Hankel contour for the squareroot of a meromorphic function, while Selberg and Delange used *complex powers* of such a function.

The associated quadratic form is $(X^2 + XY + Y^2)2/\sqrt{3}$. It is shown in [4] that of all the two-dimensional lattices of covolume 1, Σ has asymptotically the fewest distances.

In 1998 S. Schaller [6] made a general conjecture that if true would imply in particular that the hexagonal lattice is "better" than the square lattice. More precisely, he claimed that if $0 < h_1 < h_2 < \ldots$ are the integers listed in ascending order that can be written as integers as $h_i = a^2 + 3b^2$ for integers a and b and $0 < q_1 < q_2 < \ldots$ are integers listed in ascending order that can be written as $q_i = a^2 + b^2$ for integers a and b, then $q_i \leq h_i$ for every $i \geq 1$. If we let $B_3(x)$ be the counting function of the integers that can be written in the form $a^2 + 3b^2$, then the conjecture states that $B(x) \geq B_3(x)$ for every $x \geq 0$. Now already Landau could have easily shown that this conjecture is certainly asymptotically true. Namely, we have

$$B_3(x) \sim K_3 \frac{x}{\sqrt{\log x}}, \tag{16.8}$$

with

$$K_3 = \frac{1}{\sqrt{2}} \frac{1}{3^{1/4}} \prod_{p \equiv 2[3]} (1 - p^{-2})^{-1/2} \approx 0.63890940544\ldots,$$

and $K > K_3$. Using techniques from computational analytic number theory very much in the spirit of this book, the truth of Schaller's conjecture can be established, see [5].

The readers might be surprised to see the two different quadratic forms $X^2 + XY + Y^2$ and $X^2 + 3Y^2$ appear in the description of the hexagonal lattice. However, they can be shown to represent the same set of integers, the so-called *Loeschian numbers* referenced A003136 in the OEIS [3].

Exercise 16-1. Let $c(n)$ be the characteristic function of the set of those integers that can be written as a sum of two integer squares and let $F(s) = \sum_{n=1}^{\infty} c(n)n^{-s}$ be the associated L-series. Show that

$$F(s)^2 = \zeta(s)L(s, \chi_4)(1 - 2^{-s})^{-1} \prod_{p \equiv 3[4]} (1 - p^{-2s})^{-1}.$$

Note that, but for $\zeta(s)$, the factors above are analytic for $\Re s > 1/2$ and thus roughly speaking $F(s)$ around $s = 1$ behaves as a constant times $\sqrt{\zeta(s)}$.

Exercise 16-2. Show that $K_3 = 0.63890940544\ldots$

See OEIS [3] sequence A301429 for more information. Here is the general mechanism used (see [10, Eq. (15)] and [4, Eq. (3.2)]).

Exercise 16-3.

1 ⋄ Let $\mathcal{P}$ be a set of primes and let f be a function on $\mathcal{P}$ with values in $\{\pm 1\}$ Show that, for every s with $\Re s > 1$, we have

$$\prod_{\substack{p \in \mathcal{P}, \\ f(p)=-1}} (1 - p^{-s})^2 = \frac{\prod_{p \in \mathcal{P}}(1 - p^{-s})}{\prod_{p \in \mathcal{P}}(1 - f(p)p^{-s})} \prod_{\substack{p \in \mathcal{P}, \\ f(p)=-1}} (1 - p^{-2s}).$$

2 ⋄ By using for f the non-trivial character modulo 3 defnied in (15.3), find an approximate value of $\prod_{p \equiv 2[3]} (1 - p^{-2})$.

The last constant we compute is linked with the *Loeschian constant*, cf. sequence A301429 of the OEIS [3]. The next exercise proposes another mechanism to compute very rapidly some Euler products. We borrow the idea from [2].

Exercise 16-4. We consider the four characters defined by $\chi_{0,12}(n) = \mathbb{1}_{(n,12)=1}$ (this is the principal character modulo 12) and (see Eqs. (15.3) and (15.1)) $\chi_{1,12}(n) = \mathbb{1}_{(n,12)=1}\chi_3(n)$, $\chi_{2,12}(n) = \mathbb{1}_{(n,12)=1}\chi_4(n)$ and finally $\chi_{3,12}(n) = \chi_3(n)\chi_4(n)$.

1 ⋄ Check that, for any of the four functions above, say χ, we have $\chi(mn) = \chi(m)\chi(n)$ for every integers m and n. Deduce that $L(s, \chi) = \prod_p (1 - \chi(p)p^{-s})^{-1}$.

2 ⋄ Show that

$$\prod_{\substack{p \geq 5, \\ p \equiv 1[12]}} \frac{1}{(1 - p^{-s})^4} = \frac{\prod_\chi L(s, \chi)}{((1 - 2^{-2s})(1 - 3^{-2s})\zeta(2s))^2} \prod_{\substack{p \geq 5, \\ p \equiv 1[12]}} \frac{1}{(1 - p^{-2s})^2},$$

where the product is over the four characters above. Deduce from this formula an approximate value of $\prod_{p \equiv 1[12]}(1 - 1/p^2)^{-1}$ certified at 10^{-100}.

3 ⋄ Show that

$$\prod_{\substack{p \geq 5, \\ p \equiv 11[12]}} \frac{1}{(1 - p^{-s})^2} = \frac{L(s, \chi_{0,12})L(s, \chi_{3,12})}{L(s, \chi_{1,12})L(s, \chi_{2,12})} \prod_{\substack{p \geq 5, \\ p \equiv 11[12]}} \frac{1}{(1 - p^{-2s})^2}$$

and produce an approximate value of $\prod_{p \equiv 11[12]}(1 - 1/p^2)^{-1}$ certified at 10^{-100}.

See sequence A301430 of the OEIS [3]. The readers will find yet other formulae that are similarly used in the papers [8], [9] and [11] by Shanks.

References

[1] B.C. Berndt. *Ramanujan's notebooks. Part IV.* Springer-Verlag, New York, 1994, pp. xii+451. https://doi.org/10.1007/978-1-4612-0879-2 (cit. on p. 168).

[2] S. Ettahri, O. Ramaré, and L. Surel. "Fast multi-precision computation of some Euler products". In: *Math. Comp.* 90 (331) (2021), pp. 2247–2265. ISSN: 0025-5718, https://doi.org/10.1090/mcom/3630, (cit. on p. 171).

[3]   OEIS Foundation Inc. *The On-Line Encyclopedia of Integer Sequence.* http://oeis.org/. 2019 (cit. on pp. 169–171).

[4]   P. Moree and R. Osburn. "Two-dimensional lattices with few distances". In: *Enseign. Math. (2)* 52.3-4 (2006), pp. 361–380 (cit. on p. 170).

[5]   P. Moree and H. J. J. te Riele. "The hexagonal versus the square lattice". In: *Math. Comp.* 73.245 (2004), pp. 451–473. ISSN: 0025-5718. https://doi.org/10.1090/S0025-5718-03-01556-4 (cit. on p. 170).

[6]   P. Schmutz Schaller. "Geometry of Riemann surfaces based on closed geodesics". In: *Bull. Amer. Math. Soc. (N.S.)* 35.3 (1998), pp. 193–214. https://doi.org/10.1090/S0273-0979-98-00750-2 (cit. on p. 170).

[7]   J.-P. Serre. "Divisibilité de certaines fonctions arithmétiques". In: *Enseignement Math. (2)* 22.3-4 (1976), pp. 227–260 (cit. on p. 169).

[8]   D. Shanks. "On the conjecture of Hardy & Littlewood concerning the number of primes of the form $n^2 a$". In: *Math. Comp.* 14 (1960), pp. 320–332 (cit. on p. 171).

[9]   D. Shanks. "On numbers of the form n^4 1". In: *Math. Comput.* 15 (1961), pp. 186–189 (cit. on p. 171).

[10] D. Shanks, "The second-order term in the asymptotic expansion of $B(x)$". In: *Math. Comp.* 18 (1964), pp. 75–86. https://doi.org/10.2307/2003407 (cit. on pp. 168–170).

[11] D. Shanks,"Lal's constant and generalizations". In: *Math. Comp.* 21 (1967), pp. 705–707. https://doi.org/10.2307/2005014 (cit. on p. 171).

Chapter 17
Euler Products with Primes in Progressions

As we have seen in Chap. 16, in some Euler products, primes in a union of certain arithmetic progressions, instead of all primes, show up. The identity (16.4) used by Shanks calls for generalizations. We discuss in this chapter some generalizations of the cyclotomic identity that enables to evaluate such products. We advise the readers to go through the material of Chap. 9 before reading this one, though the present chapter is formally independent from it.

Most of the proofs will take place in one of the rings of *formal series* $\mathbb{Z}[[z]]$ or $\mathbb{C}[[z]]$ which are similar, in nature, to the ring of formal Dirichlet series: it consists simply of formal series $f(z) = \sum_{k \geq 0} a_k z^k$; such a series may be infinite and the coefficients a_k are taken from the integers, i.e. from $\mathbb{Z}$, for $\mathbb{Z}[[z]]$ and from the complex numbers for $\mathbb{C}[[z]]$. The readers should consider these series as a Taylor development "in the vicinity of $z = 0$", so that we may say $f(0)$ to refer to a_0. Taking the product of two formal series is mechanical and is essentially the same as taking the product of two polynomials. Indeed, the product of $\sum_{k \geq 0} a_k z^k$ times $\sum_{\ell \geq 0} b_k z^\ell$ is defined by

$$\sum_{m \geq 0} \left(\sum_{0 \leq k \leq m} a_k b_{m-k} \right) z^m .$$

17.1 The Witt Transform

Definition 17.1

For $f(z) \in \mathbb{C}[[z]]$ and $r \geq 1$ any integer, let

$$\mathcal{W}_f^{(r)}(z) = \frac{1}{r} \sum_{d \mid r} \mu(d) f(z^d)^{r/d} = \sum_{j=0}^{\infty} m_f(j, r) z^j .$$

The readers will prove the two fundamental properties of this transform in the next exercise.

© The Author(s), under exclusive license to Springer Nature Switzerland AG 2022
O. Ramaré, *Excursions in Multiplicative Number Theory*, Birkhäuser Advanced Texts Basler Lehrbücher, https://doi.org/10.1007/978-3-030-73169-4_17

Exercise 17-1.

1⋄ Show that $\sum_{\delta|r} \delta W_f^{(\delta)}(z^{r/\delta}) = f(z)^r$.

2⋄ Deduce from the first question that, when f has integer coefficients, i.e belongs to $\mathbb{Z}[[z]]$, then so does $W_f^{(r)}(z)$.

Exercise 17-2. In this exercise, we get bounds for the coefficients $m_f(j, r)$ defined in Definition 17.1 when $f(z) = F(z)/G(z)$ is the quotient of two polynomials such that $|G(0)| \geq 1$. We conclude with two other properties of the $m_f(j, r)$'s.

1⋄ Show that $1/G(z) = \sum_{\ell \geq 0} b_\ell z^\ell$ with $|b_\ell| \leq \gamma(\ell, d)B_0^\ell$, where $d = \deg G$, B_0 is the maximum of the absolute values of the inverse of the roots of G and $\gamma(\ell, d)$ is defined by $1/(1 - z)^d = \sum_{\ell \geq 0} \gamma(\ell, d)z^\ell$.

2⋄ Show that $(F(z)/G(z))^r = \sum_{\ell \geq 0} u_\ell(r)z^\ell$ with $|u_\ell(r)| \leq \gamma(\ell, r(d + 1))B_0^\ell A^r$, where we have kept the previous notation and where A is the maximum of the absolute value of the coefficients of $F(z)/G(0)$.

3⋄ Show that, when $A \geq 1$, we have $|m_f(j, r)| \leq \gamma(\ell, r(d + 1))B_0^\ell A^r$.

4⋄ Show that $\gamma(\ell, j) = \binom{j-1+\ell}{j-1}$.

5⋄ Use Exer. 9-6 to show that B is not more than the sum of the absolute values of the coefficients of $G(z)$.

6⋄ Show that, when F and G have integer coefficients and $G(0) = \pm 1$, then $m_f(j, r)$ is an integer.

7⋄ Show that, when 0 is a root of $F(z)$ order $j_0 \geq 1$, we have $m_f(j, r) = 0$ whenever $j < j_0 r$.

With this definition the cyclotomic identity is generalized as follows in [12, Theorem 8].

Theorem 17.1

Suppose that $f(z) \in \mathbb{Z}[[z]]$. Then, as formal power series in y and z, we have

$$1 - yf(z) = \prod_{j=0}^{\infty} \prod_{k=1}^{\infty} (1 - z^j y^k)^{m_f(j,k)}, \qquad (17.1)$$

where the $m_f(j, k)$ are the integers defined in Definition 17.1.

Note that the cyclotomic identity is recovered with the choice $f(z) = \alpha$ and that $W_\alpha^{(r)}(z) = M(\alpha; r)$. See Eq. (11) of [16] by Shanks for a precursor of this identity.

Proof By Exer. 17-1, we have

$$f(z)^r = \sum_{d|r} \frac{r}{d} \mathcal{W}_f^{(r/d)}(z^d) = \sum_{d|r} \frac{r}{d} \sum_{j=0}^{\infty} m_f(j, r/d) z^{dj} \,,$$

from which we infer that

$$\sum_{r \geq 1} y^r f(z)^r = \sum_{k \geq 1} k \sum_{j=0}^{\infty} m_f(j, k) \sum_{d \geq 1} z^{dj} y^{kd} \,.$$

The latter identity with both sides divided out by y can be rewritten as

$$\frac{f(z)}{1 - yf(z)} = \sum_{k \geq 1} \sum_{j=0}^{\infty} m_f(j, k) \frac{kz^j y^{k-1}}{1 - z^j y^k} \,.$$

Formal integration of both sides with respect to y gives

$$-\log(1 - yf(z)) = -\sum_{k \geq 1} \sum_{j=0}^{\infty} m_f(j, k) \log(1 - z^j y^k) \,,$$

whence

$$1 - yf(z) = \prod_{k \geq 1} \prod_{j \geq 0} (1 - z^j y^k)^{m_f(j,k)} \,.$$

This establishes the main formula. The integrality of the coefficients is proved in Exer. 17-1. □

In [17], Witt proved a generalization of the Necklace Identity (see Exer. 9-2) which we present in the next exercise.

Exercise 17-3. For $k \geq 1$, prove the Multivariate Cyclotomic Identity

$$1 - \sum_{i=1}^{k} z_i = \prod_{\substack{m_1, \ldots, m_k \geq 0, \\ m_1 + \ldots + m_k \geq 1}} (1 - z_1^{m_1} \cdots z_k^{m_k})^{M(m_1, \ldots, m_k)} \,, \tag{17.2}$$

where the integer $M(m_1, \ldots, m_k)$ is defined by

$$M(m_1, \ldots, m_k) = \frac{1}{N} \sum_{d \mid \gcd(m_1, m_2, \ldots, m_k)} \mu(d) \frac{(N/d)!}{(m_1/d)! \cdots (m_k/d)!}$$

with $N = m_1 + \ldots + m_k$.

Further generalizations may be found in the paper [10] by S.-J. Kang and M.-H. Kim. Another analogue of the necklace identity is the *Sherman identity*, which is a special non-trivial case of the *Feynman identity*, conjectured by R. Feynman in the setting of the Ising model in two dimensions in physics, see [4] by da Costa and Graciele Zimmermann for particulars.

Exercise 17-4. Show, as an application of (17.2), that the series $D(s)$ defined in (8.11) admits a meromorphic continuation to the half-plane $\Re s > 1/2$.

Exercise 17-5. Show, as an application of (17.2), that the series $H(s)$ defined in (8.1) admits a meromorphic continuation to the half-plane $\Re s > 1$.

17.2 Euler Products with Characters

Just as the cyclotomic identity can be used to evaluate standard Euler products, so can Theorem 17.1 deal with certain Euler products involving Dirichlet characters. Recall that the Dirichlet L-series $L(s, \chi^k)$, is defined, for $\Re s > 1$, by $\sum_{n=1}^{\infty} \chi^k(n)n^{-s}$. The function χ^k is multiplicative, hence

$$L(s, \chi^k) = \prod_{p \geq 2} \left(1 - \frac{\chi(p)^k}{p^s}\right)^{-1} \tag{17.3}$$

and, generalizing (9.12), we further define:

$$L_P(s, \chi^k) = \prod_{p > P} \left(1 - \frac{\chi(p)^k}{p^s}\right)^{-1}. \tag{17.4}$$

Theorem 17.2

Let $f(z) = F(z)/G(z)$ is the quotient of two integer polynomials such that $|G(0)| = 1$ and 0 is a root of $F(z)$ order $j_0 \geq 1$. Let B_0 be the maximum of the inverse of the roots of G and A be the maximum of the absolute value of the coefficients of $F(z)$, let χ be a Dirichlet character, let s be a complex number of real part $\sigma > 1/j_0$ and define $C = 2^{(\deg G+1+j_0)/j_0} A^{1/j_0} B_0$. When the real parameter P satisfies $P^{\sigma} \geq 2C$, we have

$$\prod_{p > P} \left(1 - \chi(p)\frac{F(1/p^s)}{G(1/p^s)}\right) = \prod_{1 \leq k \leq J/j_0} \prod_{k j_0 \leq j \leq J}^{\infty} L_P(js, \chi^k)^{-m_f(j,k)} \times I,$$

where the coefficients $m_f(j, k)$ are the integers defined in Definition 17.1 and where I satisfies

$$|\log|I|| \leq 4P^{\sigma} (C/P^{\sigma})^J.$$

When the condition $|G(0)| = 1$ is not met, but χ is real valued and s is a real number, the same approximation holds.

The character χ may well be the principal character to some modulus. We recall that B_0 is easily bounded above by appealing to Exer. 9-6.

Proof Exer. 17-2 shows that the coefficients $m_f(j, k)$ are integers. It also shows that they vanish when $j > j_0 k$. To prove our formula, we first use Theorem 17.2 to infer formally that we have

$$\prod_{p > P} \left(1 - \frac{\chi(p)F(1/p^s)}{G(1/p^s)}\right) = \prod_{p > P} \prod_{k \geq 1} \prod_{j \geq j_0 k} \left(1 - \frac{\chi(p)^k}{p^{sj}}\right)^{m_f(j, k)}.$$

We now use Theorem 3.3; the possibility that some $m_f(j, k)$ may be negative introduces a slight difficulty. Indeed, when $m_f(j, k) \geq 0$ we take $u_{j,k,p} = -\chi(p)^k/p^{sj}$ that we repeat $m_f(j, k)$ times. When $m_f(j, k) < 0$ we take

$$u_{j,k,p} = \frac{\chi(p)^k}{p^{sj} - \chi(p)^k}.$$

So what we have to bound is

$$M(T) = \sum_{p > P} \sum_{j \geq T} \sum_{k \leq j/j_0} |m_f(j, k)| \frac{1}{p^{\sigma j} - 1}.$$

We employ the bound for $|m_f(j, k)|$ given in the third and fourth question of Exer. 17-2 (remember that $d = \deg G$) to show that

$$M(T) \leq \sum_{p > P} \sum_{j \geq T} \sum_{k \leq j/j_0} \binom{k(d + 1) - 1 + j}{k(d + 1) - 1} B_0^j A^k \frac{1}{p^{\sigma j} - 1}.$$

We infer from this inequality that

$$M(T) \leq \sum_{p > P} \sum_{j \geq T} 2^{\frac{j}{j_0}(d+1) - 1 + j} B_0^j A^{j/j_0} \frac{2}{p^{\sigma j}}$$

$$\leq \sum_{j \geq T} 2^{\frac{j}{j_0}(d+1) + j} B_0^j A^{j/j_0} \frac{2}{P^{\sigma(j-1)}}$$

by using $\sum_{n > P} 1/n^{\sigma j} \leq 2/P^{\sigma(j-1)}$ that we encourage the readers to prove. By our assumption on P, this upper bound for M is finite. On selecting $T = j_0$, we deduce from this discussion that our initial product is absolutely convergent in the sense of Godement. We may permute the factors as we want. By using $T = J$, we control the tail. With $D = C/P^\sigma$, we find that

$$M(J) \leq 2P^\sigma \frac{D^J}{1 - D} \leq 4P^\sigma D^J.$$

By Theorem 3.3, this offers of bound for the remaining product, say I, and since we have used only the absolute value of $m_f(j, k)$, this bound is also valid for $1/|I|$. This ends the first part of the proof. When we stick to real values, it is not anymore required to assume that $m_f(j, k)$ is an integer. We only need to notice that $(1 - \chi(p)^k/p^{js})^{m_f(j,k)} = 1 + u$ with $|u| \leq m_f(j, k)/p^{j\sigma}$ when $m_f(j, k) \geq 0$ and $|u| \leq 2|m_f(j, k)|/p^{j\sigma}$ when $m_f(j, k) < 0$. The proof then takes the same tracks as the previous one. $\qquad \square$

Since Dirichlet L-series are easily evaluated with high decimal precision, this result allows one, when $G(0) = \pm 1$, to evaluate the constants appearing with high precision. The readers will find in [11] some usage of the above theorem. However, in general, the constraint $G(0) = \pm 1$ is too strong, so that the theory is incomplete, see [13] for more on this issue. We now detail an example, partially in the form of an exercise.

Exercise 17-6. Let S be the set of integers all of whose prime factors are congruent to 2 modulo 3.

1 ◇ Show that, as x goes to infinity, we have $\displaystyle\sum_{\substack{n \leq x, \\ n \in S}} \frac{1}{\varphi(n)} \sim A\sqrt{\log x}$, where

$A = 3\sqrt{2/\pi} \prod_{p \geq 3} A_p$ with $A_p = (1 + p/(p-1)^2)\sqrt{1 - p^{-1}}$ when $p \equiv 2 \pmod 3$ and $A_p = \sqrt{1 - p^{-1}}$ when $p \equiv 1 \pmod 3$.

2 ◇ Show that, when $p > 2$ is a prime, we have $A_p = B_p C_p$ with $B_p = \left(1 + \right.$

$\left.\dfrac{p}{2(p-1)^2}\right)\left(1 - \dfrac{1}{p}\right)^{1/2}$, and $C_p = 1 - \dfrac{\chi_3(p)}{2(p-1)^2 + p}$.

3 ◇ Show that $z^2/(2 - z + 2z^2) = \sum_{\ell \geq 2} v_\ell z^\ell$ where the coefficients $w_\ell = 2^{\ell-1} v_\ell$ are integers satisfying $w_1 = 0$, $w_2 = 1$ and, when $\ell \geq 1$, $w_{\ell+2} - w_{\ell+1} + 4w_\ell = 0$.

Here is a Sage script that creates a list of lists containing the $u_\ell^*(r) = 2^{\ell-1} u_\ell(r)$, where the $u_r(\ell)$ are defined by

$$\left(\frac{z^2}{2 - z + z^2}\right)^r = \sum_{\ell \geq 0} u_\ell(r) z^\ell .$$

```
def CreateWOfEll (ellmax):
  global WOfEll;
  WOfEll = [0, 0, 1] #(w0, w1,w2)
  for ell in range(3, ellmax):
      WOfEll = WOfEll + [WOfEll[ell-1]-4*WOfEll[ell-2],]

def GetCreateUstar (ellmax, rmax):
  global Ustar, WOfEll;
  CreateWOfEll(ellmax)
  Ustar = list(list(0 for ell in range(ellmax+1))
                 for r in range(rmax + 1))
  Ustar[0] = WOfEll
  for r in range(1, rmax + 1):
      for ell in range(ellmax + 1):
          res = 0
          for k in range(ell + 1):
              res += Ustar[r-1][ell-k]*WOfEll[k]
          Ustar[r][ell] = res
```

The readers may check the values with Pari/GP and, for instance, compute

```
(z^2/(2-z+2*z^2))^2+0(z^10)
```

4⋄ Write a Sage script to compute the corresponding $m_f(k, \ell)$.

5⋄ Compute the first fifty digits of $\prod_{p \geq 3} C_p$ and check the result with Pari/GP.

6⋄ By following the above process and Theorem 17.2, compute a high precision value of $\prod_{p \geq 3} B_p^2$.

7⋄ Compute the first fifty digits of the constant A defined at the beginning.
We extract the next exercise from Paragraph 6 of the paper [5] by Flajolet and Vardi.

Exercise 17-7. For $s > 1$, we set $\ell(s) = \log\left((1 - 2^{-s})\dfrac{\zeta(s)}{L(s, \chi_4)}\right)$.

1⋄ Show that $\ell(s) = \displaystyle\sum_{p \equiv 3[4]} \log \dfrac{1 + p^{-s}}{1 - p^s}$.

2⋄ Find the Taylor expansion of the function $x \mapsto \log \dfrac{1 + x}{1 - x}$ around $x = 0$.

3⋄ Set $R(s) = \displaystyle\sum_{p \equiv 3[4]} 1/p^s$. Find a functional relation between $R(s)$ and $\ell(s)$.

4⋄ On using Exer. 6.8, express the function R in terms of the function ℓ and deduce from it an efficient way of computing $R(s)$.

We leave to the readers to find how to compute with high accuracy the two sums

$$\sum_{\substack{p \geq 2, \\ p \equiv 3[4]}} \frac{\log p}{p^2 - p + 1} \quad \text{and} \quad \sum_{\substack{p \geq 2, \\ p \equiv 3[4]}} \frac{1}{p^2 - p + 1} .$$

17.3 On the Bateman–Horn Conjecture

Suppose that $f_1, \ldots, f_k$ are irreducible polynomials in $\mathbb{Z}[x]$ with positive leading coefficients, their degrees being $h_1, \ldots, h_k$, respectively. Let $Q(f_1, f_2, \ldots, f_k; x)$ denote the number of positive integers $n \leq x$ such that $f_1(n), f_2(n), \ldots, f_k(n)$ are simultaneously prime. Even in the case $k = 1$, determining whether $Q(f_1, f_2, \ldots, f_k; x) > 0$ is a widely open question! That it should be the case under natural conditions is the topic of the V. Bunyakovsky conjecture in [3] which was later generalized in [15] by A. Schinzel and W. Sierpinski to become A. Schinzel's hypothesis H. These conjectures are qualitative. From a quantitative viewpoint and by following the path opened by Hardy and Littlewood in the series of papers *Partitio Numerorum**, [7–9], P.T. Bateman and R. Horn proposed in [2] that

$$Q(f_1, f_2, \ldots, f_k; x) \sim \frac{C(f_1, \ldots, f_k)}{h_1 h_2 \cdots h_k} \int_2^x \frac{dt}{(\log t)^k},$$

where

$$C(f_1, \ldots, f_k) = \prod_p \frac{1 - \omega(p)/p}{(1 - 1/p)^k},$$

the function $\omega(p)$ being the number of solutions, without multiplicity, of

$$f_1(x) f_2(x) \cdots f_k(x) \equiv 0 \bmod p.$$

As an example let us take $f_1(x)$ and $f_2(x) = x + 2$. We obtain that $T(x)$, the number of primes $p \le x$ such that $p + 2$ is also a prime, should satisfy

$$T(x) \sim 2 \prod_{p>2} \frac{1 - 2/p}{(1 - 1/p)^2} \int_2^x \frac{dt}{(\log t)^k} \sim 2T \int_2^x \frac{dt}{(\log t)^k},$$

with T the twin prime constant defined in (9.1). This does not apply to the Goldbach problem which is to count the number of solutions to $N = p_1 + p_2$, where both p_1 and p_2 are primes, and N is assumed to be even; A. Schinzel in [14] proposes a conjecture that treats such cases.

Generalizing the twin prime problem, we may consider k linear forms $f_1, \ldots, f_k$ satisfying $\omega(p) < p$ for every prime p (a finite test is enough to verify this). Then $C(f_1, \ldots, f_k)$ is a rational number times

$$C_k = \prod_{p>k} \frac{1 - k/p}{(1 - 1/p)^k}.$$

Each factor satisfies $1 + O_k(p^{-2})$ and so the product converges. Using the cyclotomic identity we obtain

$$C_k = \prod_{j=2}^\infty \zeta_k(j)^{-M(k;j)}.$$

Noting that $C_2 = T$, the twin prime constant, we infer that

$$T = \prod_{j=2}^\infty \zeta_2(j)^{-M(2;j)}.$$

The results concerning the Bateman–Horn conjecture are scarce and only some linear systems have been properly treated. Let us mention two spectacular results in this context: A. Balog proved in [1] that, given any k, there are infinitely many chains of

---

* E. Landau wrote in Zentralblatt in 1921 the following comment about these papers: "The new method marks a new epoch in analytic number theory and places the authors in the rank of the greatest arithmeticians of all time".

primes $p_1, p_2, \ldots, p_k$ such that all the mid-points $(p_i + p_j)/2$ are also prime; and B. Green and T. Tao proved in [6] that, given any k, there are infinitely chains of primes $p_1, p_2, \ldots, p_k$ such that all the differences $p_{i+1} - p_i$ are the same.

Exercise 17-8. A *prime magic square* of order k is a $k \times k$ array of distinct prime numbers such that the sum of every column, the sum of every lign and the sum of the two diagonals are equal to a same integer. One may ask that the intervening primes should be the first k^2 odd primes, but we wave this condition here. Find the number of 3×3 prime magic squares whose common sum is not more than 2000. Can you find 3×3 prime magic squares whose common sum is also a prime number? The readers may similarly investigate larger prime magic squares.

See Exer. 29-5 for more on prime magic squares and generally Chap. 29 for a sieve approach to these questions.

References

[1]   A. Balog, "Linear equations in primes". In: *Mathematika* 39.2 (1992), pp. 367–378. https://doi.org/10.1112/S0025579300015096 (cit. on p. 180).

[2]   P.T. Bateman and R.A. Horn, "A heuristic asymptotic formula concerning the distribution of prime numbers". In: *Math. Comp.* 16 (1962), pp. 363–367. https://doi.org/10.2307/2004056 (cit. on p. 179).

[3]   V. Bouniakowsky. "Nouveaux théorèmes relatifs à la distinction des nombres premiers et à la de composition des entiers en facteurs". In: *Sc. Math. Phys.* 6 (1857), pp. 305–329 (cit. on p. 179).

[4]   G.A.T.F. da Costa and G.A. Zimmermann, "An analogue to the Witt identity". In: *Pacific J. Math.* 263.2 (2013), pp. 475–494. https://doi.org/10.2140/pjm.2013.263.475 (cit. on p. 175).

[5]   P. Flajolet and I. Vardi. "Zeta Function expansions of Classical Constants". In: *preprint* (1996), pp. 1–10. http://algo.inria.fr/flajolet/Publications/landau.ps (cit. on p. 179).

[6]   B. Green and T. Tao. "The primes contain arbitrarily long arithmetic progressions". In: *Ann. of Math. (2)* 167.2 (2008), pp. 481–547. https://doi.org/10.4007/annals.2008.167.481 (cit. on p. 180).

[7]   G.H. Hardy and J.E. Littlewood. "Some problems of "Partitio Numerorum" III. On the expression of a number as a sum of primes". In: *Acta Math.* 44 (1922), pp. 1–70 (cit. on p. 179).

[8]   G.H. Hardy and J.E. Littlewood. "Some problems of "Partitio Numerorum" IV. The singular series in Waring's Problem and the value of the number $G(k)$". In: *Math. Zeitschrift* 12 (1922), pp. 161–188 (cit. on p. 179).

[9]   G.H. Hardy and J.E. Littlewood. "Some problems of "Partitio Numerorum" (VI): Further researches in Waring's problem". In: *Mathematische Z.* 23 (), pp. 1–37 (cit. on p. 179).

[10] Seok-Jin Kang and Myung-Hwan Kim, "Dimension formula for graded Lie algebras and its applications". In: *Trans. Amer. Math. Soc.* 351.11 (1999), pp. 4281–4336. https://doi.org/10.1090/S0002-9947-99-02239-4 (cit. on p. 175).

[11] P. Moree. "Convoluted convolved Fibonacci numbers". In: *J. Integer Seq.* 7.2 (2004), Article 04.2.2, 14 (cit. on p. 177).

[12] P. Moree, "On the average number of elements in a finite field with order or index in a prescribed residue class". In: *Finite Fields Appl.* 10.3 (2004), pp. 438–463. https://doi.org/10.1016/j.ffa.2003.10.001 (cit. on p. 174).

[13]   O. Ramaré. "Accurate computations of Euler products over primes in arithmetic progressions". In: *Funct. Approx. Comment. Math.* 65(1) (2021), pp. 33–45. ISSN: 0208-6573, arxiv.org/pdf/10.7169/facm/1853.pdf, (cit. on p. 177).

[14]   A. Schinzel, "A remark on a paper of Bateman and Horn". In: *Math. Comp.* 17 (1963), pp. 445–447. https://doi.org/10.2307/2004008 (cit. on p. 180).

[15]   A. Schinzel and W. Sierpinski. "Sur certaines hypothèses concernant les nombres premiers". In: *Acta Arith.* 4 (1958), 185–208, erratum 5 (1958), 259. https://doi.org/10.4064/aa-4-3-185-208 (cit. on p. 179).

[16]   D. Shanks, "Lal's constant and generalizations". In: *Math. Comp.* 21 (1967), pp. 705–707. https://doi.org/10.2307/2005014 (cit. on p. 174).

[17]   E. Witt, "Treue Darstellung Liescher Ringe". In: *J. Reine Angew. Math.* 177 (1937), pp. 152–160. https://doi.org/10.1515/crll.1937.177.152 (cit. on p. 175).

Chapter 18
The Chinese Remainder Theorem and Multiplicativity

18.1 The Chinese Remainder Theorem

We presented $\mathbb{Z}/q\mathbb{Z}$ in Chaps. 5 and 15. The theme of the present section comes from a simple remark. When we study the algebraical equation $x^2 \equiv -2 (\bmod 3 \times 11)$, we want solutions x say in $\{1, \ldots, 33\}$. The condition $33 | x^2 + 2$ splits in two: we need $3 | x^2 + 2$ and $11 | x^2 + 2$. The first condition removes only the divisors of 3, leaving 10 possibilities, of which the second condition leaves $\{8, 14, 19, 25\}$:

```
for(x = 1, 33, print("x = ", x, " (x^2+2)%33 = ", (x^2+2)%33));
```

Understanding these four (why *four*?) numbers is a difficulty. The Chinese Remainder Theorem clears the situation.

> **Theorem 18.1 (Chinese Remainder Theorem)**
>
> Let q_1 and q_2 be two coprime positive integers. The map
>
> $$\mathbb{Z}/q_1 q_2 \mathbb{Z} \rightarrow \mathbb{Z}/q_1 \mathbb{Z} \times \mathbb{Z}/q_2 \mathbb{Z}$$
> $$n + q_1 q_2 \mathbb{Z} \mapsto (n + q_1 \mathbb{Z}, n + q_2 \mathbb{Z})$$
>
> is a one-to-one homomorphism of rings.

We should be more specific about what "homomorphism of rings" means. On calling this map F, it means that $F(xy + z) = F(x)F(y) + F(z)$. This is important for us, as it transforms algebraic equations into algebraic equations: $F(x^2 + 2) = F(0)$ becomes simply $y^2 + 2 = 0$ with $y = F(x)$. Hence finding solutions in $\mathbb{Z}/q_1 q_2 \mathbb{Z}$ reduces to finding solutions in $\mathbb{Z}/q_1 \mathbb{Z}$ and in $\mathbb{Z}/q_2 \mathbb{Z}$.

Proof First note that $\mathbb{Z}/q_1 q_2 \mathbb{Z}$ contains $q_1 q_2$ points, and that $\mathbb{Z}/q_1 \mathbb{Z} \times \mathbb{Z}/q_2 \mathbb{Z}$ contains also $q_1 \cdot q_2$ points. Hence, it is enough to show that any two elements in $\mathbb{Z}/q_1 q_2 \mathbb{Z}$, say n and n' cannot be sent on the same image. If so, then $n \bmod q_1 = n' \bmod q_1$,

© The Author(s), under exclusive license to Springer Nature Switzerland AG 2022
O. Ramaré, *Excursions in Multiplicative Number Theory*, Birkhäuser Advanced Texts Basler Lehrbücher, https://doi.org/10.1007/978-3-030-73169-4_18

which means that $q_1|n-n'$. Similarly $q_2|n-n'$. But since q_1 and q_2 are coprime, this equivalent to $q_1q_2|n-n'$, which, in turn, is equivalent to $n \bmod q_1q_2 = n' \bmod q_1q_2$. This proves our theorem.   □

Let us continue with our example. In $\mathbb{Z}/3\mathbb{Z}$, the equation $x^2 + 2$ has the two solutions $\{1+3\mathbb{Z}, 2+3\mathbb{Z}\}$, while in $\mathbb{Z}/11\mathbb{Z}$, it has the two solutions $\{3+11\mathbb{Z}, 8+11\mathbb{Z}\}$. The solutions modulo 33 are thus

$$F^{-1}(1+3\mathbb{Z}, 3+11\mathbb{Z}),\ F^{-1}(1+3\mathbb{Z}, 8+11\mathbb{Z}),$$
$$F^{-1}(2+3\mathbb{Z}, 3+11\mathbb{Z}),\ F^{-1}(2+3\mathbb{Z}, 8+11\mathbb{Z})$$

and these are

$$25+33\mathbb{Z},\ 19+33\mathbb{Z},\ 14+33\mathbb{Z},\ 8+33\mathbb{Z}.$$

We have thus recovered our *four* solutions and found some order in them. The readers will find very detailed explanations on how to solve similar congruences in Chap. 2 of the book [6] by I. Niven, H. Zuckerman and H.L. Montgomery. Here is the most important corollary for us.

Corollary 18.1

Let $P(X)$ be some polynomial with integer coefficients. The function $\rho_P(d)$ that, to any positive integer d, associates the number of solutions of $P(x)=0$ with a variable x in $\mathbb{Z}/d\mathbb{Z}$ is multiplicative.

Exercise 18-1.

1 ◇ Find all the solutions to the equation $x^5 - 2x + 1 \equiv 0[100]$.

2 ◇ Find all the solutions to the equation $x^5 + 31x - 30 \equiv 0[100]$.

3 ◇ Find all the solutions to the equation $x^5 - 5x + 4 \equiv 0[100]$.

Exercise 18-2. Show that $30|n^5 - n$ for every integer n.

Exercise 18-3. Let n be an integer coprime to 6. Show that 24 divides $n^2 + 47$.

Exercise 18-4. For any modulus $r \geq 1$, let us write

$$\Xi_r = \left\{ \frac{u}{r},\ 1 \leq u \leq r, \gcd(u,r) = 1 \right\}.$$

Let q and q' be two coprime integers. Prove that the function

$$\Psi : \Xi_q \times \Xi_{q'} \to \Xi_{qq'}$$
$$(a/q, a'/q') \mapsto (aq' + a'q)/(qq')$$

is a one-to-one map.

Exercise 18-5.

1 ◇ Prove that, for each n, the Ramanujan sum $c_d(n)$ defined in (1.9) is multiplicative as a function of d.

2 ◇ Show that $c_p(n) = p - 1$ when $p|n$ and $c_p(n) = -1$ otherwise, p being a prime number.

3 ◇ Show that, when m is prime to d, we have $c_d(nm) = c_d(n)$.

4 ◇ Determine the value of $c_{p^k}(n)$ and prove the Kluyver Identity (1.10).

18.2 On the Number of Divisors of $n^2 + 1$

Our aim here is to prove the following theorem.

Theorem 18.2

We have, when $x \geq 2$, $\displaystyle\sum_{n \leq x} d(n^2 + 1) = \frac{3}{\pi} x \log x + O(x)$.

The proof takes three steps in the course of which the readers will see the richness of the tools that are at their disposal now. One can trace this result to Eira Scourfield in [8] though it may have been proved earlier. With this same error term and with explicit bounds, it has been recently reproved by Kostadinka Lapkova in [5]. We first need some input from modular arithmetic.

Lemma 18.1

Let d be some positive integer. The number $\rho(d)$ of solutions x in $\mathbb{Z}/d\mathbb{Z}$ to the equation $x^2 + 1 \equiv 0 \pmod{d}$ is the multiplicative function defined by

$$\rho(p^k) = \begin{cases} 1 & \text{when } p = 2 \text{ and } k = 1, \\ 0 & \text{when } p = 2 \text{ and } k \geq 2, \\ 2 & \text{when } p \equiv 1 \bmod 4, k \geq 1, \\ 0 & \text{when } p \equiv 3 \bmod 4, k \geq 1. \end{cases}$$

Proof Cor. 18.1 proves the multiplicativity. It is easy to check that $\rho(2) = 1$. When $p \equiv 1 \pmod 4$, we have $(-1)^{\frac{p-1}{2}} = 1$, which by (5.7) means that -1 is a square modulo p, say $-1 \equiv a^2 \pmod p$. Whence equation $x^2 + 1$ has the two roots a and $-a$, and they are distinct, for otherwise $0 = a - (-a) = 2a \pmod p$, and this would imply that $p = 2$. The same argument shows that the equation $x^2 + 1 = 0 \pmod{p}$ has no solution when $p \equiv 3 \pmod 4$. This proves the lemma when $k = 1$.

Let us proceed to see what happens when k is larger. When p is congruent to 3 modulo 4, the situation is easy: there cannot be any solution modulo p^k as it would yield a solution modulo p. When p is congruent to 1 modulo 4, we show by recursion on k that $\rho(p^k) = 2$. Indeed, any solution modulo p^{k+1}, say A, gives a solution modulo p^k, say a. We thus have $A = a + p^k b$ for some chosen a in $\{1, \ldots, p^k\}$. Since $A^2 = a^2 + 2bp^k + b^2 p^{2k} \equiv -1 \pmod{p}^{k+1}$, this gives us

$$b \equiv \frac{-1 - a^2}{p^k} 2^{-1} \bmod p$$

hence determining b modulo p, hence A modulo p^{k+1}. The neat outcome is that a root modulo p^k *lifts* in a single root modulo p^{k+1}. The process we are using is called *Hensel's Lift* after the mathematician K. Hensel who introduced it. The situation modulo 2 could be more tricky as the above argument fails because 2 is not invertible, but by chance, we check that the equation has no solution in $\mathbb{Z}/4\mathbb{Z}$ and thus none in $\mathbb{Z}/2^k\mathbb{Z}$ when k is at least 2.   □

Exercise 18-6.

1 ⋄ Show that $\sum_{d \leq x} \rho(d)/d = c \log x + O(1)$, for some constant c, where ρ is the function defined in Lemma 18.1 and the constant c is given by

$$c = \frac{3}{2} \prod_{p \geq 3} F_p, \quad F_p = \begin{cases} (p+1)/p & \text{when } p \equiv 1 \bmod 4, \\ (p-1)/p & \text{when } p \equiv 3 \bmod 4. \end{cases}$$

2 ⋄ Deduce that $\sum_{d \leq x} \rho(d) \ll x$.

3 ⋄ Recall the character χ_4 is defined in (15.2). Recall (15.7), i.e.

$$L(1, \chi_4) = \prod_{p \geq 3} G_p, \quad G_p = \begin{cases} p/(p-1) & \text{when } p \equiv 1 \bmod 4, \\ p/(p+1) & \text{when } p \equiv 3 \bmod 4. \end{cases}$$

Show that $(2/3)c/L(1, \chi_4) = (4/3)/\zeta(2)$ (Beware of the Euler factor at 2) and deduce that $c = 3/\pi$.

Proof (of Theorem 18.2) We first note that

$$d(n^2 + 1) = 2 \sum_{d \leq \sqrt{n^2+1}} 1 .$$

Indeed, we can pair the divisors by $d \leftrightarrow (n^2 + 1)/d$. One such pair could consist of a single element: we would then have $d^2 = n^2 + 1$, i.e. $(d - n)(d + n) = 1$ which is impossible. Hence the expression above. We thus get

$$\sum_{n \le x} d(n^2 + 1) = 2 \sum_{d \le \sqrt{x^2+1}} \sum_{\substack{d \mid n^2+1, \\ \sqrt{d^2-1} \le n \le x}} 1 \, .$$

In the last sum, each of the $\rho(d)$ solutions to the equation $x^2 + 1 = 0$ mod d gives rise to an arithmetic progression of solutions, i.e. to

$$\frac{x - \sqrt{d^2 - 1}}{d} + O(1)$$

integers n. Therefore

$$\sum_{n \le x} d(n^2 + 1) = 2 \sum_{d \le \sqrt{x^2+1}} \rho(d) \frac{x - \sqrt{d^2 - 1}}{d} + O\left(\sum_{d \le \sqrt{x^2+1}} \rho(d) \right) .$$

The error term is $\ll x$ by Exer. 18-6. We also have

$$\sum_{d \le \sqrt{x^2+1}} \rho(d) \frac{\sqrt{d^2 - 1}}{d} \le \sum_{d \le \sqrt{x^2+1}} \rho(d) \ll x \, .$$

The other sum is evaluated in Exer. 18-6. $\qquad\qquad\square$

One of the classical papers concerning the average number of divisors of a polynomial is [2], due to Erdos.

Exercise 18-7.

1 ⋄ Show that the integers k and $k + 1$ are always coprime.

2 ⋄ Prove that there exists a constant C such that, for every $x \ge 2$, we have $\sum_{k \le x} d(k(k + 1)) = Cx \log^2 x + O(x \log x)$.

3 ⋄ Prove that there exists a constant C' such that, for every $x \ge 2$, we have
$$\sum_{k \le x} \frac{\varphi(k(k + 1))}{k(k + 1)} = C'x + O(\log^2 x).$$

The above line of thought developped into the very active area called *correlation of multiplicative functions*, see, for instance, the paper [4] due to O. Klurman or [1] with more elementary techniques, due to P. Darbar.

Exercise 18-8. Let f be a non constant polynomial with integer coefficients. We define $\mathcal{P}(f)$ to be the set of primes for which there exists an integer n such that $p \mid f(n)$.

1 ⋄ Let $a \ne 0$ and b be two integers. Show that $\mathcal{P}(ax + b) = \{\text{all primes}\} \setminus \{p \mid a, p \nmid b\}$.

2 ⋄ Prove the following fact, due to I. Schur in [7, p. 41]: the set $\mathcal{P}(f)$ is always infinite.

3 ⋄ By considering the polynomial $f(X) = 9X^2 + 3X + 1$, show that there are infinitely many primes congruent to 1 modulo 3.

Further Reading

The readers interested in the algebraic aspect can study the book [3] by K. Ireland and M. Rosen.

References

[1]   P. Darbar. "Triple correlations of multiplicative functions". In: *Acta Arith.* 180.1 (2017), pp. 63–88. https://doi.org/10.4064/aa8605-4-2017 (cit. on p. 187).

[2]   P. Erdös. "On the sum $\sum_{k=1}^{x} d(f(k))$". In: *J. London Math. Soc.* 27 (1952), pp. 7–15. https://doi.org/10.1112/jlms/s1-27.1.7 (cit. on p. 187).

[3]   K. Ireland and M. Rosen. *A classical introduction to modern number theory*. Second. Vol. 84. Graduate Texts in Mathematics. Springer-Verlag, New York, 1990, pp. xiv+389. https://doi.org/10.1007/978-1-4757-2103-4 (cit. on p. 187).

[4]   O. Klurman. "Correlations of multiplicative functions and applications". In: *Compos. Math.* 153.8 (2017), pp. 1622–1657. https://doi.org/10.1112/S0010437X17007163 (cit. on p. 187).

[5]   K. Lapkova. "Explicit upper bound for the average number of divisors of irreducible quadratic polynomials". In: *Monatsh. Math.* 186.4 (2018), pp. 663–673. https://doi.org/10.1007/s00605-017-1061-y (cit. on p. 185).

[6]   I. Niven, H.S. Zuckerman, and H.L. Montgomery. *An introduction to the theory of numbers*. Fifth edition. John Wiley & Sons, Inc., New York, 1991, pp. xiv+529 (cit. on p. 184).

[7]   I Schur. *Über die Existenz unendlich vieler Primzahlen in einigen speziellen arithmetischen Progressionen*. German. Sitzungsber. Berl. Math. Ges. 11, 40-50 (1912). 1912 (cit. on p. 187).

[8]   E.J. Scourfield. "The divisors of a quadratic polynomial". In: *Proc. Glasgow Math. Assoc.* 5 (1961), 8–20 (1961) (cit. on p. 185).

Chapter 19
The Riemann Zeta Function

The Dirichlet series $D(\mathbb{1}, s)$ associated to the function $\mathbb{1}$ is of fundamental importance and is called *the Riemann ζ-function*. We have started our investigations on this function in Sect. 3.2. Let us recall its definition: when $\Re s > 1$, we have

$$\zeta(s) = \sum_{n \geq 1} 1/n^s .$$

This central object has been introduced by Euler very early, but it is Riemann who considered it not only as a generating series but as a full-fledged function on the complex plane. By the Euler product formula proved in Exer. 3-18, this function offers a link between the integers and the prime numbers. Though this is the simplest Dirichlet series one can think of, many of its properties are still shrouded in mystery. Here is a proof of the beginning of the Taylor expansion of $\zeta(s)$ around 1.

Exercise 19-1. We define the series of functions $\sum f_n$ by

$$f_n(x) = \frac{1}{n^x} - \int_n^{n+1} \frac{dt}{t^x} .$$

1 ◇ Show that the sequence of general term $u_n = \sum_{k=1}^n k^{-1} - \log(n + 1)$ is convergent. Denote its limit by γ and call it the *Euler constant*.

2 ◇ Show that when $n \geq 1$ and $x > 0$, one has: $0 \leq f_n(x) \leq n^{-x} - (n + 1)^{-x}$.

3 ◇ Prove that the series of functions $\sum f_n$ is convergent on $(0, \infty)$.

4 ◇ Let S be the sum of the series $\sum f_n$ over $\mathbb{R}_+^\star$. Show that $S(1) = \gamma$ and give an expression of $S(x)$ when $x > 1$.

5 ◇ Show that the convergence of the series $\sum f_n$ is uniform on $[1, \infty)$.

6 ◇ Conclude that, when x tends to 1, the quantity $\zeta(x) - \frac{1}{x-1}$ tends to γ.

© The Author(s), under exclusive license to Springer Nature Switzerland AG 2022
O. Ramaré, *Excursions in Multiplicative Number Theory*, Birkhäuser Advanced Texts
Basler Lehrbücher, https://doi.org/10.1007/978-3-030-73169-4_19

19.1 Upper Bounds in the Critical Strip

Let us write

$$
\begin{aligned}
\zeta(s) &= \sum_{n \le N} n^{-s} + s \int_N^\infty \sum_{N < n \le u} 1 \, \frac{du}{u^{s+1}} \\
&= \sum_{n \le N} n^{-s} + s \int_N^\infty (u - \{u\} - N) \frac{du}{u^{s+1}} \\
&= \sum_{n \le N} n^{-s} - \frac{N^{1-s}}{1-s} - s \int_N^\infty \{u\} \frac{du}{u^{s+1}}
\end{aligned}
\tag{19.1}
$$

for an integer $N \ge 0$ to be chosen later. With $N = 1$, we obtain

$$
\zeta(s) = \frac{s}{s-1} - s \int_1^\infty \{u\} \frac{du}{u^{s+1}} \ .
\tag{19.2}
$$

Theorem 19.1

The function ζ can be extended to a meromorphic function on the half-plane $\Re s > 0$ with a unique pole at $s = 1$ with residue 1.

Actually, the Riemann ζ-function can be extended to the whole complex plane with a unique pole at $s = 1$. See Exers. 19-4 and 20-10.

Proof Eq. (19.1) gives the extension for $\Re s > 0$.   □

Exercise 19-2.

1◇ Show that, when $\Re s \in (0, 1)$, we have $\dfrac{\zeta(s)}{s} = \displaystyle\int_0^\infty \{1/t\} t^{s-1} dt$.

2◇ Show that, when $\Re s \in (0, 1)$ and for any two sequences $(c_k)_{k \le K}$ and $(\theta_k)_{k \le K}$ of real numbers such that $\theta_k \ge 1$, we have

$$
1 - \zeta(s) \sum_{k \le K} \frac{c_k}{\theta_k^s} = s \int_0^\infty \left(\mathbb{1}_{[0,1]} + \sum_{k \le K} c_k \left\{ \frac{1}{\theta_k t} \right\} \right) t^{s-1} dt \ .
$$

This last equation can be seen as the heart of the Nyman–Beurling criterium. See [1] by L. Báez-Duarte, M. Balazard, B. Landreau and É. Saias.

Exercise 19-3. Let $g(u) = \int_1^u (\{v\} - \frac{1}{2}) dv$ for $u \ge 1$.

1◇ Show that $g(u)$ is bounded.

2◇ Show that $\zeta(s) = \dfrac{s}{s-1} - \dfrac{1}{2} - s(s+1) \displaystyle\int_1^\infty g(u) \frac{du}{u^{s+2}}$ and deduce that the Riemann zeta function extends to $\Re s > -1$. Establish also that $\zeta(0) = -1/2$.

The next exercise proves the existence of a meromorphic continuation of $\zeta(s)$ by employing an idea of S. Ramanujan in [6] (let r go to zero in Eq. (20) therein) as modified by J. Écalle in Sect. 6 of [2]. More on this idea can be read in [7] by B. Saha and in [4] by B. Saha and Sanoli Gun.

Exercise 19-4.

1 ◇ Show that, when $\Re s > 1$ and $k \geq 0$, we have $\zeta(s + k) - 1 \ll 1/2^k$.

2 ◇ Prove that, for any real number $x \in (-1, 1)$ and any complex number α. the following generalization of the binomial theorem holds true: $(1 - x)^{-\alpha} = 1 + \sum_{k\geq 0} (\alpha)_k x^{k+1}$, where $(\alpha)_k = \frac{\alpha(\alpha+1)\cdots(\alpha+k)}{(k+1)!}$.

3 ◇ On using the above formula for $n \geq 2$ to develop $(n - 1)^{1-s} - n^{1-s}$, establish the *translation formula*: $1 = \sum_{k\geq 0} (s - 1)_k \big(\zeta(s + k) - 1\big)$ for $\Re s > 1$.

4 ◇ Deduce from the *translation formula* that the Riemann ζ-function may be extended to the whole complex plane in a meromorphic function.

Proposition 19.1

For $s = \sigma + it$ with $\sigma > 0$ we have

$$\begin{cases} |\zeta(s)| \leq |s| \left(\dfrac{1}{|1 - \sigma|} + \dfrac{1}{\sigma} \right) & \text{if } |t| \leq 2, \\ |\zeta(s)| \leq 4 + \log |t| & \text{if } |t| \geq 2 \text{ and } 2 \geq \sigma \geq 1, \\ |\zeta(s)| \leq 6\sigma^{-1} |t|^{1-\sigma} \log |t| & \text{if } |t| \geq 2 \text{ and } 1 \geq \sigma \geq 0. \end{cases}$$

Proof We start with Eq. (19.2), which gives us for $|t| \leq 2$ that

$$|\zeta(s)| \leq \frac{|s|}{|\sigma - 1|} + |s| \int_1^\infty \frac{du}{u^{1+\sigma}}.$$

Further, for $|t| \geq 2$ and $\sigma \geq 1$, we take $N = 1 + [|t|]$ (it is 1 plus the integer part of the absolute value of t) in Eq. (19.1) and obtain

$$|\zeta(s)| \leq \sum_{n\leq N} n^{-\sigma} + \frac{N^{1-\sigma}}{|1 - s|} + (2 + |t|) \int_N^\infty \frac{du}{u^{\sigma+1}}$$

$$\leq 1 + \log N + \frac{1}{2} + \frac{2 + |t|}{N} \leq 4 + \log |t|.$$

Finally suppose that $|t| \geq 2$ and $\sigma \leq 1$. We have

$$|\zeta(s)| \le \sum_{n \le N} n^{-\sigma} + \frac{N^{1-\sigma}}{|1-s|} + (2+|t|) \int_N^\infty \frac{du}{u^{\sigma+1}}$$

$$\le 1 + \int_1^N \frac{du}{u^\sigma} + \frac{N^{1-\sigma}}{1+|t|} + \frac{2+|t|}{\sigma N^\sigma}$$

$$\le 1 + \frac{N^{1-\sigma} - 1}{1-\sigma} + \frac{N^{1-\sigma}}{1+|t|} + \frac{2+|t|}{\sigma(1+|t|)} N^{1-\sigma} .$$

Since the derivative (with respect to x) of $N^x - 1 - xN^x \log N$ is negative and this function vanishes at $x = 0$, we find that

$$\frac{N^{1-\sigma} - 1}{1-\sigma} \le N^{1-\sigma} \log N \quad (0 \le \sigma \le 1). \tag{19.3}$$

Consequently, we have

$$|\zeta(s)| \le 1 + N^{1-\sigma} \log N + \frac{N^{1-\sigma}}{1+|t|} + \frac{2+|t|}{\sigma(1+|t|)} N^{1-\sigma} .$$

We finally bound N by $1 + |t|$ and since $|t| \ge 2$, we obtain

$$\left(1 + \frac{\log(|t|+1)}{3} + \frac{4}{3\sigma}\right) \frac{(|t|+1)^{1-\sigma}}{|t|^{1-\sigma} \log |t|} \le 3(1+\sigma^{-1}) \le 6/\sigma .$$

Exercise 19-5. Show that the $\zeta(s)$ is convex for real $s \in (1, +\infty)$.

Exercise 19-6. Let s be a complex number satisfying $\Re s > 1$.

1 ◇ Show that $\zeta(s) = s \int_1^\infty \frac{[u]}{u^{s+1}} du$ and that $\frac{1}{\zeta(s)} = s \int_1^\infty \frac{M(u)}{u^{s+1}} du$, where $M(t) = \sum_{n \le t} \mu(n)$.

2 ◇ Show that $-\frac{\zeta'(s)}{\zeta(s)} = s \int_1^\infty \frac{\psi(u)}{u^{s+1}} du$, where $\psi(u) = \sum_{n \le u} \Lambda(n)$.

Exercise 19-7. Prove that, when $\Re s > 1$, we have $\log \zeta(s) = \sum_{n \ge 2} \frac{\Lambda(n)}{n^s \log n}$.

19.2 Computing $\zeta(s)$

Equation (19.1) provides us with a first tool to compute values of $\zeta(s)$ in case $\Re s \in (0, 1]$. Indeed, for a positive integer N to be selected at our convenience we have

$$\zeta(s) = \sum_{n \le N} n^{-s} - \frac{N^{1-s}}{1-s} + O^*\left(\frac{|s|}{\sigma N^\sigma}\right) . \tag{19.4}$$

This gives us the following script to evaluate $\zeta(s)$.

```
{myzeta(s, prec) =
  my(sig = real(s), N = ceil((abs(s)/sig/prec)^(1/sig)));
  return(sum(n = 1, N, n^(-s), -N^(1-s)/(1-s)));
}
myzeta(1 + I, 0.001)
zeta(1 + I)
```

The above compares the value given by (19.4) with one given by the inbuilt `zeta`: the readers may safely rely on Pari/GP for this task, but it is always a better idea to know how things can be computed.

It is also important to control the local variations.

Lemma 19.1

For $s = \sigma + it$ with $\sigma \geq 3/4$ and $|t| \geq 2$ we have

$$|\zeta'(s)| \leq 8\,|s|^{\max(1-\sigma,0)}(\log|s|)^2 .$$

Furthermore, when $|t| \leq 2$, $\sigma \geq 3/4$ and $|s-1| \geq 1/2$, we have $|\zeta'(s)| \leq 7$.

We presented in Sect. 3.2 several ways to compute $\zeta'(s)$ in Pari/GP as well as in Sage, and a further one is included in the proof below.

Proof By differentiating (19.1), we get

$$\zeta'(s) = -\sum_{n \leq N} \frac{\log n}{n^s} - \frac{N^{1-s}}{(1-s)^2}$$

$$+ \frac{N^{1-s}\log N}{1-s} - \int_N^\infty \{u\}\frac{du}{u^{s+1}} - s\int_N^\infty \{u\}(\log u)\frac{du}{u^{s+2}}$$

and thus

$$|\zeta'(s)| \leq \sum_{n \leq N} \frac{\log n}{n^\sigma} + \frac{N^{1-\sigma}(1+\log N)}{|1-s|} + \frac{1}{\sigma N^\sigma} + \frac{(1+\log N)|s|}{(1+\sigma)N^{1+\sigma}}$$

since

$$\int_N^\infty (\log u)\frac{du}{u^{\sigma+2}} = \frac{\log N}{(1+\sigma)N^{1+\sigma}} + \frac{1}{1+\sigma}\int_N^\infty \frac{du}{u^{\sigma+2}} \leq \frac{1+\log N}{(1+\sigma)N^{1+\sigma}} .$$

Let us start with the case $\sigma \in [3/4, 1]$. Since $\sigma \geq 3/4$, the function $n \mapsto (\log n)/n^\sigma$ is non-increasing from $n = 4$ onwards, whence

$$\sum_{n \leq N} \frac{\log n}{n^\sigma} \leq \frac{\log 2}{2^{3/4}} + \frac{\log 3}{3^{3/4}} + \frac{\log 4}{4^{3/4}} + \int_4^N (\log t)\frac{dt}{t^\sigma} \leq 1.4 + N^{1-\sigma}(\log N)^2$$

by using (19.3). In that case, we get

$$|\zeta'(s)| \le 1.4 + N^{1-\sigma}(\log N)^2 + \frac{N^{1-\sigma}(1 + \log N)}{|1 - s|}$$

$$+ \frac{4}{3N^\sigma} + \frac{4(1 + \log N)|s|}{7N^{1+\sigma}} \qquad (19.5)$$

i.e. when $2 \le N \le |s|$ and $|s| \ge \sqrt{4 + 9/16} = 2.13\ldots$

$$\frac{|\zeta'(s)|}{|s|^{1-\sigma}(\log |s|)^2} \le \frac{1.4}{(\log 2.13)^2} + 1 + \frac{1 + 1/\log 2}{2 \log 2.13}$$

$$+ \frac{4}{3 \cdot 2^{3/4}(\log 2.13)^2} + \frac{4(1 + 1/\log 2)|s|^\sigma}{7N^{1+\sigma} \log 2.13},$$

which we simplify into

$$\frac{|\zeta'(s)|}{|s|^{1-\sigma}(\log |s|)^2} \le 6.5 + \frac{2|s|^\sigma}{N^{1+\sigma}}.$$

When $|s| \le 3$, we select $N = 2$, getting that is above is at most 8. When $|s| \ge 3$, we select $N = [|s|] \ge 2$ and get the same bound.

Let us now consider the case when $\sigma \ge 1$. We find that

$$\sum_{n \le N} \frac{\log n}{n} \le \frac{\log 2}{2} + \frac{\log 3}{3} + \frac{\log 4}{4} + \frac{1}{2}\left(\log^2 N - \log^2 4\right) \le \frac{1}{10} + \frac{(\log N)^2}{2}$$

whence

$$|\zeta'(s)| \le \frac{1}{10} + \frac{(\log N)^2}{2} + \frac{(1 + \log N)}{|1 - s|} + \frac{1}{N} + \frac{(1 + \log N)|s|}{2N^2}$$

and now the same reasoning as before yields in fact a better bound.

When $|t| \le 2$, $\sigma \in [3/4, 3/2]$ and $|s - 1| \ge 1/2$, we employ (19.5) with $N = 1$ and get

$$|\zeta'(s)| \le 1.4 + \frac{1}{|1 - s|} + \frac{4}{3} + \frac{4|s|}{7} \le 7.$$

Finally, when $\sigma \ge 3/2$, we find that, by looking at the Dirichlet series expansion, $|\zeta'(s)| \le -\zeta'(3/2) \le 4$. This last inequality gives us the opportunity to use yet another Pari/GP function:

```
sumpos( n = 2, log(n)/n^(3/2))
```

Note that we have shown another way to compute $\zeta'(s)$ at the end of Sect. 3.2. The proof is complete. $\qquad\square$

Exercise 19-8. Write a Pari/GP script to establish that, when $\Re s \in [3/4, 1]$ and $|\Im s| \le 100$ and $|s - 1| \ge 1$, we have $3/20 \le |\zeta(s)| \le 7/2$.

The paper [5] of two of the authors together with Izabela Petrykiewicz proposes a pleasant stroll around computations linked with the Riemann zeta function, and reading it a good (though trying!) exercise.

We conclude this chapter with a Sage plot of $|\zeta(\sigma + it)|$ when $1/10 \le \sigma \le 9/10$ and $-40 \le t \le 40$ in Fig. 19.1. This plot shows that the Riemann zeta function already varies quite a bit in this domain! We have obtained this plot with the script:

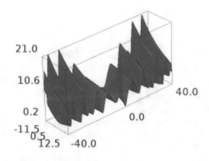

Fig. 19.1: Plot of $5|\zeta(\sigma + it)|$ when $1/10 \le \sigma \le 9/10$ and $-40 \le t \le 40$

```
(nbx, nby) = (100, 130)

def fx(x):
    return(0.5 + x/2.5)

def fy(y):
    return(40*y)

zetaPlot = list_plot3d([[(30*(fx(x/nbx)-1/2)+1/2 , fy(y/nby),
                       5*abs(zeta(fx(x/nbx) + I*fy(y/nby))))
                       for x in range(-nbx, nbx+1)
                       for y in range(-nby, nby+1)],
                       interpolation_type = 'spline')

zetaPlot.show()
```

Further Reading

The theory of the Riemann zeta function is very developed. A first step for the readers would be the two books [8] by E. C. Titchmarsh and revised by D.R. Heath-Brown, and [3] by H.M. Edwards. This latter monograph also emphasizes the computational aspects of the theory.

References

[1]   L. Báez-Duarte et al. "Notes sur la fonction ζ de Riemann. III". In: *Adv. Math.* 149.1 (2000), pp. 130–144. https://doi.org/10.1006/aima.1999.1861 (cit. on p. 192).

[2]   J. Ecalle. "ARI/GARI, la dimorphie et l'arithmétique des multizêtas: un premier bilan". In: *J. Théor. Nombres Bordeaux* 15.2 (2003), pp. 411–478. http://jtnb.cedram.org/item/JTNB_2003__15_2_411_0/ (cit. on p. 192).

[3]   H.M. Edwards. *Riemann's zeta function*. Pure and Applied Mathematics, Vol. 58. Academic Press [A subsidiary of Harcourt Brace Jovanovich, Publishers], New York-London, 1974, pp. xiii+315 (cit. on p. 197).

[4]   Sanoli Gun and Biswajyoti Saha. "Multiple Lerch zeta functions and an idea of Ramanujan". In: *Michigan Math. J.* 67.2 (2018), pp. 267–287. https://doi.org/10.1307/mmj/1516330974 (cit. on p. 192).

[5]   P. Moree, I. Petrykiewicz, and A. Sedunova. "A computational history of prime numbers and Riemann zeros". In: arxiv.org/abs/1810.05244 (2020), 29pp (cit. on p. 196).

[6]   S. Ramanujan. *A series for Euler's constant γ*. English. Messenger 46, 73-80 (1916). 1916 (cit. on p. 192).

[7]   Biswajyoti Saha. "An elementary approach to the meromorphic continuation of some classical Dirichlet series". In: *Proc. Indian Acad. Sci. Math. Sci.* 127.2 (2017), pp. 225–233. https://doi.org/10.1007/s12044-017-0327-6 (cit. on p. 192).

[8]   E.C. Titchmarsh. *The theory of the Riemann zeta-function*. Second. Edited and with a preface by D.R. Heath-Brown. The Clarendon Press, Oxford University Press, New York, 1986, pp. x+412 (cit. on p. 197).

Part IV
The Mellin Walk

Chapter 20
The Mellin Transform

Between 1880 and 1920, H. Mellin, a student of Weierstrass, studied the pair of transforms

$$\begin{cases} \Phi(x) = \dfrac{1}{2i\pi} \displaystyle\int_{a-i\infty}^{a+i\infty} F(z)x^{-z}dz\,, & (x > 0) \\[2ex] F(z) = \displaystyle\int_{0}^{\infty} \Phi(x)x^{z-1}dx\,. \end{cases}$$

Each of these integrals of course requires proper hypotheses to exist, see Theorem 20.1. In the favourable cases, they are mutually inverse. The function F is said to be the *Mellin transform* of Φ and we write $F = \check{\Phi}$ (that's a capital ϕ with a caron˘ on top).

20.1 Some Examples

Before entering the theory, let us compute some complex integrals.

Lemma 20.1

When $x > 0$ and $a > 0$ are positive real numbers, we have

$$\frac{1}{2i\pi} \int_{a-i\infty}^{a+i\infty} \frac{x^{-s}ds}{s(s+1)} = \begin{cases} 1 - x & \text{when } x < 1, \\ 0 & \text{when } x \geq 1. \end{cases}$$

Lemma 20.2

When $x > 0$ and $a > 0$ are positive real numbers, we have

© The Author(s), under exclusive license to Springer Nature Switzerland AG 2022
O. Ramaré, *Excursions in Multiplicative Number Theory*, Birkhäuser Advanced Texts
Basler Lehrbücher, https://doi.org/10.1007/978-3-030-73169-4_20

$$\frac{1}{2i\pi} \int_{a-i\infty}^{a+i\infty} \frac{x^{-s}ds}{s^2} = \begin{cases} -\log x & \text{when } x < 1, \\ 0 & \text{when } x \geq 1. \end{cases}$$

Lemma 20.3

When $x > 0$ and $a > 0$ are positive real numbers, we have

$$\frac{1}{2i\pi} \int_{a-i\infty}^{a+i\infty} \frac{2x^{-s}ds}{s(s+1)(s+2)} = \begin{cases} (1-x)^2 & \text{when } x < 1, \\ 0 & \text{when } x \geq 1. \end{cases}$$

Proof Indeed, when $x \geq 1$, it is enough to shift the line of integration to the right-hand side, noticing that the contribution from the poles should be affected with a minus sign and that the horizontal segments do not contribute. Such a sentence may sound mysterious, hence we give *one* full proof here. We start this investigation by noticing that the integral over the whole line is the limit of the integral over a finite path:

$$\frac{1}{2i\pi} \int_{a-i\infty}^{a+i\infty} \frac{x^{-s}ds}{s(s+1)} = \lim_{T,T' \to \infty} \frac{1}{2i\pi} \int_{a-iT}^{a+iT'} \frac{x^{-s}ds}{s(s+1)},$$

where the variable s ranges the vertical segment from the bottom to the top, and where T and T' tend independently to infinity. We then compare the integral to

$$\frac{-1}{2i\pi} \int_{A-iT}^{A+iT'} \frac{x^{-s}ds}{s(s+1)}$$

for some $A \geq a$ that we shall select very large but that is at first fixed. We realize this comparison by applying the Cauchy Residue Theorem to the rectangle of vertices $a + iT'$, $a - iT$, $A - iT$ and $A + iT'$. The integrals over the horizontal segments are bounded above, in absolute value, respectively, by A/T^2 and A/T'^2; both quantities tend to 0 when T and T' tend to infinity. Notice that no residue of the summand belongs to the enclosed region, hence

$$\frac{1}{2i\pi} \int_{a-iT}^{a+iT'} \frac{x^{-s}ds}{s(s+1)} = \frac{1}{2i\pi} \int_{A-iT}^{A+iT'} \frac{x^{-s}ds}{s(s+1)} + o(1),$$

where this $o(1)$ represents a function that tends to 0 when T and T' both go to infinity. We next notice that the integral over $\Re s = A$ is $O(x^{-A})$ since

$$\frac{1}{2\pi} \int_{A-iT}^{A+iT'} \left| \frac{ds}{s(s+1)} \right|$$

is bounded above independently of T, T' and $A \geq a$. It suffices then to let A go to infinity. For brevity we will merely write "we shift the line of integration to the right-hand side" to indicate this procedure.

When $x < 1$, we shift the line of integration to the left-hand side and meet polar contributions at $s = 0$ and $s = -1$ giving the value $1 - x$ as expected. The second and third lemma are proved in a similar fashion. $\qquad\square$

Exercise 20-1. When $x > 0$, we have

$$\frac{1}{2i\pi} \int_{1/2-i\infty}^{1/2+i\infty} \frac{2x^s\, ds}{s^2(1-s)(2-s)} = \begin{cases} 2x - x^2/2 & \text{when } x < 1, \\ \log x + 3/2 & \text{when } x \geq 1. \end{cases}$$

Exercise 20-2. When $x > 0$, we have

$$\frac{1}{2i\pi} \int_{1/2-i\infty}^{1/2+i\infty} \frac{x^s\, ds}{s(1-s)(2-s)} = \begin{cases} x - x^2/2 & \text{when } x < 1, \\ 1/2 & \text{when } x \geq 1. \end{cases}$$

The readers should note that moving the line of integration from $\Re s = 1/2$ to $\Re s = 3$ would change the functions in the last two exercises.

Exercise 20-3. When $x > 0$ and $n \geq 1$, show that

$$\frac{1}{2\pi i} \int_{1/2-i\infty}^{1/2+i\infty} \frac{x^{-z}}{z^{n+1}}\, dz = \begin{cases} \frac{(-1)^n}{n!} \log^n x, & \text{when } x \leq 1, \\ 0, & \text{when } x > 1. \end{cases}$$

The readers may have a look at Exer. 23-6 where a fancier kernel is being used.

20.2 The complex Stirling Formula and the Cahen-Millen Formula

The Euler Gamma-function is a classical object. We recall some of its properties in the exercises below. On the historical side, we mention that the notation Γ for this function is due to A.-M. Legendre.

Exercise 20-4. Recall that the Euler Gamma-function is defined by

$$\Gamma(s) = \int_0^\infty e^{-t} t^s \frac{dt}{t}.$$

This function is holomorphic when $\Re(s) > 0$.

$1 \diamond$ Show that the Γ-function satisfies the functional equation $\Gamma(s+1) = s\Gamma(s)$ (this enables one to extend it as a meromorphic function over $\mathbb{C}$).

$2 \diamond$ When $\Re(s) > 1$, show that $\zeta(s) = \dfrac{1}{\Gamma(s)} \displaystyle\int_0^\infty \frac{1}{e^t - 1} t^s \frac{dt}{t}$.

The Stirling Formula has been extensively studied in classical textbooks. The version we use comes from the book [7, Part A, chapter 2] with C. Gudermann's improvement from the theorem p. 61 of the same book. We said it is classical analysis: Gudermann's improvement on the classical formula dates from 1845! It leads to the constant $1/12$, where earlier investigations only had $1/8$. For any $\delta \in (0, \pi)$, let us define the domain

$$W_\delta = \left\{ |z|e^{i\phi} \in \mathbb{C}^* : |\phi| \le \pi - \delta \right\} . \tag{20.1}$$

Our estimates will be expressed in terms of a function that is traditionally denoted by $\mu(z)$ though it has nothing to do with the Möbius μ-function.

Lemma 20.4 (Complex Stirling Formula)

Let $\delta \in (0, \pi)$. Let $z \in W_\delta$ be some complex number. We have

$$\Gamma(z) = \sqrt{2\pi} e^{-z} z^{z-1/2} e^{\mu(z)} ,$$

with

$$|\mu(z)| \le \frac{1}{12 \sin^2(\delta/2)} \frac{1}{|z|} .$$

In particular, when $\delta = \pi/2$ we have $|\mu(z)| \le 1/(6|z|)$.

This lemma uses complex exponentiation: we have to raise a complex number to some complex power and some caution is required. The expression z^z is defined as $\exp(z \log z)$, where we have just shifted the problem from defining the exponential to the one of defining the logarithm! The logarithm we use (called *the principal branch of the logarithm*) is defined by:

$$\log z = \log r + i\theta, \qquad (z = re^{i\theta}, \, r \ge 0, \, \theta \in (-\pi, \pi]) . \tag{20.2}$$

In our applications, we will only have to bound the modulus of $\Gamma(z)$. Here is a lemma that does just that.

Lemma 20.5

When $\sigma > 0$, we have

$$|\Gamma(\sigma + it)| \le \sqrt{2\pi} (\sigma^2 + t^2)^{(2\sigma-1)/4} e^{-\pi|t|/2} \exp\left(\frac{1}{6\sqrt{\sigma^2 + t^2}} \right) .$$

Proof We first note that $\Gamma(\sigma - it) = \overline{\Gamma(\sigma + it)}$, showing that we only have to worry about the non-negative values of t. On writing $\sigma + it = \sqrt{\sigma^2 + t^2} \exp(i\theta)$, we readily see that $\arctan(t/\sigma) = \theta$ and since $\arctan x + \arctan(1/x) = \pi/2$ for real positive x, we deduce that $\theta = \pi/2 - \arctan(\sigma/t)$. We note also that $\arctan(\sigma/t) \leq \sigma/t$. Let us proceed to put these elements together:

$$\Re\left\{-(\sigma + it) + (\sigma - \tfrac{1}{2} + it)(\log \sqrt{\sigma^2 + t^2} + i\theta)\right\}$$

$$= -\sigma + (\sigma - \tfrac{1}{2})\log \sqrt{\sigma^2 + t^2} - t\theta$$

$$= (\sigma - \tfrac{1}{2})\log \sqrt{\sigma^2 + t^2} - \tfrac{\pi}{2}t + t\arctan\frac{\sigma}{t} - \sigma,$$

and our bound follows. $\square$

Exercise 20-5. Show that, when $u \geq 0$, we have $\arctan u \geq u - u^3$ and deduce that

$$|\Gamma(\sigma + it)| \geq \sqrt{2\pi}(\sigma^2 + t^2)^{(2\sigma-1)/4}e^{-\pi|t|/2}\exp\left(-\frac{\sigma^3}{t^2}\right)$$

when both σ and t are positive.

This detour involving the Γ-function was to prepare the way for using another classical formula, often referred to as the *Cahen-Millen Formula*.

Lemma 20.6

When $x > 0$ and $a > 0$ are two positive parameters, we have

$$e^{-x} = \frac{1}{2\pi i}\int_{a-i\infty}^{a+i\infty} \Gamma(s)x^{-s}\,ds.$$

The convergence of the integral is guaranteed by the Complex Stirling Formula, which ensures that $\Gamma(1/2 + it)$ decreases exponentially in $|t|$.

Proof (of Lemma 20.6) Recall that the poles of Γ are located at the negative integers $-n$. We shift the line of integration further and further to the left. In each non-positive integer $-n$ that we pass in this way the Γ-function has a simple pole with residue $(-1)^n/n!$, as is shown, for instance, by employing the *Complement Formula*

$$\frac{1}{\Gamma(s)\Gamma(1 - s)} = \frac{\sin \pi s}{\pi}.$$

Adding the various contributions yields $\sum_{n\geq 0}(-x)^n/n! = e^{-x}$, as expected. $\square$

Exercise 20-6. (Mellin–Barnes formula).

1 ◇ Define the Beta function by $B(s, \xi) = \int_0^1 x^{s-1}(1-x)^{\xi-1}dx$ for $\Re s > 0$ and $\Re \xi > 0$. Of which function of s is it the Mellin transform?

2 ◇ Express the Beta function in terms of the Gamma-function.

3 ◇ Show that $B(s, \xi) = \int_0^\infty \dfrac{x^{s-1}}{(1+x)^{s+\xi}}dx$.

4 ◇ Prove the Mellin–Barnes formula: when x and λ are positive, we have
$$\frac{1}{(1+x)^\lambda} = \frac{1}{2i\pi} \int_{c-i\infty}^{c+i\infty} \frac{\Gamma(s)\Gamma(\lambda-s)}{\Gamma(\lambda)} x^{-s}ds \text{ for any } c \in (0, \lambda).$$ This formula is also valid when λ is a complex number with a positive real part and $c \in (0, \Re\lambda)$.

E.W. Barnes introduced such formulas in [1].

20.3 Mellin Transforms/Fourier Transforms

Let us first illustrate how one can use the preceding lemmas. For instance, we may write

$$\sum_{q \leq Q} f_0(q)(1 - q/Q) = \sum_{q \geq 1} f_0(q) \frac{1}{2\pi i} \int_{3-\infty}^{3+i\infty} \frac{(q/Q)^{-s}ds}{s(s+1)}$$

$$= \frac{1}{2\pi i} \int_{3-i\infty}^{3+i\infty} \sum_{q \geq 1} \frac{f_0(q)}{q^s} \frac{Q^s ds}{s(s+1)} .$$

This links the initial study to the one of the Dirichlet series $D(f_0, s)$, which we completed in Step 3 of Sect. 8.1. Formally speaking, we start from a function w defined on $(0, \infty)$ (we had $w(x) = \max(0, 1 - x)$ in our opening example) with the aim of writing it in the form

$$w(x) = \frac{1}{2\pi i} \int_{c-i\infty}^{c+i\infty} \check{w}(s)x^{-s}ds . \tag{20.3}$$

With the change of variables $s = c + 2i\pi y$ and $x = e^u$, this becomes

$$w(e^u) = \int_{-\infty}^\infty \check{w}(c + 2\pi iy)e^{-uc}e^{-2\pi iuy}dy,$$

i.e. $\check{w}(c+2i\pi y)$ is the *Fourier transform* of $e^{uc}w(e^u)$, which, under suitable hypotheses, can be expressed in the form

$$\check{w}(c + 2\pi iy) = \int_{-\infty}^\infty w(e^u)e^{uc}e^{2\pi iuy}du .$$

On going back to our initial variables, we get

$$\check{w}(s) = \int_0^\infty w(x)x^{s-1}dx .$$

(20.4)

The function $\check{w}$ is called the *Mellin transform* of w. This is a formal argument and convergence problems may arise that should be addressed seriously: the examples given in Exers. 20-1 and 20-2 show that the situation may be delicate. This book is not a treatise on these issues, but the above shows how to reach (20.3) and to *guess* $\check{w}$. We mention for convenience the next theorem which is proved by converting the problem to a Fourier inversion question.

Theorem 20.1

Assume that w is a piecewise continuous function on $(0, \infty)$ that takes a value halfway between the limit values at any jump discontinuities. Assume further that $\int_0^\infty |w(x)|x^\kappa dx/x < \infty$ for some real parameter κ. Then we have

$$w(x) = \frac{1}{2\pi i} \int_{c-i\infty}^{c+i\infty} \check{w}(s)x^{-s}ds$$

for every $x > 0$ as soon as $c \geq \kappa$.

Exercise 20-7. Compute the Mellin transform of the function

$$w_0(t) = \begin{cases} t-1 & \text{when } 1 \leq t \leq 2, \\ 3-t & \text{when } 2 \leq t \leq 3, \\ 0 & \text{otherwise.} \end{cases}$$

20.4 Truncated Transform

Let Y be the function that vanishes on $(0, 1)$, takes the value $1/2$ at 1 and 1 afterwards. Its Mellin transform is $1/s$, which tends rather slowly to zero in vertical strips, giving rise to convergence problems. A way out is to employ a truncated transform. We start with a basic lemma.

Lemma 20.7

For $\kappa > 0, T > 0$ and $x > 0$, we have

$$\left| Y(x) - \frac{1}{2\pi i} \int_{\kappa-iT}^{\kappa+iT} \frac{x^s ds}{s} \right| \leq \frac{x^\kappa}{\pi} \min\left(\frac{7}{2}, \frac{1}{T|\log x|}\right) .$$

It is clear from the proof that if we had selected any real value between 0 and 1 for $Y(1)$, the very same estimate would have resulted.

Proof When $x < 1$, we write for any $K > \kappa$ tending to infinity:

$$\left(\int_{\kappa-iT}^{\kappa+iT} + \int_{\kappa+iT}^{K+iT} + \int_{K+iT}^{K-iT} + \int_{K-iT}^{\kappa-iT} \right) \frac{x^s \, ds}{s} = 0 \, .$$

The third integral tends to 0 when K goes to infinity. The two integrals over the horizontal segments are each bounded above (in absolute value) by $x^\kappa / (T |\log x|)$, whence

$$\left| Y(x) - \frac{1}{2\pi i} \int_{\kappa-iT}^{\kappa+iT} \frac{x^s \, ds}{s} \right| \leq \frac{x^\kappa}{\pi T |\log x|} \qquad (0 < x < 1).$$

The same estimate holds true when $x > 1$, which we prove this time by sending the line of integration to the left-hand side. These bounds are efficient when $T |\log x|$ is large enough; if not, we write

$$\int_{\kappa-iT}^{\kappa+iT} \frac{x^s \, ds}{s} = x^\kappa \int_{\kappa-iT}^{\kappa+iT} \frac{ds}{s} + x^\kappa \int_{-T}^{T} \frac{(x^{it} - 1) i \, dt}{\kappa + it} \, .$$

We check that the first integral equals $2 \arctan(T/\kappa) \leq \pi$ while, to handle the second one, we first notice that

$$\left| \frac{x^{it} - 1}{it \log x} \right| = \left| \int_0^1 e^{iut \log x} \, du \right| \leq 1 \, .$$

This leads to the upper bound $2T |\log x|$ (even when $x = 1$), whence

$$\left| \frac{1}{2\pi i} \int_{\kappa-iT}^{\kappa+iT} \frac{x^s \, ds}{s} \right| \leq \frac{x^\kappa}{\pi} \left(\frac{\pi}{2} + T |\log x| \right) \, .$$

This is enough when $x < 1$. When $x > 1$, we note that

$$1 - \frac{x^\kappa}{2\pi i} \int_{\kappa-iT}^{\kappa+iT} \frac{ds}{s} = 1 - \frac{x^\kappa}{\pi} \arctan(T/\kappa) \, ,$$

which is $\geq -x^\kappa / 2$ and $\leq 1 \leq x^\kappa$. Finally, we have obtained

$$\left| Y(x) - \frac{1}{2i\pi} \int_{\kappa-iT}^{\kappa+iT} \frac{x^s \, ds}{s} \right| \leq \frac{x^\kappa}{\pi} \min \left(\pi + T |\log x|, \frac{1}{T |\log x|} \right) \, .$$

We simplify this bound by noticing that

$$\min(\pi + u, 1/u) \leq \min(\alpha, 1/u) \, ,$$

where $\alpha = 1/u_0 = \pi + u_0$. Since this gives us $\alpha \leq 7/2$, the lemma follows readily. $\square$

The memoir [2] of Perron has been a high point in the theory of Dirichlet series, but inversion formulae appeared earlier, most notably in the famous 1859-memoir of Riemann. We now deduce from Lemma 20.7 the following rather classical and very useful *truncated* summation formula that avoids the delicate problems of convergence arising in the un-truncated formula.

Theorem 20.2 (Truncated Perron Summation Formula)

Let $F(s) = \sum_n \dfrac{a_n}{n^s}$ be some Dirichlet series that is absolutely converging when $\Re s > \kappa_a$, and let $\kappa > 0$ be strictly greater than κ_a. For $x \geq 1$ and $T \geq 1$, we have

$$\sum_{n \leq x} a_n = \frac{1}{2\pi i} \int_{\kappa-iT}^{\kappa+iT} F(s) \frac{x^s \, ds}{s} + O^* \left(\int_{1/T}^{\infty} \sum_{|\log(x/n)| \leq u} \frac{|a_n|}{n^\kappa} \frac{2 x^\kappa \, du}{T u^2} \right).$$

The error term in this form is taken from [5, Theorem 7.1] and is further studied in [6]. It depends on the short sums

$$\sum_{|\log(x/n)| \leq u} |a_n|/n^\kappa,$$

where the summation is over n and can be alternatively written as $e^{-u} x \leq n \leq e^u x$. In [4], it is shown that instead of $|a_n|$ we may put a_n at the price of a small technical complication. Returning to our problem, when $u \geq 1$, it usually suffices to bound the latter sum by the complete sum $\sum_{n \geq 1} |a_n|/n^\kappa$. When u gets smaller, we appeal most of the time to bounds of the shape $u x^{\kappa_a} B/x^\kappa$ for some *decent* B (typically a constant time $\log x$), and this leads to the error term

$$O \left(\frac{B x^{\kappa_a} \log T}{T} + \frac{x^\kappa}{T} \sum_{n \geq 1} \frac{|a_n|}{n^\kappa} \right).$$

The shortest sums we shall have to consider are of size $\simeq x/T$.

Proof We start from

$$\sum_{n \leq x} a_n = \sum_{n \geq 1} a_n Y(x/n) = \sum_{n \geq 1} a_n \frac{1}{2i\pi} \int_{\kappa-iT}^{\kappa+iT} \frac{(x/n)^s \, ds}{s}$$

$$+ O^* \left(\sum_{n \geq 1} \frac{|a_n| x^\kappa}{\pi n^\kappa} \min \left(\frac{7}{2}, \frac{1}{T |\log(x/n)|} \right) \right)$$

by appealing to Lemma 20.7. Let us set $\varepsilon = 1/T$. In order to study the error term, let us first isolate the contribution of those integers n satisfying $|\log(x/n)| \leq \varepsilon$, which we keep unchanged. Otherwise, we write

$$\sum_{\varepsilon \le |\log(x/n)|} \frac{|a_n|x^\kappa}{n^\kappa |\log(x/n)|} = \sum_{\varepsilon \le |\log(x/n)|} \frac{|a_n|x^\kappa}{n^\kappa} \int_{|\log(x/n)|}^\infty \frac{du}{u^2}$$

$$= \int_\varepsilon^\infty \sum_{|\log(x/n)| \le u} \frac{|a_n|x^\kappa}{n^\kappa} \frac{du}{u^2} - \int_\varepsilon^\infty \sum_{|\log(x/n)| \le \varepsilon} \frac{|a_n|x^\kappa}{n^\kappa} \frac{du}{u^2}$$

and these are enough details allowing the readers to complete the proof. □

To illustrate as well as to measure the relative strength of this theorem, let us try to evaluate the number of integers below x. The relevant Dirichlet series is of course the Riemann ζ-function, for which $\kappa_a = 1$. We select $\kappa = 1 + 1/\log x$ and readily get an error term of size $O(\log(xT)/T)$, provided that $T \le x$. We next shift the integral to the line $\kappa = 0$ (but avoid the point $s = 0$), where the Riemann ζ-function is of size $O(\sqrt{|t| + 1} \log(|t| + 2))$ by Proposition 19.1. Combining all this gives

$$\sum_{n \le x} 1 = x + O\big(\sqrt{T}\log^2 T + x\log(xT)/T\big) .$$

We select $T = x^{2/3}$ and obtain an error term of size $O(x^{1/3}\log^2 x)$. This shows that this technique may incur some loss: the error term in this problem could be . . . $O(1)$! However, when we use this summation formula, the loss is usually compensated by the fact that we incorporate information on the Dirichlet series in the proof.

Lemma 20.8

When $x > 0$, we have $\dfrac{1}{2\pi i} \displaystyle\int_{1/2-i\infty}^{1/2+i\infty} \frac{x^s ds}{s \sin \pi s} = \log(1 + x)$.

Proof We distinguish again according to whether $x < 1$ or not. When it is, we shift the line of integration to the right and recover what we recognize to be the Taylor series of $\log(1 + x)$. Otherwise, we shift the line of integration to the left, take into account the double pole at $s = 0$ which contributes $\log x$, whereas the other poles contribute $\log(1 + x^{-1})$. When $x = 1$, we use the Dominated Convergence Theorem. □

20.5 Smoothed Formulae

The readers have now all the material and experience to prove the next two summation formulae.

Theorem 20.3

Let $F(s) = \sum_n a_n/n^s$ be a Dirichlet series that converges absolutely for $\Re s > \kappa_a$, and let $\kappa > 0$ be a parameter strictly larger than κ_a. When $x \geq 1$, we have

$$\sum_{n \leq x} a_n(1 - n/x)^2 = \frac{1}{2\pi i} \int_{\kappa-i\infty}^{\kappa+i\infty} F(s)\frac{2x^s \, ds}{s(s+1)(s+2)}.$$

Theorem 20.4

Let $F(s) = \sum_n a_n/n^s$ be a Dirichlet series that converges absolutely for $\Re s > \kappa_a$, and let $\kappa > 0$ be a parameter strictly larger than κ_a. When $x \geq 1$, we have

$$\sum_{n \geq 1} a_n e^{-n/x} = \frac{1}{2\pi i} \int_{\kappa-i\infty}^{\kappa+i\infty} F(s)\Gamma(s)x^s \, ds.$$

Exercise 20-8. Let k and x be two real positive parameters. Generalize Theorem 20.4 to $\sum_{n \leq 1} a_n e^{-(n/x)^k} = \dfrac{1}{2i\pi} \int_{\kappa-i\infty}^{\kappa+i\infty} F(s)\Gamma\left(1 + \dfrac{s}{k}\right)x^s \dfrac{ds}{s}$, where we have kept notation and hypotheses from Theorem 20.4.

The function Γ decreases in vertical strips exponentially fast. The next formula uses a Mellin transform that decreases even faster.

Exercise 20-9. Let $F(s) = \sum_n a_n/n^s$ be a Dirichlet series that converges absolutely for $\Re s > \kappa_a$, and let $\kappa > 0$ be a parameter strictly larger than κ_a. Prove that when $y > 0$ we have

$$\sum_{n \geq 1} a_n e^{-\frac{\log^2 n}{4y}} = \frac{1}{i\sqrt{\pi}} \int_{\kappa-i\infty}^{\kappa+i\infty} F(s)e^{s^2 y} \, ds.$$

20.6 Smoothed Representations

The formula $F(s) = s \int_0^\infty \left(\sum_{n \leq x} a_n\right) dt/t^{s+1}$ of (3.9) may be too rough to infer properties on $F(s)$ because the error term due to the approximation of $\sum_{n \leq x} a_n$ is not small enough*. The next theorem proposes some easier-to-control representations.

---

* Though this is outside the scope of this monograph, it is typically the case for the Dedekind zeta function.

Theorem 20.5

Let σ be some real parameter. Let $F(s) = \sum_{n\geq 1} a_n/n^s$ be a Dirichlet series absolutely convergent for $\Re s \geq \sigma$ and let w be an integrable function over $[0, \infty)$. Assume that $\sum_{n\geq 1} |a_n| \int_0^\infty |w(n/u)| du/u^{1+\sigma} < \infty$. Then, when $\Re s \geq \sigma$, we have

$$\check{w}(s)F(s) = \int_0^\infty \sum_{n\geq 1} a_n w\left(\frac{n}{u}\right) \frac{du}{u^{s+1}} \, .$$

Proof It is enough to notice that

$$\check{w}(s) = \int_0^\infty n^s \int_0^\infty w(n/u) \frac{du}{u^{s+1}} \, .$$

The reader will easily complete the argument from there.   □

By using $w(t) = \mathbb{1}_{t\leq 1}$ whose Mellin transform is $1/s$, we recover the usual representation (3.9). By selecting $w(t) = \max(1 - t, 0)$, we obtain the less known formula:

$$F(s) = s(s + 1) \int_0^\infty \left(\sum_{n\leq u} a_n(u - n)\right) \frac{du}{u^{s+2}} \, . \tag{20.5}$$

Exercise 20-10.

1 ◇ By using (20.5) and Exer. 7-6, show that the Riemann zeta function admits a meromorphic continuation to the half-plane $\Re s > -1$.

2 ◇ Show that the polynomial $B_3(X) = X^3 - \frac{3}{2}X^2 + \frac{1}{2}X$ satisfies the identity $B_3(X + 1) - B_3(X) = 3X^2$ and that $3 \sum_{n\leq u}(1 - n/u)^2 = (B_3(u) - B_3(\{u\}))/u^2$ for any positive u. (See also Exer. 7-6)

3 ◇ By selecting $w(t) = (1 - t)^2 \mathbb{1}_{t\leq 1}$ in Theorem 20.5, show that

$$\frac{2\zeta(s)}{s(s + 1)(s + 2)} = \frac{1}{3(s - 1)} - \frac{1}{2s} + \frac{1}{6(s + 1)} - \int_1^\infty B_3(\{u\}) \frac{du}{u^{s+3}} \, .$$

Deduce that $\zeta(s)$ admits a meromorphic continuation to the half-plane $\Re s > -2$ and that this extension satisfies $|\zeta(-1 + it)| \leq \frac{2}{3}(t^2 + 1)^{3/2}$ for every t.

The bound proved in the last question is largely improved in Lemma 1 of [3] by Pintz: the exponent $3/2$ becomes $3/4$ there. This approach can be generalized by using $w(t) = (1 - t)^k \mathbb{1}_{t\leq 1}$ and the *Bernoulli Polynomials*.

Exercise 20-11. Let f be a C^∞-function over $\mathbb{R}^+$ that is of rapid decay at infinity.

1 ⋄ Show that, when $\Re(s) > 0$, the function $B(f, s) = \dfrac{1}{\Gamma(s)} \displaystyle\int_0^\infty f(t)\, t^s \dfrac{dt}{t}$ admits a holomorphic continuation to the whole complex plane.

2 ⋄ Show that, for any $n \in \mathbb{N}$, we have $B(f, -n) = (-1)^n f^{(n)}(0)$.

3 ⋄ Consider $f_0(t) = \dfrac{t}{e^t - 1} = \displaystyle\sum_{n=0}^\infty \dfrac{b_n}{n!} t^n$. The numbers b_n are easily shown to be rational numbers; they are called the Bernoulli numbers. We have in particular $b_0 = 1$, $b_1 = -\frac{1}{2}$, $b_2 = \frac{1}{6}$, $b_4 = -\frac{1}{30}, \ldots,$ $b_{12} = -\frac{691}{2730}$, and $b_{2k+1} = 0$ for $k \geq 1$. By using Theorem 20.5 with $w(t) = e^{-t}$, show that

a. When $n \in \mathbb{N}$, we have $\zeta(-n) = (-1)^n \frac{b_{n+1}}{n+1}$.

b. For $n \geq 1$, we have $\zeta(2n) = (-1)^{n-1} 2^{2n-1} \dfrac{b_{2n}}{(2n)!} \pi^{2n}$.

Exer. 20-11 shows that $\zeta(2) = \pi^2/6$. Evaluating $\zeta(2)$ was for many years an open problem and when Euler finally managed to do so (and in the process also determined $\zeta(2n)$ for $n \in \mathbb{N}$), it made him overnight into a celebrity.

The formula under a. is also famous. Taking $n = 1$ it shows how to make sense of Euler's claim that $1 + 2 + 3 + \ldots = -1/12$.

Exercise 20-12. Let $F(s) = \sum_{n\geq 1} a_n/n^s$ be a Dirichlet series absolutely convergent for $\Re s > 1$ and let w be an integrable function over $[0, \infty)$. Assume that $\int_0^\infty |w(u)| du/u^{1+\sigma'} < \infty$ for every $\sigma' \geq \sigma$ for some $\sigma < 1$ and that

$$\sum_{n\geq 1} a_n w(n/u) \ll u^\sigma$$

for every $u \geq 1$. Prove that the Dirichlet series $F(s)$ admits a meromorphic continuation to the half-plane $\Re s > \sigma$; the only possible poles are located at the zeros of $\check{w}(s)$.

References

[1] E.W. Barnes. "A new development of the theory of the hypergeometric functions". In: *Proc. Lond. Math. Soc.* (2) 6 (1908), pp. 141–177 (cit. on p. 203).

[2] O. Perron. "Zur Theorie der Dirichletschen Reihen". German. In: *J. Reine Angew. Math.* 134 (1908), pp. 95–143. https://doi.org/10.1515/crll.1908.134.95 (cit. on p. 206).

[3] J. Pintz. "An effective disproof of the Mertens conjecture". In: 147-148. Journées arithmétiques de Besançon (Besançon, 1985). 1987, pp. 325–333, 346 (cit. on p. 209).

[4] D.S. Ramana and O. Ramaré. "Variant of the truncated Perron formula and primes in polynomial sets". In: *Int. J. Number Theory* (2) 16 (2020), pp. 309–323. ISSN 1793-0421, https://doi.org/10.1142/S1793042120500165 (cit. on p. 206).

[5] O. Ramaré. "Eigenvalues in the large sieve inequality". In: *Funct. Approximatio, Comment. Math.* 37 (2007), pp. 7–35 (cit. on p. 206).

[6] O. Ramaré. "Modified truncated Perron formulae". In: *Ann. Blaise Pascal* 23.1 (2016), pp. 109–128 (cit. on p. 206).

[7] R. Remmert. *Classical topics in complex function theory*. Vol. 172. Graduate Texts in Mathematics. Translated from the German by Leslie Kay. Springer-Verlag, New York, 1998, pp. xx+349. https://doi.org/10.1007/978-1-4757-2956-6 (cit. on p. 201).

Chapter 21
Proof of Theorem $\mathscr{C}$

We go back to the function f_0 defined by formula (0.1) and proceed to prove Theorem $\mathscr{C}$. Our aim is to facilitate a thorough understanding by the readers and for this reason we present a very detailed proof. This appears to be rather long. The interested readers are invited to shorten the proof on their own!

Keeping in mind steps 1, 2 and 3 of Sect. 8.1 we now proceed further.

Claim (STEP 4) The Dirichlet series $D(f_0, s)$ has a simple pole at $s = 2$ with residue

$$H(2) = \mathscr{C}_0 = \prod_{p \geq 2}\left(1 - \frac{3}{p(p+1)}\right) .$$

For $1 + 3/4 \leq \sigma \leq 2$ and real t with $|t| \geq 2$, we have

$$|D(f_0, s)| \leq 160|t|^{2-\sigma} \log|t| .$$

When $\sigma \geq 7/4$ and t is real, we have $|D(f_0, s)| \leq 20|\zeta(s-1)|$. When $\sigma = 7/4$ and $|t| \leq 2$, we have $|D(f_0, s)| \leq 203$.

In the considered region, the function $H(s)$ is bounded in absolute value by (here $s = \sigma + it$)

$$\prod_{p \geq 2}\left(1 + \frac{3p+3}{(p^\sigma - 1)p^{\sigma-1}}\right) \leq \prod_{p \geq 2}\left(1 + \frac{3p+3}{(p^{7/4} - 1)p^{3/4}}\right) \leq 20 ,$$

where the upper bound is found using Pari/GP and the small script:

```
P0 = 300000;
{f(p) = 3*(p+1)/(p^(7/4)-1)/p^(3/4);}
prodeuler(p = 2, P0, 1.0 + f(p))*exp(7/P0^(1/2))
```

© The Author(s), under exclusive license to Springer Nature Switzerland AG 2022
O. Ramaré, *Excursions in Multiplicative Number Theory*, Birkhäuser Advanced Texts
Basler Lehrbücher, https://doi.org/10.1007/978-3-030-73169-4_21

Proof We majorize the tail via

$$\sum_{p>P_0} \log\left(1 + \frac{3(p+1)}{(p^{7/4}-1)p^{3/4}}\right) \le \int_{P_0}^{\infty} \frac{3(t+1)dt}{(t^{7/4}-1)t^{3/4}} \quad \le \frac{1+P_0^{-1}}{1-P_0^{-7/4}} \int_{P_0}^{\infty} \frac{3dt}{t^{3/2}} \ .$$

Thus, Lemma 19.1 gives us that

$$|D(f_0, s)| \le \frac{6 \cdot 20}{\sigma-1} |t|^{2-\sigma} \log|t| \ .$$

The last inequality comes from using part 1 of Lemma 19.1 and noticing that

$$20\sqrt{(3/4)^2+4}(4+3/4) \le 203 \ .$$

Then we can easily conclude since $\sigma - 1 \ge 3/4$. $\qquad\qquad\qquad\qquad\qquad\square$

The preparation being over, we can invoke Theorem 20.4 with $\kappa = 3$.

Claim (STEP 5) For $x \ge 10$ we have

$$\sum_{n\ge 1} f_0(n)e^{-n/x} = \frac{1}{2\pi i} \int_{3-i\infty}^{3+i\infty} D(f_0, s)\Gamma(s)x^s \, ds \ .$$

We are quite familiar with the Dirichlet series $D(f_0, s)$ from the previous discussion, while the Γ-function is controlled by the complex Stirling formula (Lemma 20.4). In our region, we have $\Gamma(\sigma + it) \ll (1 + |t|)^{5/8} \exp{-(\pi|t|/2)}$.

Claim (STEP 6) For $x \ge 10$, we have

$$\sum_{n\ge 1} f_0(n)e^{-n/x} = \mathscr{C}_0 x^2 + \frac{1}{2i\pi} \int_{7/4-i\infty}^{7/4+i\infty} D(f_0, s)\Gamma(s)x^s \, ds \ .$$

The proof proceeds by shifting the contour. By Cauchy's theorem we have, for real $T \ge 2$, that

$$\frac{1}{2\pi i} \int_{2-iT}^{2+iT} D(f_0, s)\Gamma(s)x^s \, ds + \frac{1}{2\pi i} \int_{2+iT}^{7/4+iT} D(f_0, s)\Gamma(s)x^s \, ds$$

$$-\frac{1}{2\pi i} \int_{7/4-iT}^{7/4+iT} D(f_0, s)\Gamma(s)x^s \, ds + \frac{1}{2\pi i} \int_{7/4-iT}^{2-iT} D(f_0, s)\Gamma(s)x^s \, ds = H(2)x^2 \ ,$$

since the only pole of the integrand $D(f_0, s)\Gamma(s)x^s$ is at $s = 2$; this pole is simple and has residue $H(2)\Gamma(2)x^2 = H(2)x^2$. The integrals over the two horizontal segments are easy to deal with:

$$\int_{3+iT}^{7/4+iT} \left| D(f_0, s)\Gamma(s)x^s ds \right| \leq 20 \int_{3/4}^{3} |\zeta(u+iT)||\Gamma(1+u+iT)|x^3 du$$

$$\ll \left(4 + \log T + \frac{6}{3/4}T^{1/4}\log T\right)x^3 T^{5/2}e^{-\pi T/2} .$$

We bounded $\zeta(u+iT)$ using Lemma 19.1: we added the second and the third bounds to cover all the cases for u (either ≥ 1 or ≤ 1). The same bound works for $\int_{7/4-iT}^{2-iT}$ (note that we have to divide by 2π). All of these tend to zero when T tends to infinity.

The last step of this procedure is the following:

Claim (STEP 7) We have

$$\frac{1}{2\pi}\left|\int_{7/4-i\infty}^{7/4+i\infty} D(f_0, s)\Gamma(s)x^s ds\right| \leq 833 \cdot x^{7/4} .$$

It is enough to estimate the integral as follows:

$$\left|\int_{7/4-i\infty}^{7/4+i\infty} D(f_0, s)\Gamma(s)x^s ds\right| \leq x^{7/4}\int_{7/4-i\infty}^{7/4+i\infty} |D(f_0, s)\Gamma(s)||ds|$$

$$\leq 2x^{7/4}\left(\int_0^2 203|\Gamma(7/4 + it)|dt + 160\int_2^\infty t^{1/4}\log t \cdot \sqrt{2\pi}(4 + t^2)^{5/8}e^{-\pi t/2}e^{1/12}dt\right) .$$

We remark that the last two integrals are finite. For the first one, the integral representation of Γ ensures us that $|\Gamma(7/4 + it)| \leq |\Gamma(7/4)| \leq 0.92$ as given by

```
gamma(7/4)
```

We thus find that, on majorizing $(4 + t^2)^{5/8}$ by $(2t)^{5/8}$ when $t \geq 2$,

$$\left|\int_{7/4-i\infty}^{7/4+i\infty} D(f_0, s)\Gamma(s)x^s ds\right| \leq 2x^{7/4}\left(374 + 673\int_2^\infty t^{7/8}\log t\, e^{-\pi t/2}dt\right) .$$

We have now gone far enough in the analysis of the problem to finish by numerical integration:

```
aux = intnum( t = 2, [[+1], Pi/2], t^(7/8)*log(t)*exp(-Pi/2*t));
2*(374 + 673 * aux)
```

On collecting all the pieces (the readers should not forget the division by 2π for the remainder term) we establish Theorem $\mathscr{C}$:

Claim (STEP 8) We have

$$\sum_{n \le x} f_0(n) e^{-n/x} = \mathscr{C}_0 x^2 + O^*\left(133 \cdot x^{7/4}\right) .$$

Exercise 21-1. Using the same technique, show that

$$\sum_{n \ge 1} \varphi(n) e^{-n/x} = (1 + o(1)) \pi^2 x^2 / 12 .$$

Exercise 21-2. Using Theorem 20.3, prove that

$$\sum_{n \le x} f_0(n)(1 - n/x)^2 = (1 + o(1)) \mathscr{C}_0 x^2 / 12 .$$

Here is a fully detailed exercise following the same pattern as the proof of Theorem $\mathscr{C}$.

Exercise 21-3. Consider the function f_1 defined by

$$f_1(n) = \prod_{p | n} \frac{p + 1}{p + 2} . \tag{21.1}$$

1 ⬦ Check that the function f_1 is multiplicative.

2 ⬦ Check that the Dirichlet series $D(f_1, s)$ can be written as

$$D(f_1, s) = \prod_{p \ge 2} \left(1 + \frac{p + 1}{(p + 2)(p^s - 1)}\right) .$$

3 ⬦ Check that $D(f_1, s) = K(s) \zeta(s)$, where

$$K(s) = \prod_{p \ge 2} \left(1 - \frac{1}{(p + 2)p^s}\right) . \tag{21.2}$$

This product converges absolutely in the sense of Godement when $\Re s > 0$.

4 ⬦ Check that the Dirichlet series $D(f_1, s)$ has a simple pole at $s = 1$ with residue

$$K(1) = \mathscr{C}_2 = \prod_{p \ge 2} \left(1 - \frac{1}{p(p + 2)}\right) .$$

Check also that, for $1/4 \le \sigma \le 2$ and real t such that $|t| \ge 2$, we have $|D(f_1, s)| \le 3|\zeta(s)|$. Check also that we have $|D(f_1, s)| \le 36$ when $s = 1/4 + it$ and $|t| \le 2$.

5 ⬦ Check that, for $x \ge 10$, we have

$$\sum_{n \ge 1} f_1(n)\left(1 - \frac{n}{x}\right) = \frac{1}{2\pi i} \int_{2-i\infty}^{2+i\infty} D(f_1, s) \frac{x^s \, ds}{s(s + 1)} .$$

6◇ Check that $\dfrac{1}{2\pi}\left|\displaystyle\int_{1/4-i\infty}^{1/4+i\infty} D(f_1,s)\dfrac{x^s\,ds}{s(s+1)}\right| \le 436\,x^{1/4}$.

7◇ Show that

$$\sum_{n\le x} f_1(n)\left(1-\frac{n}{x}\right) = \tfrac{1}{2}\mathscr{C}_2 x + O^*\left(436\cdot x^{1/4}\right)$$

where $\mathscr{C}_2$ is that constant that is given by

$$\mathscr{C}_2 = \prod_{p\ge 2}\left(1-\frac{1}{p(p+2)}\right) = 0.75947\ 92316\ 30837\ 16720\ldots \qquad (21.3)$$

Chapter 22
Roughing up: Removing a Smoothening

The readers should by now be able to prove the next theorem by using the convolution method. We also expect them to be sufficiently fluent with the material of the previous chapters to establish it on their own by using the truncated Perron summation formula! If an example is needed, this truncated formula is used in the proof of Theorem 23.4.

Theorem 22.1

Let x be a positive real number. We have

$$\sum_{n \le x} f_0(n) = (1 + o(1)) \tfrac{1}{2} \mathscr{C}_0 x^2 .$$

The question we address in this chapter is whether one can *deduce* Theorem 22.1 from Theorem $\mathscr{C}$. We analyse two cases of smoothening removal: removing a factor $e^{-n/x}$ and removing a factor $\max(0, 1 - n/x)$. Two different techniques will be used, the first one being by far the trickier. In both cases, we use the proven asymptotic with different values of x. We start with the second one which we expose in the next exercise.

Exercise 22-1. We start from the result of Exer. 21-3 with the aim of establishing that

$$\sum_{n \le x} f_1(n) = \mathscr{C}_1 x + O(x^{5/8}) \tag{22.1}$$

from the result of Question 7 of Exer. 21-3.

1 ⋄ Show that

$$(x + L) \sum_{n \le x+L} f_1(n)\left(1 - \frac{n}{x+L}\right) - x \sum_{n \le x} f_1(n)\left(1 - \frac{n}{x+L}\right)$$

$$= L \sum_{n \le x} f_1(n) + O^*\left(L \sum_{x < n \le x+L} f_1(n)\right).$$

© The Author(s), under exclusive license to Springer Nature Switzerland AG 2022
O. Ramaré, *Excursions in Multiplicative Number Theory*, Birkhäuser Advanced Texts
Basler Lehrbücher, https://doi.org/10.1007/978-3-030-73169-4_22

2 ⋄ Show that we have $\sum_{n \leq x} f_1(n) = \mathscr{C}_1 x + O(x^{5/4} L^{-1} + L)$ for $L \leq x$.

3 ⋄ Conclude.

In Theorem $\mathscr{C}$, a very strong smoothing is used: the function $n \mapsto e^{-n/x}$ is only a pale approximation of the characteristic function of the interval $[1, x]$. The advantage of such a smoothing is that the corresponding Mellin transform, namely, $\Gamma(s)$, decreases exponentially fast in vertical strips, simplifying largely our analysis. It can be proved that deducing a result like Theorem 22.1 with a condition $n \leq x$, i.e. with a weight $\mathbb{1}_{n \leq x}$, from a similar one but with a weight $e^{-n/x}$ is impossible in general. However, we have an additional hypothesis at our disposal that will prove to be enough: the function f_0 is non-negative. The proof we present belongs to the field of *Tauberian theorems* and follows the path described in the paper [6] by J. Karamata.

We first need two lemmas from usual calculus.

Lemma 22.1

Let h be a piecewise continuous function on $[0, 1]$. For every positive ε, there exists two polynomials q and Q such that

$$\forall u \in [0, 1], \quad q(u) \leq h(u) \leq Q(u), \quad \int_0^1 (Q(u) - q(u))du \leq \varepsilon .$$

Better theorems are true, with weaker hypotheses and stronger conclusions! We just want to show the general idea.

Proof The function h is bounded in absolute value, say by B. It has a finite number of discontinuities, say C. They are apart from each other and from the endpoints 0 and 1 by at least δ say. We can modify h in C intervals of length $\min(\delta, \varepsilon/(4BC))$ around these discontinuities into two continuous functions h^+ and h^- that verify $h^-(u) \leq h(u) \leq h^+(u)$ for every $u \in [0, 1]$ and $\int_0^1 (h^+ - h^-)(u)du \leq \varepsilon/4$. We next appeal to the Stone–Weierstrass approximation theorem to find two polynomials q_1 and Q_1 such that

$$\max_{0 \leq u \leq 1} |h^+(u) - Q_1(u)| \leq \varepsilon/4, \quad \max_{0 \leq u \leq 1} |h^-(u) - q_1(u)| \leq \varepsilon/4 .$$

We set $Q(u) = Q_1(u) + \varepsilon/4$ and $q(u) = q_1(u) - \varepsilon/4$. These two polynomials satisfy our conditions. $\qquad\Box$

Lemma 22.2

Let g be a non-negative Riemann-integrable function over $[0, 1]$. We have

$$\limsup_{n \to \infty} \frac{1}{N^2} \sum_{n \geq 1} n e^{-n/N} g(e^{-n/N}) \leq \int_0^1 |\log t| g(t) dt .$$

Proof Let us first notice that, for $A \geq 1$ and $N \geq 1$, we have

$$\sum_{n > AN} (n/N) e^{-n/N} \leq 3NA^2 e^{-A} .$$

Indeed, we can use a comparison to an integral to derive that

$$\sum_{n > AN} \frac{n}{N} e^{-n/N} \leq A e^{-A} + \int_{NA}^{\infty} \frac{t}{N} e^{-t/N} dt = A e^{-A} + (A+1) e^{-A} N$$

which is not more than $3NAe^{-A}$.

The second remark is that since g is Riemann-integrable, it is bounded, say by G, and thus

$$\left(\int_0^1 |\log t| g(t) dt \right)^2 \leq \int_0^1 (\log t)^2 dt \int_0^1 g(t)^2 dt \leq 2G \int_0^1 g(t) dt < \infty$$

since a change of variables shows that $\int_0^1 (\log t)^2 dt = \int_0^\infty u^2 e^{-u} du = 2$.

Let $\epsilon > 0$. Choose A such that $3AGe^{-A} \leq \epsilon$. Over $[e^{-A}, 1]$, the function $|\log t| g(t)$ is bounded and Riemann-integrable. The points $e^{-n/N}$ and $e^{-(n+1)/N}$ are spaced by $e^{-n/N}(1 - e^{-1/N})$. Hence

$$\frac{1}{N^2} \sum_{n \leq AN} n e^{-\frac{n}{N}} g(e^{-\frac{n}{N}}) = \frac{N/N^2}{1 - e^{-\frac{1}{N}}} \sum_{n \leq AN} (e^{-\frac{n}{N}} - e^{-\frac{n+1}{N}})(|\log| \cdot g)(e^{-\frac{n}{N}})$$

behaves like a Riemann sum. When N goes to infinity, it converges to $\int_{e^{-A}}^1 |\log t| g(t) dt$, which is bounded above by $\int_0^1 |\log t| g(t) dt$. We have thus shown, in particular, that

$$\limsup_{n \to \infty} (1/N^2) \sum_{n \geq 1} n e^{-n/N} g(e^{-n/N}) \leq \int_0^1 |\log t| g(t) dt + \epsilon$$

for every positive $\epsilon > 0$. We let ϵ go to zero to conclude. □

Proof (of Theorem 22.1 from Theorem $\mathscr{C}$) Let us define $S_n = \sum_{m \leq n} f_0(m)$. The first step of the argument is to set $S_n^* = S_n - \mathscr{C}_0 n^2 / 2$ and to check, by using, for instance, Exer. 7-5, that

$$\left| \sum_{n \geq 1} (S_n^* - S_{n-1}^*) e^{-n/x} \right| \leq C x^{7/4}$$

for some positive constant C. The second step of the proof, and one of its main feature, is to write S_N^* as a telescopic sum in a somewhat elaborate manner, i.e. to write that, when $N \geq 1$, we have

$$S_N^* = \sum_{n \geq 1} (S_n^* - S_{n-1}^*) e^{-n/N} h(e^{-n/N}),$$

where the function h is defined by

$$h(u) = \begin{cases} 1/u & \text{when } 1/e < u \leq 1, \\ 0 & \text{when } 0 \leq u \leq 1/e. \end{cases}$$

In the third step, we choose some $\varepsilon \in (0, 1]$ and apply Lemma 22.1 to the function h. This yields a polynomial $q(u) = \sum_k q_k u^k$ that verifies $q(u) \leq h(u)$ throughout the interval $[0, 1]$ and also $\int_0^1 (h - q)(u) du \leq \varepsilon$. We then write

$$S_N^* = \sum_{n \geq 1} (S_n^* - S_{n-1}^*) e^{-n/N} (h - q)(e^{-n/N}) + \sum_{n \geq 1} (S_n^* - S_{n-1}^*) e^{-n/N} q(e^{-n/N}).$$

Note that $S_n^* - S_{n-1}^* = S_n - S_{n-1} - \frac{1}{2}\mathscr{C}_0(n^2 - (n-1)^2) \geq -\mathscr{C}_0 n$ since the function f_0 is non-negative. The proof then runs smoothly:

$$S_N^* \geq \sum_{n \geq 1} -\mathscr{C}_0 n e^{-n/N} (h - q)(e^{-n/N}) + \sum_k q_k \sum_{n \geq 1} (S_n^* - S_{n-1}^*) e^{-n(k+1)/N}$$

$$\geq -\mathscr{C}_0 (1 + \varepsilon) N^2 \int_0^1 |\log t| (h - q)(t) dt - C \sum_k |q_k| \left(\frac{N}{k+1}\right)^{7/4}$$

by Lemma 22.2 for the first summand and provided $N \geq N_0(\varepsilon)$, and by using our hypothesis for the second one. We find that

$$\left(\int_0^1 |\log t| (h - q)(t) dt\right)^2 \leq \int_0^1 |\log t| dt \int_0^1 (h - q)^2(t) dt$$

$$\leq 2e \int_0^1 (h - q)(t) dt \leq 2e\varepsilon$$

since $0 \leq (h - q)(t) \leq h(t) \leq e$. As a conclusion, we obtain that

$$S_N^*/N^2 \geq -6\sqrt{\varepsilon}\mathscr{C}_0 - CN^{-1/4} \sum_k \frac{|q_k|}{(k+1)^{7/4}}.$$

The coefficients q_k depend only on ε, so we can take N large enough to make this last summand not more than ε say. We have thus proved that $S_N^* \geq N^2 o(1)$. We can proceed similarly for the upper bound and prove that $S_N^* \leq N^2 o(1)$ which results in $S_N^* = o(N^2)$. The proof is complete. $\square$

The above argument is not quantitative and the question remains to know how good an error term one can infer. Such a question has been treated by several authors, and the best solution is in [2, 3] by G. Freud: we can replace the $o(1)$ by a $O(1/\log x)$. The removal of the smoothing $\max(0, 1 - n/x)$ in Exer. 22-1 led to a much stronger outcome since we were able to save a power of x there.

Removing a smoothing usually involves estimating the summand in short interval. The next exercise proposes some material in this direction.

Exercise 22-2. Let $d_r(n)$ be the number of r-tuples of integers $(n_1, \cdots, n_r)$ such that $n_1 n_2 \cdots n_r = n$. The function $d_2 = d$ is the usual divisor function.

1 ◇ Show that $d_r(n) \leq r \displaystyle\sum_{d|n, d \leq n^{1/r}} d_{r-1}(n/d)$.

2 ◇ Prove that $\displaystyle\sum_{x-y < n \leq x} d(n) \leq y \log x + 2(y + \sqrt{x})$ when $x \geq y \geq 1$.

3 ◇ Show that $\displaystyle\sum_{x-y < n \leq x} d_3(n) \leq y(\log^2 x + 5 \log x + 6) + 12x^{2/3}$ when $x \geq y \geq 1$

The readers may use Exer. 4-14 to generalize the above inequalities, and to get estimates that are more acute when y is smaller. Here is an example.

Exercise 22-3.

1 ◇ Use Exer. 4-14 to prove that $\displaystyle\sum_{x-y < n \leq x} d(n) \leq 3^6(y + x^{1/3}) \sum_{\delta \leq x^{1/3}} \frac{d^3(\delta)}{\delta}$ under the conditions $x \geq y \geq 1$.

2 ◇ Show that $\displaystyle\sum_{\delta \leq D} \frac{d^3(\delta)}{\delta} \leq \left(\sum_{\delta \leq D} \frac{d^2(\delta)}{\delta}\right)^2$ and that $\displaystyle\sum_{\delta \leq D} \frac{d^2(\delta)}{\delta} \leq \left(\sum_{\delta \leq D} \frac{d(\delta)}{\delta}\right)^2$.

3 ◇ Conclude that $\displaystyle\sum_{x-y < n \leq x} d(n) \leq \frac{1}{9}(y + x^{1/3})(\log x + 3)^8$ when $x \geq y \geq 1$.

P. Shiu treats in [10] a more general situation. We continue with an algorithmical approach borrowed from [1, Proposition 4.2] by J.-M. Deshouillers and F. Dress.

Exercise 22-4. Let g be the multiplicative function defined on the prime powers p^k by

$$g(p^k) = \begin{cases} 5/2 & \text{when } k = 1, \\ (5k-1)/2 & \text{when } k \geq 2. \end{cases}$$

Our aim is to show that $d_3(n) \leq \dfrac{243}{76} \displaystyle\sum_{\substack{\delta|n, \\ \delta \leq \sqrt{n}}} g(\delta)$ $(\star)$.

We define $G^*(n) = \displaystyle\sum_{\delta|n} \min\left(g(\delta), g\left(\frac{n}{\delta}\right)\right)$.

1 ◇   Show that G^* is *super-multiplicative*, i.e. that $G^*(mn) \geq G^*(m)G^*(n)$ whenever m and n are coprime integers.

2 ◇   Show that (⋆) follows from the inequality $d_3(n) \leq \frac{243}{152}G^*(n)$ and that the function $\Phi(n) = d_3(n)/G^*(n)$ is *sub-multiplicative*, i.e. that $\Phi(mn) \leq \Phi(m)\Phi(n)$ whenever m and n are coprime integers.

3 ◇   If $n = p_1^{\alpha_1} \cdots p_k^{\alpha_k}$ is the prime decomposition of n in prime powers written in such a way that $\alpha_1 \geq \cdots \geq \alpha_k$, notice that $\Phi(n)$ depends only on the k-tuple $(\alpha_1, \ldots, \alpha_k)$. We write $\Phi(n) = \Phi((\alpha_1, \ldots, \alpha_k))$.

4 ◇   Show that $\Phi(p^\alpha) \leq 1$ when $\alpha \geq 15$ and deduce that it is enough to establish (⋆) when $\alpha_1 \leq 14$.

5 ◇   Order the set $\mathscr{S}$ of non-increasing integer sequences $(\alpha_1, \ldots, \alpha_k)$ by using the *lexicographical order*. So $(\alpha_1, \ldots, \alpha_k) > (\beta_1, \ldots, \beta_\ell)$ if and only if there is an index $j \geq 0$ such that $\alpha_i = \beta_i$ for $i \leq j$ and $\alpha_{j+1} > \beta_{j+1}$. When $j = 0$, the first half of the condition is empty, and if $j + 1 > \ell$, we understand that $\beta_{j+1} = 0$ Show that the successor of $(\alpha_1, \ldots, \alpha_k)$ is $(\alpha_1, \ldots, \alpha_k, 1)$.

6 ◇   If $(\alpha_1, \ldots, \alpha_k)$ satisfies $\alpha_\ell > \alpha_{\ell+1} = \alpha_{\ell+2} = \cdots = \alpha_k$, we define $\sigma((\alpha_1, \ldots, \alpha_k)) = (\alpha_1, \ldots, \alpha_\ell, \alpha_{\ell+1} + 1)$. Let also $\mathscr{S}(\alpha_1, \ldots, \alpha_k)$ be the subset of all the finite non-increasing integer sequences that start by $(\alpha_1, \ldots, \alpha_k)$ including this sequence. Show that $\mathscr{S}(\alpha_1, \ldots, \alpha_k) = \{s \in \mathscr{S} / s < \sigma((\alpha_1, \ldots, \alpha_k))\}$.

7 ◇   Let $C(\alpha_1, \ldots, \alpha_k) = \max\{\Phi(s), s < (\alpha_1, \ldots, \alpha_k)\} \geq 1$. Show that when $\Phi(\alpha_1, \ldots, \alpha_k)) \leq 1$, we have $C(\alpha_1, \ldots, \alpha_k) = \max\{\Phi(s), s \in \mathscr{S}(\alpha_1, \ldots, \alpha_k)\}$.

8 ◇   Let $\alpha_1 = \alpha_2 = \cdots = \alpha_k = 1$. Show that, when k is odd, the value of $G^*((\alpha_1, \ldots, \alpha_k))$ is $2 \displaystyle\sum_{i=0}^{(k-1)/2} \binom{k}{i}(5/2)^i$ and find a similar formula to cover the case k even. Deduce that the lowest value we could set for $g(p)$ is $9/4$.

9 ◇   Use the above properties to write a Sage script to prove (⋆).

10 ◇   Use (⋆) to prove the inequality $\displaystyle\sum_{x-y<n\leq x} d_3(n) \leq C(y + \sqrt{x})(\log x)^{5/2}$ valid when $x \geq y \geq 3$ and provide a possible numerical value for the constant C.

11 ◇   Adapt the previous idea to show that $\mu^2(n)d_3(n) \leq 4 \displaystyle\sum_{\substack{d|n, \\ d\leq\sqrt{n}}} (47/20)^{\omega(d)}$.

22.1 Exploring a Mean Value with Sage

We want to examine the error term in asymptotic (22.1). To this end, here is the Sage script we use. An accurate value of $\mathscr{C}_1$ is required; though it is provided to us by the result of Exer. 9-9, we propose now a shorter version, which we simply write

$$\mathscr{C}_1 = \prod_{p \geq 2}\left(1 - \frac{1}{p(p+2)}\right) = \frac{1}{\zeta(2)} \prod_{p \geq 2}\left(1 + \frac{2}{(p^2-1)(p+2)}\right).$$

```
R = RealIntervalField(100)

def g(n):
    res = 1
    l = factor(n)
    for p in l:
        res *= (p[0]+1)/(p[0]+2)
    return(R(res))


P = 1000000
aaa = R(1)
p = 2
while p <= P:
    aaa *= R(1 + 2/(p+1)/(p-1)/(p+2))
    p = next_prime (p)
x = 2/(R(P)-1)^2
x = exp(x)
aaa = aaa * x.union(R(1)) * 6 / R(pi)^2

def model(z):
    return(aaa * R(z))

def getbounds (zmin, zmax):
    zmin = max (0, floor (zmin))
    zmax = ceil (zmax)
    res = R(0)
    output = []
    for n in range (1, zmin + 1):
        res += g(n)
    maxi = abs(res - model (zmin)).upper()
    maxiall = maxi
    for n in range (zmin + 1, zmax + 1):
        m = model (n)
        maxi = max (maxi, abs(res - m).upper())
        res += g(n)
        maxi = max (maxi, abs(res - m).upper())
        if n % 100000 == 0:
            print("Upto ", n, " : ", maxi, cputime())
            output.append(maxi)
            maxiall = max (maxiall, maxi)
```

```
        maxi = R(-1000).upper()
    maxi = max (maxi, abs (res - model (zmax)).upper())
    maxiall = max (maxiall, maxi)
    print("When ", zmin, " <= z <= ", zmax)
    print("|sum_{n <= z}f1(n) - C1 z| <= ", maxiall)
    return([maxiall, output])
```

The readers may put this script in the file "Myscript.sage" and load it in Sage:

```
sage: load("Myscript.sage")
sage: myres = getbounds( 2, 10000000)
sage: list_plot(myres[1], plotjoined = True)
```

The output is impressive as it shows that

$$\left| \sum_{n \le x} f_1(n) - \mathscr{C}_1 x \right| \le 0.72 \quad (2 \le x \le 10^7).$$

We have also checked this inequality in the larger interval $[2, 10^8]$. There is surely some explanation for such a regular behaviour... On the related problem of the distribution of $\varphi(n)/n$ around its average, the readers may consult the paper [9] by S.S. Pillai and S.D. Chowla, as well as the two papers [7, 8] by Y.-F.S. Pétermann. If there would indeed be a strong analogy between $\varphi(n)/n$ and $f_1(n)$, then these papers suggest that the above error term is indeed very small, but may take arbitrary large values (in absolute value).

A very different behaviour occurs when considering f_0 rather than f_1.

Exercise 22-5. Examine numerically the error term in Theorem 22.1. First notice that a precise manner of computing $\mathscr{C}_0$ is given in Exer. 9-8. Show that the maximal value of

$$\left| \sum_{n \le x} f_0(n) - \mathscr{C}_0 x^2 \right| / (x^{3/2} \log^2 x)$$

on $[2, y]$ when y ranges $[2, 10^7]$ tends to slowly increase, from about 0.11 to about 1.78.

The next exercise is connected with the material of Sect. 10.5.

Exercise 22-6. Define $S(x) = \sum_{n \le x} \dfrac{\mu^2(n)}{\varphi(n)} - \log x - c$, where c is the constant described in (10.7). Find numerically $\alpha \in [0, 1]$ such that $|S(x)|/x^\alpha$ remains of unit size for all $x \le 10^9$. Proceed similarly for $\sum_{n \le x} 1/\varphi(n)$.

Further Reading

Tauberian theorems have been studied rather intensively. The book [4] and the very clear paper [5] by T. Ganelius are good sources on the subject. The interested readers may also consult the book [11] by G. Tenenbaum.

References

[1] J.-M. Deshouillers and F. Dress. "Sommes de diviseurs et structure multiplicative des entiers". In: *Acta Arith.* 49.4 (1988), pp. 341–375. https://doi.org/10.4064/aa-49-4-341-375 (cit. on p. 221).

[2] G. Freud. "Restglied eines Tauberschen Satzes. I". In: *Acta Math. Acad. Sci. Hungar.* 2 (1951), pp. 299–308. https://doi.org/10.1007/BF02020734 (cit. on p. 220).

[3] G. Freud. "Restglied eines Tauberschen Satzes. II". In: *Acta Math. Acad. Sci. Hungar.* 3 (1952), 299–307 (1953). https://doi.org/10.1007/BF02027829 (cit. on p. 220).

[4] T.H. Ganelius, *Tauberian remainder theorems*, Lecture Notes in Mathematics, Vol. 232. Springer-Verlag, Berlin-New York, 1971, pp. vi+75 (cit. on p. 224).

[5] T.H. Ganelius. "Géza Freud's work on Tauberian remainder theorems". In: vol. 46. 1. Papers dedicated to the memory of Géza Freud. 1986, pp. 42–50. https://doi.org/10.1016/0021-9045(86)90085-7 (cit. on p. 224)

[6] J. Karamata. "Neuer Beweis und Verallgemeinerung der Tauberschen Sätze, welche die Laplacesche und Stieltjessche Transformation betreffen". In: *J. Reine Angew. Math.* 164 (1931), pp. 27–39. https://doi.org/10.1515/crll.1931.164.27 (cit. on p. 218).

[7] Y.-F.S. Pétermann, "About a theorem of Paolo Codecà's and Ω-estimates for arithmetical convolutions". In: *J. Number Theory* 30.1 (1988), pp. 71–85. https://doi.org/10.1016/0022-314X(88)90026-1 (cit. on p. 224)

[8] Y.-F.S. Pétermann. "On an estimate of Walfisz and Saltykov for an error term related to the Euler function". In: *J. Théor. Nombres Bordx.* 10.1 (1998), pp. 203–236. https://doi.org/10.5802/jtnb.225 (cit. on p. 224).

[9] S.S. Pillai and S.D. S.D. Chowla. "On the error terms in some asymptotic formulae in the theory of numbers. I." English. In: *Journal L. M. S.* 5 (1930), pp. 95–101. https://doi.org/10.1112/jlms/s1-5.2.95 (cit. on p. 224).

[10] P. Shiu. "A Brun-Titchmarsh theorem for multiplicative functions". In: *Journal für die reine und angewandte Mathematik* 313 (1980), pp. 161–170. https://eudml.org/doc/152201 (cit. on p. 221).

[11] G. Tenenbaum. *Introduction à la théorie analytique et probabiliste des nombres.* Second. Vol. 1. Cours Spécialisés. Paris: Société Mathématique de France, 1995 (cit. on p. 224).

Chapter 23
Proving the Prime Number Theorem

In 1896, J. Hadamard in [4] and Ch. de la Vallée-Poussin in [2]* independently proved the following long-sought result:

> **Theorem 23.1 (Prime Number Theorem)**
>
> For any constant A and any $x \geq 2$, we have
>
> $$\sum_{p \leq X} \log p = X + O_A(X/(\log X)^A) \,.$$
>
> The constant implied in the O-symbol depends on the choice of A.

This is slightly off our main road, but we have in fact all the ingredients at our disposal to prove it! This will give us the opportunity to show a different route than the one that most authors take, and the opportunity to show how the Truncated Perron Summation Formula, i.e. Theorem 20.2, may be used.

The proof we chose is unusual (we propose the more classical one in an exercise). We first prove an estimate pertaining to the Möbius function which leads to the explicit Theorem 23.4. Explicit result of this nature is scarce in the literature.

The first step of our argument is to show that the Riemann zeta function does not vanish in the vicinity of the line $\Re s = 1$. This crucial fact has been discovered independently in 1896 by Hadamard and de la Vallée-Poussin and is the key to these results.

We expect that the readers have by now enough experience on the subject, so our proof is going to be slightly less detailed.

---

* Ch. de la Vallée-Poussin was only 29 years old when he published this proof!

© The Author(s), under exclusive license to Springer Nature Switzerland AG 2022
O. Ramaré, *Excursions in Multiplicative Number Theory*, Birkhäuser Advanced Texts Basler Lehrbücher, https://doi.org/10.1007/978-3-030-73169-4_23

23.1 The Riemann zeta function in the vicinity of the line $\Re s = 1$

The subject we discuss in this section has a long and fruitful history and it would take us too far afield to explain it all to the readers. We shall just brush the matter and show that a simple approach already goes rather far, establishing some essential facts. The computations that we complete here are also meant to convince the readers that it is difficult, even *extremely* difficult, and an active research area for more than 60 years, to get decent constants. This part may be disregarded for a first reading and the readers may only retain Theorem 23.3.

Lemma 23.1

When $s = \sigma + it$ with $\sigma \in (1, 2]$ and $|t| \geq 10$, we have $|\zeta(\sigma + it)| \geq \dfrac{2(\sigma - 1)^{3/4}}{5(\log |t|)^{1/4}}$.

Stronger results are accessible with more efforts; the readers may consult [10] by T. Trudgian.

Proof The trigonometric inequality $3 + 4\cos t + \cos(2t) = 2(1 + \cos t)^2 \geq 0$ implies that

$$\sum_{n \geq 1} \frac{\Lambda(n)}{n^\sigma \log n} \left(3 + 2n^{-it} + 2n^{it} + \frac{1}{2}n^{-2it} + \frac{1}{2}n^{2it} \right) \geq 0 .$$

On recalling the Dirichlet series expansion of $\log \zeta(s)$ when $\Re s > 1$ (proved in Exer. 19-7), we get

$$3 \log \zeta(\sigma) + 4\Re \log \zeta(\sigma + it) + \Re \log \zeta(\sigma + 2it) \geq 0 .$$

The real part of a logarithm being the logarithm of the modulus, the above simplifies into $|\zeta(\sigma)|^3 |\zeta(\sigma + it)|^4 |\zeta(\sigma + 2it)| \geq 0$. Since $\zeta(\sigma) \leq \sigma/(\sigma - 1)$ by Exer. 3-6 and $|\zeta(\sigma + 2it)| \leq 5 + \log |t|$ by Proposition 19.1, this gives us

$$|\zeta(\sigma + it)|^4 \geq \frac{(\sigma - 1)^3}{\sigma^3(5 + \log |t|)} \geq \frac{(\sigma - 1)^3}{8(1 + 5/\log(10)) \log |t|}$$

and the lemma readily follows.   □

The method used, via trigonometric inequalities, has been devised by F. Mertens in [7]. Lemma 23.1 already tells us that the Riemann zeta function does not vanish on the line $\Re s = 1$. Indeed, if $1 + it_0$ were a zero of ζ with $t_0 \geq 10$, we would have $\zeta(\sigma + it_0) = O(\sigma - 1)$, contradicting Lemma 23.1 when σ is close enough to 1. Notice that ζ does not vanish on the half-plane $\Re s > 1$ because its Euler product representation is absolutely convergent in the sense of Godement: if it were to vanish, one of the factors would vanish and this does not happen. Other trigonometrical inequalities are possible, see [8] by M. Mossinghoff and T. Trudgian as well as the second half of Chap. 4 of [5] by H. Iwaniec. The next exercise proposes a related but distinct path.

Exercise 23-1.

1 ◇ Let $D(t) = \sum_{n \geq 1} a_n n^{it}$ and $D^*(t) = \sum_{n \geq 1} a_n^* n^{it}$ be two Dirichlet series, both absolutely convergent for $\Re s \geq 0$. Assume that $|a_n| \leq a_n^*$. Establish the inequality $2|\Re D(t)|^2 \leq D^*(0)(D^*(0) + \Re D^*(2t))$, valid for any real number t.

2 ◇ Use the above inequality to prove that the Riemann zeta function does not vanish in the vicinity of the line $\Re s = 1$.

This exercise is continued in Exer. 23-4.

Theorem 23.2

When $|t| \leq 10$ and $\sigma \geq 3/4$, we have $|1/\zeta(\sigma + it)| \leq 10$.

Proof When $|s - 1| \leq 3/5$, we use Eq. (19.2) and $|\{u\} - 1/2| \leq 1/2$ to get

$$\zeta(s) = \frac{s}{s-1} - \frac{1}{2} + O^*\left(\frac{|s|}{2\sigma}\right) = \frac{s+1}{2(s-1)} + O^*\left(\frac{|s|}{2\sigma}\right).$$

We next notice that $|s + 1| \geq \dfrac{1}{10} + \dfrac{|s|}{2 \times 3/4}$, since it is implied by

$$2 - |s - 1| \geq \frac{1}{10} + \frac{2}{3}(1 + |s - 1|),$$

i.e. $37/50 \geq |s - 1|$, which holds true.

When $\sigma \in [3/4, 5/4]$, $|t| \leq 10$ and $|s - 1| \geq 1/2$, we invoke Lemma 19.1 to write the next script.

```
{Check(uplimit = 10, lowsig = 3/4, upsig = 5/4) =
  my(value, mymax = 0, s, nbstep = 10^3);
  for( m = 0, nbstep,
    for( k = 0, nbstep,
      s = lowsig + ( m*(upsig - lowsig) + I*k*uplimit)/nbstep;
      if(abs(s-1)>=3/5,
        value = abs(zeta(s));
        if(value <= 77*1.5/nbstep,
          print("!! Problems! Increase parameter nbstep !");
          return,);
        mymax = max(1/(value - 77*1.5/nbstep) , mymax),)));
  print("When |t| <= ", uplimit);
  print(" and sigma in [ ", lowsig, ", ", upsig, "]");
  print("we have 1/|zeta(sigma+it)| <= ", mymax);
}
Check();
```

We have used 1.5 as an upper bound for $\sqrt{2}$ in this script. A run of the program ensures that $1/|\zeta(s)| \leq 2.14$ in the concerned region.

Finally, when $\sigma \geq 5/4$, we use $|1/\zeta(s)| \leq \zeta(\sigma) \leq \zeta(5/4)$ which we obtain by comparing the Dirichlet series of $1/\zeta(s)$ and of $\zeta(s)$.   □

Exercise 23-2. Show that, when $\Re s = \sigma > 1$, we have $|1/\zeta(s)| \leq \zeta(\sigma)/\zeta(2\sigma)$.

Theorem 23.3

When $s = \sigma + it$ satisfies $\sigma \geq 1 - \dfrac{1}{(6 \log \max(10, |t|))^9}$, the Riemann zeta function does not vanish at $\sigma + it$ and we further have $|\zeta(\sigma + it)| \geq 1/(9 \log \max(10, |t|))^7$.

This result yields a *zero-free region* for the Riemann zeta function, by which we mean a region on the left-hand side of the line $\Re s = 1$ where $\zeta(s)$ does not vanish. Better explicit zero-free regions than the above one have been given, for instance, by Habiba Kadiri in [6] (slightly improved by Mossinghoff and Trudgian in [8]) or by K. Ford in [3]. The widely shared belief is that all the non-real zeroes of zeta lie on the line $\Re s = 1/2$, a conjecture known as the (world famous) *Riemann hypothesis* and often shortened as *RH*.

Proof When $|t| \leq 10$, the result follows from Theorem 23.2. Let us assume from now on that $|t| \geq 10$. We use the mean value theorem together with Lemma 19.1 to write, for any $\sigma' \in (1, 2]$,

$$|\zeta(\sigma + it) - \zeta(\sigma' + it)| \leq (\sigma' - \sigma) \cdot 2(4 + t^2)^{(1-\sigma)/2}(\log(4 + t^2))^2,$$

and thus, by appealing to Lemma 23.1,

$$|\zeta(\sigma + it)| \geq \frac{2(\sigma' - 1)^{3/4}}{5(\log |t|)^{1/4}} - (\sigma' - \sigma) \cdot 2(4 + t^2)^{(1-\sigma)/2}(\log(4 + t^2))^2 .$$

We notice that

$$(4 + t^2)^{(2(\log |t|)^9)^{-1}} (\log(4 + t^2))^2 \leq \tfrac{9}{2}(\log |t|)^2 . \tag{23.1}$$

We set $\rho = 1/(\log |t|)^9$, $\sigma' - 1 = x\rho$ and assume that $1 - \sigma \leq y\rho$. We get

$$|\zeta(\sigma + it)| \geq \frac{2x^{3/4}}{5}\rho^{\frac{3}{4} + \frac{1}{36}} - 9(x + y)\rho^{1 - \frac{2}{9}} = \left(\frac{2x^{3/4}}{5} - 9(x + y)\right)\rho^{7/9} .$$

We now take $x = y = 1/5^9$.   □

Exercise 23-3. Show that, in the region given in Theorem 23.3 and when $|t| \geq 1$, we also have $|(\zeta'/\zeta)(\sigma + it)| \leq (9 \log |s|)^9$.

Exercise 23-4. Use Exer. 23-1 to prove that there exist two positive constants c_1 and c_2 such that $|\zeta(\sigma + it)| \geq c_1/(\log |t|)^6$ when $|t| \geq 2$ and $\sigma \geq 1 - \dfrac{c_2}{(\log |t|)^8}$.

23.2 The Prime Number Theorem for the Möbius Function

Theorem 23.4 (Prime Number Theorem for the Möbius Function)

When $x \geq 10$, we have $\left| \displaystyle\sum_{n \leq x} \mu(n) \right| \leq 70 \, x (\log x) \exp\left\{ -\left(\dfrac{\log x}{6} \right)^{1/10} \right\}$.

The proof that $\sum_{n \leq x} \mu(n) = o(x)$ appeared for the first time in the paper [11] by H.C.F. von Mangoldt. The name "Prime Number Theorem for the Möbius Function" may sound strange to the readers and indeed it is, since no prime is involved! We know however since Landau's doctoral thesis that proving such a theorem has very similar consequences for the Λ-function. We shall deduce rather easily Theorem 23.1 from Theorem 23.4. The name "Prime Number Theorem" is so famous that it stuck to the corresponding result for the Möbius function!

Proof We apply Theorem 20.2 with $a_n = \mu(n)$, $F(s) = 1/\zeta(s)$ and $\kappa = 1 + 1/\log x$. We assume the parameter T lies in $[10, x]$. We first have to take care of the error term that arises in Theorem 20.2.

- When u lies in $[1/T, 1]$, we use

$$\sum_{|\log(x/n)| \leq u} \frac{|\mu(n)|}{n^\kappa} \leq (e^u/x)(e^u x - e^{-u} x + 1)$$

$$\leq e(2 \, \mathrm{sh}\, u + 1/x) \ .$$

We note that the function $\mathrm{sh}(u)/u$ is non-decreasing for positive u, and that $1/x \leq 1/T \leq u$. Hence, the total contribution is at most $9u$.
- When $u \geq 1$, we simply majorize the sum over n by $\zeta(\kappa)$, which we subsequently bound by $\kappa/(\kappa - 1)$ by Exer. 3-6 and then by $2 \log x$.

As a conclusion, we find that

$$\int_{1/T}^{\infty} \sum_{|\log(x/n)| \leq u} \frac{|\mu(n)|}{n^\kappa} \frac{2x^\kappa \, du}{Tu^2} \leq 2ex\left(9 \frac{\log T}{T} + \frac{2 \log x}{T}\right) \leq \frac{60x \log x}{T} \ .$$

We now shift our line of integration to the line

$$\mathcal{L}: \quad \sigma = 1 - \frac{1}{(6 \log \max(10, |t|))^9} \ . \tag{23.2}$$

Theorem 23.3 tells us that on this line $|\zeta(\sigma+it)| \geq 1/(9 \log \max(10, |t|))^7$. During this shifting, we do not encounter any pole of the integrand. The two horizontal segments contribute equally in absolute value. Concerning the upper one, its contribution is bounded above in absolute value by

$$\int_{1-\eta}^{\kappa} \frac{x^{\kappa}(9 \log T)^7 d\sigma}{T} \quad \left(\eta = \frac{1}{(6 \log T)^9}\right),$$

which we bound above by

$$\frac{3x(9 \log T)^7}{T}.$$

The main contribution comes from the integral over $\mathcal{L}$. In the horizontal segments, our saving was coming from the s in the denominator, but now s can become small. However we save from the exponent of x which is this time always strictly less than 1. This more-or-less vertical line contributes to

$$2x^{1-\eta} \int_0^T \frac{(9 \log T)^7 |ds|}{|s|},$$

where s lies on $\mathcal{L}$. We next check that

$$\int_0^T \frac{|ds|}{|s|} \leq \int_0^T \frac{2dt}{1+t} \leq 2 \log(1+T) \leq 2 \log T \left(1 + \frac{1}{T \log T}\right).$$
$$\leq 3 \log T.$$

We have thus proved that

$$\left|\sum_{n \leq x} \mu(n)\right|/x \leq \frac{60 \log x}{T} + \frac{6(9 \log T)^7}{T} + x^{-\eta}(9 \log T)^8.$$

We select T by

$$\log T = \big((\log x)/6\big)^{1/10}, \tag{23.3}$$

and first assume that $\log x \leq 6 \cdot 10^{10}$, so that $T \geq 10$. With this choice, the readers will check that $(9 \log T)^8 \leq \log x$, while $x^{-\eta} = 1/T^6$, so that our upper bound becomes

$$\left|\sum_{n \leq x} \mu(n)\right|/x \leq \frac{67 \log x}{T}.$$

We next check that, when $y \leq 10^{10}$, we have

$$70 \cdot 6 \cdot y \exp\left(-y^{1/10}\right) \geq 1.$$

Indeed, this is equivalent with $10 \log z - z \geq -\log(420)$ for $z = y^{1/10} \leq 10$ (and $z \geq 1$), which is swiftly verified. $\qquad\square$

We have given a completely explicit proof, albeit with poor numerical constants. Though a better version [1] has been worked out by Kirsty A. Chalker in her master's term paper, there is still ground to cover before reaching a decent result. With our treatment, we have to ask for x to be at least as large as $\exp 10^{17}$ for the bound to be less than x, a bound that is astronomically large! A fair number of improvements are available, which we did not carry out. Further the readers may use more recent results like the zero-free region of Habiba Kadiri as improved by M. Mossinghoff and T. Trudgian in [8] and consequently the lower bound for $|\zeta(s)|$, replacing our Theorem 23.3, see [10]. We encourage the readers to try them out as this subject is a current field of inquiry!

What about the Prime Number Theorem, by the way? We simply relate the distribution of the Λ-function to one of the Möbius function by the convolution identity $\Lambda = \mu \star \log$ and use the Hyperbola formula. Since the computations somewhat obscure the clarity of the argument, we first deduce a simpler bound from Theorem 23.4.

Corollary 23.1

Let $A \geq 1$ be arbitrary. Then $\displaystyle\sum_{n \leq x} \mu(n) = O_A\left(x/(\log x)^A\right)$. The constant implied in O_A-symbol depends on the choice of A.

Exercise 23-5. *Irregular numbers* have been defined in Exer. 3-15. They are the integers that have an odd number of prime factors. Show that the number of such irregular numbers up to x is asymptotic to $x/2$.

Here is an exercise inspired by the paper [9] of J. Pintz.

Exercise 23-6. In this exercise, we use the notation $M(t) = \sum_{n \leq t} \mu(n)$. Let $\rho = \beta + i\gamma$ be a zero of the Riemann zeta function in the strip $\frac{1}{2} \leq \beta \leq 1$, if any We consider the function

$$w(u) = \frac{1}{2i\pi} \int_{2-i\infty}^{2+i\infty} \frac{s(s-1)\zeta(s)}{(s+2)^7(s-\rho)} u^{s+1} du, \quad I(Y) = \int_1^\infty M(t)w(Y/t)dt .$$

1 ⋄ Show that $w(u) = 0$ when $u < 1$.

2 ⋄ By using Exer. 20-10, show that $|w(u)| \leq 1/3$ for every positive u.

3 ⋄ Show that $I(Y) \leq (2/3) \int_1^Y |M(u)|du$.

4 ⋄ Show that $I(Y) = \dfrac{1}{2i\pi} \displaystyle\int_{2-i\infty}^{2+i\infty} \frac{(s-1)Y^{s+1}}{(s+2)^7(s-\rho)}ds$ and deduce from this expression that

$$\frac{|I(Y)|}{Y} \geq \frac{Y^\beta}{|\gamma|^6} - \frac{2}{3} .$$

5 ⋄ Show that, for every $Y \geq |\gamma|^{12}$, there exists $u_0 \in [1, Y]$ such that $|M(u_0)| \geq Y^\beta/(3|\gamma|^6)$.

In order to prove Cor. 23.1, we have used the non-vanishing of the Riemann zeta function on the line $\Re s = 1$. A consequence of the above exercise is that, conversely, we can *infer* this non-vanishing *from* the conclusion of Cor. 23.1.

23.3 Proof of Theorem 23.1

Proof As announced, we start with

$$\sum_{n \le x} \Lambda(n) = \sum_{\ell \le \sqrt{x}} \mu(\ell) \sum_{m \le x/\ell} \log m + \sum_{m \le \sqrt{x}} \log m \sum_{\sqrt{x} < \ell \le x/m} \mu(\ell) .$$

On invoking Cor. 23.1, we see that the second sum is bounded is absolute value by

$$O\left(\frac{x}{(\log x)^A}\right) \sum_{m \le \sqrt{x}} \frac{\log m}{m} \ll_A \frac{x}{(\log x)^{A-2}} .$$

As for the first part, we use Lemma 7.1 to evaluate the sum over m and get

$$\sum_{\ell \le \sqrt{x}} \mu(\ell) \sum_{m \le x/\ell} \log m = x \sum_{\ell \le \sqrt{x}} \frac{\mu(\ell)}{\ell} \left(\log \frac{x}{\ell} - 1 \right) + O(\sqrt{x} \log x) .$$

We thus need information on the two sums $\sum_{\ell \le \sqrt{x}} \mu(\ell)(\log \ell)^k / \ell$ for $k \in \{0, 1\}$. Both converge, the first one to $F(1) = 0$ where $F(s) = 1/\zeta(s)$ and the second one to $-F'(1) = -1$. Indeed, we have $\zeta(s) = 1/(s-1) + O(1)$ when s is close to 1, which implies that $F(s) = (s-1) + O((s-1)^2)$ and the claimed value fo $F'(1)$. The rate of convergence of the series to their limits is readily obtained, for instance, when $k = 1$:

$$\sum_{n \le y} \frac{\mu(n)}{n} \log n = \sum_{n \ge 1} \frac{\mu(n)}{n} \log n - \int_y^\infty \sum_{y < n \le t} \mu(n) \frac{(\log t - 1) \, dt}{t^2} .$$

The second term is $O_A(1/(\log y)^A)$ for any positive parameter A by Cor. 23.1. The readers will swiftly complete the proof from this point onwards. □

Exercise 23-7. Show that the number of integers below x that have an even number of prime factors is asymptotically equal to $x/2$.

Exercise 23-8. Prove Theorem 23.1 directly by using Theorem 20.2 with $F(s) = -(\zeta'/\zeta)(s)$ and Exer. 23-3.

Exercise 23-9. Show that $\sum_{n \le x} \sqrt{d(n)} = \mathscr{C}_1 x (\log x)^{\sqrt{2}-1} (1 + o(1))$.

References

[1]   K.A. Chalker. "Perron's formula and resulting explicit bounds on sums". MA thesis. Mathematics, 2019. https://opus.uleth.ca/handle/10133/5441 (cit. on p. 238).

[2]   C.-J.G.N.B. de la Vallée-Poussin. "Sur la fonction $\zeta(s)$ de Riemann et le nombre des nombres premiers inférieurs à une limite donnée". In: *Belg. Mém. cour. in 8°* LIX (1899), 74pp (cit. on p. 227).

[3]   K. Ford. "Zero-free regions for the Riemann zeta function". In: *Number theory for the millennium, II (Urbana, IL, 2000)*. A K Peters, Natick, MA, pp. 25–56 (cit. on p. 236).

[4]   J. Hadamard. "Sur la distribution des zéros de la fonction $\zeta(s)$ et ses conséquences arithmétiques". In: *Bull. S.M.F.* 24 (1896), pp. 199–220 (cit. on p. 227).

[5]   H. Iwaniec. *Lectures on the Riemann zeta function*. Vol. 62. University Lecture Series. American Mathematical Society, Providence, RI, 2014, pp. viii+119. https://doi.org/10.1090/ulect/062 (cit. on p. 228).

[6]   H. Kadiri. "Une région explicite sans zéros pour la fonction ζ de Riemann". In: *Acta Arith.* 117.4 (2005), pp. 303–339 (cit. on p. 230).

[7]   F. Mertens. "Ueber eine Eigenschaft der Riemann'schen ζ-Function". In: *Wien. Ber.* 107 (1898), pp. 1429–1434 (cit. on p. 228).

[8]   M.J. Mossinghoff and T.S. Trudgian. "Nonnegative trigonometric polynomials and a zero-free region for the Riemann zeta-function". In: *J. Number Theory* 157 (2015), pp. 329–349. https://doi.org/10.1016/j.jnt.2015.05.010 (cit. on pp. 228, 230, 232).

[9]   J. Pintz. "An effective disproof of the Mertens conjecture". In: 147-148. Journées arithmétiques de Besançon (Besançon, 1985). 1987, pp. 325–333, 346 (cit. on p. 233).

[10] T.S. Trudgian. "Explicit bounds on the logarithmic derivative and the reciprocal of the Riemann zeta-function". In: *Funct. Approx. Comment. Math.* 52.2 (2015), pp. 253–261. https://doi.org/10.7169/facm/2015.52.2.5 (cit. on pp. 228, 232).

[11] H.C.F. von Mangoldt. "Beweis der Gleichung $\sum_{k=1}^{\infty} \frac{\mu(k)}{k} = 0$". German. In: *Berl. Ber.* 1897 (1897), pp. 835–852 (cit. on p. 231).

Chapter 24
The Selberg Formula

While striving to understand sieves and a surprising phenomenon he called the *parity principle* in [12] (see also [13]), A. Selberg found a formula that eventually led to an elementary proof of the Prime Number Theorem. This formula has some interesting consequences and follow up which we brush in this chapter.

24.1 The Iseki–Tatuzawa Formula

We start with a theorem from [14] due to the K. Iseki and T. Tatuzawa.

Theorem 24.1

Let F be a complex-valued function of the real variable. We define
$$G(x) = (\log x) \sum_{n \le x} F(x/n) .$$
For any $x \ge 1$, we have $F(x) \log x + \sum_{n \le x} F(x/n) \Lambda(n) = \sum_{d \le x} \mu(d) G(x/d)$.

Proof An easy way to proceed is to start from
$$F(x) \log x = \sum_{n \le x} F(x/n) \sum_{d|n} \mu(d) \log \frac{x}{n}$$

and to use $\sum_{d|n} \mu(d) \log \frac{n}{d} = \Lambda(n)$. $\qquad\square$

In this way, we arrive at the following result of Selberg.

© The Author(s), under exclusive license to Springer Nature Switzerland AG 2022
O. Ramaré, *Excursions in Multiplicative Number Theory*, Birkhäuser Advanced Texts
Basler Lehrbücher, https://doi.org/10.1007/978-3-030-73169-4_24

Corollary 24.1

For any $x \geq 1$, we have $\psi(x) \log x + \sum_{n \leq x} \Lambda(n)\psi(x/n) = 2x \log x + O(x)$.

Proof Use the Iseki–Tatuzawa formula once with $F_1(x) = \psi(x)$ and once with $F_2(x) = x - \gamma$. $\qquad\qquad\square$

Exercise 24-1. Use again the Iseki–Tatuzawa formula to prove that $M(x) \log x + \sum_{n \leq x} \Lambda(n)M(x/n) = O(x)$, where $M(x)$ is the summatory function of the Möbius function.

24.2 A Different Proof

Corollary 24.1 was the first step of the first elementary proof of the Prime Number Theorem, and as such, it deserves closer scrutiny. At the core of the machine, we find the identity

$$\Lambda \log + \Lambda \star \Lambda = \mu \star \log^2 . \tag{24.1}$$

Several proofs of it are available, the analytical one being surely the more illuminating. A first one is obtained by simply writing

$$D(-\Lambda \log, s) = \left(-\frac{\zeta'(s)}{\zeta(s)} \right)' = -\frac{\zeta''(s)}{\zeta(s)} + \left(-\frac{\zeta'(s)}{\zeta(s)} \right)^2$$
$$= -D(\log^2, s)D(\mu, s) + D(\Lambda, s)^2 .$$

The difficulty of evaluating the average of the right-hand side of (24.1) is thus transferred to its left-hand side. The evaluation of this part can be achieved in two different manners: either by using the Iseki–Tatuzawa formula or by evaluating

$$\sum_{n \leq x} \frac{\mu(n)}{n} \left(\log \frac{x}{n} \right)^2 .$$

The results of Chap. 6 apply here.

Proposition 24.1

For every $x \geq 1$, we have $\sum_{n \leq x} \frac{\mu(n)}{n} \log^2 \frac{x}{n} = 2 \log x + O(1)$.

Proof We use Theorem 6.2 with $h(t) = (-\log t)/t$. On recalling Lemma 7.1 as well as Eq. (7.1), we find that $\sum_{n \le t} \dfrac{\log(t/n)}{n} = \dfrac{1}{2} \log^2 t - \gamma \log t + \gamma_1 + O\left(\dfrac{\log(2t)}{t}\right)$. This leads to our choice of H, as the main term above should be $H'(t)$. Since

$$(at \log^2 t + bt \log t + ct)' = a \log^2 t + (2a + b) \log t + b + c$$

we select $H(t) = \frac{1}{2} t \log^2 t + (\gamma - 1) t \log t + (\gamma_1 - \gamma + 1) t$. Theorem 6.2 with these choices gives us

$$\sum_{n \le x} \frac{\mu(n)}{n} \left(\frac{1}{2} \log^2 \frac{x}{n} + (\gamma - 1) \log \frac{x}{n} + (\gamma_1 - \gamma + 1) \right) = (\gamma_1 - \gamma + 1) \frac{M(x)}{x} + \log x$$

$$+ 1 - \frac{1}{x} + \frac{1}{x} \int_1^x M(x/t) \left(\frac{1}{2} \log^2 t - \gamma \log t + \gamma_1 - \sum_{n \le t} \frac{\log(t/n)}{n} \right) dt \ .$$

We then plug in the estimates $|M(t)| \le t$, $|\sum_{n \le x} \mu(n)/n| \le 1$ as well as the von Mangoldt estimate

$$\sum_{n \le x} \frac{\mu(n)}{n} \log \frac{x}{n} \ll 1$$

to conclude the proof.   □

Exercise 24-2. Prove the following generalization of Selberg's identity:

$$\tfrac{1}{6} \Lambda \log^3 + \Lambda \log \star \Lambda \log = \tfrac{1}{6} \mu \star \log^4 - \tfrac{2}{3} \mu \star \mu \star \log^3 \star \log + \tfrac{1}{2} \mu \star \mu \star \log^2 \star \log^2 \ .$$

Such identities originate from the paper [6] by H. Diamond and J. Steinig.

Exercise 24-3.

1 ◇ Show that $\mu \log^2 = \mu \star (\Lambda \star \Lambda - \Lambda \log)$, where $\mu \log^2$ is a function that associates to n the value $\mu(n)(\log n)^2$.

2 ◇ Similarly show that $\mu \log^3 = \mu \star (\Lambda \log^2 - 3\Lambda \star (\Lambda \log) + \Lambda \star \Lambda \star \Lambda)$.

In [11, Theorem 2.1], the readers will find a series of identities in which the above fits, as well as an application of it. The strength of these identities comes from the fact that the functions $\Lambda \star \Lambda - \Lambda \log$ and $\Lambda \log^2 - 3\Lambda \star (\Lambda \log) + \Lambda \star \Lambda \star \Lambda$ are zero on average, by the Prime Number Theorem.

24.3 A Glimpse at the Bombieri Asymptotic Sieve and P_k-Numbers

A different look at the Selberg identity (24.1) arises by considering the numbers that have at most two prime factors, the so-called P_2-numbers. The readers will guess that P_k-*numbers* are integers with at most k prime factors. Another way of writing Cor. 24.1 is

$$\sum_{n \le x} \left(\Lambda(n) \log n + (\Lambda \star \Lambda)(n) \right) = 2x \log x + O(x) . \tag{24.2}$$

We are thus able to determine easily how many P_2-numbers there are below x, though with a well-defined weight. The situation is much simpler than with the primes. It should be noted that the Prime Number Theorem implies that

$$\sum_{n \le x} \Lambda(n) \log n \sim x \log x \sim \sum_{n \le x} (\Lambda \star \Lambda)(n)$$

so that the numbers having one prime factor and the one having two prime factors contribute equally in (24.2). The situation has been further explored by E. Bombieri in [2] (see also [3]) who considered

$$\Lambda_k = \mu \star \log^k , \tag{24.3}$$

so that $\Lambda_1 = \Lambda$, $\Lambda_2 = \Lambda \log + \Lambda \star \Lambda$ and in general

$$\Lambda_{k+1} = \Lambda_k \log + \Lambda \star \Lambda_k .$$

Let us prove the validity of this last expression.

Proof The readers by now should have guessed that an argument using Dirichlet series will be easier. Indeed, the Dirichlet series corresponding to Λ_k is $(-1)^k \zeta^{(k)}(s)/\zeta(s)$ and we find that

$$\left((-1)^k \frac{\zeta^{(k)}(s)}{\zeta(s)} \right)' = (-1)^k \frac{\zeta^{(k+1)}(s)}{\zeta(s)} - (-1)^k \frac{\zeta'(s)}{\zeta(s)} \frac{\zeta^{(k)}(s)}{\zeta(s)} ,$$

from which the readers will readily be able to finish the argument. $\qquad\square$

The above shows by recursion that $\Lambda_k(n)$ vanishes on integers having more than $k+1$ prime factors. Told in another manner: $(\Lambda_k(n))$ is a weighted version of the sequence of P_k-numbers. A further generalization may be found in [7] by H. Iwaniec and J. Friedlander (see also [8]), and a strengthening in [10].

Exercise 24-4. (An Iseki–Tatuzawa Formula with P_2-numbers). Let F be a complex-valued function of a real variable. We define for $i \in \{1, 2\}$,

$$H_i(x) = \sum_{d \le x} \mu(d) \left(\log \frac{x}{d} \right)^i \sum_{n \le x/d} F\left(\frac{x}{dn} \right) .$$

1 ◇ Show that

$$F(x)(\log x)^2 + 2 \sum_{n \le x} \Lambda(n) \left(\log \frac{x}{n} \right) F\left(\frac{x}{dn} \right) + \sum_{n \le x} \Lambda_2(n) F\left(\frac{x}{dn} \right) = H_2(x) . \tag{24.4}$$

2 ◇ Show that, with $M(x)$ being the summatory function of the Möbius function, we have

$$M(x)(\log x)^2 + 2 \sum_{n \leq x} \Lambda(n)\left(\log \frac{x}{n}\right)M(x/n) + \sum_{n \leq x} \Lambda_2(n)M(x/n) = O(x) .$$

3 ◇ Use Cor. 24.1 to prove that $\sum_{n \leq x} \Lambda_2(n) = 2x \log x + O(x)$ and deduce that

$$M(x)\log x + 2 \sum_{n \leq x} \Lambda(n)\frac{\log(x/n)}{\log x}M(x/n) = O(x) .$$

4 ◇ Show that $|F(x)|(\log x)^2 \leq 2|H_1(x)| + |H_2(x)| + 3 \sum_{n \leq x} \Lambda_2(n)\left|F\left(\frac{x}{dn}\right)\right|.$

Identity (24.4) is the case $k = 2$ of Eq. (20) from the paper [1] by A. Balog.

Exercise 24-5. Get an asymptotic for $\sum_{n \leq x} \frac{\mu(n)}{n}\left(\log \frac{x}{n}\right)^3$ as in the proof of Proposition 24.1 and deduce an asymptotic for $\sum_{n \leq x} \Lambda_3(n)$.

Exercise 24-6. K. Chandrasekharan gives in the book [4] the asymptotics $\sum_{n \leq x} \frac{\mu(n)}{n}\left(\log \frac{x}{n}\right)^3 = 3 \log^3 x - 6\gamma \log x + O(1)$. Write a Pari/GP script that computes the maximum of the left-hand side minus the main term of the right-hand side for integer values of x between two bounds. Run this script for x in $[10^3, 10^6]$, giving partial results at every multiple of 10^5. What is your conclusion? Extend this script to cover all the *real* values of x in the prescribed interval.

Exercise 24-7. Use the Iseki–Tatuzawa formula with $F(x) = x \log x$ and deduce that $\sum_{d \leq x} \frac{\Lambda(d)}{d} \log \frac{x}{d} = \frac{1}{2} \log^2 x + \log x + O(1)$.

The Bombieri asymptotic sieve studies quantities of the shape $\sum_n f(n)\Lambda_k(n)$ for some *well-distributed* and usually non-negative arithmetical function f. It is, for instance, possible to take for f the characteristic function of an arithmetic progression. Deeper examples are outside the scope of the present monograph, but the readers may consult [9, Chap. 3] by Friedlander and Iwaniec and [5] by Nathalie Debouzy.

Exercise 24-8. Give an asymptotic for $\sum_{\substack{n \leq x, \\ n \equiv 2[7]}} \big(\Lambda(n)\log n + (\Lambda \star \Lambda)(n)\big)$. The readers may remember Exer. 6-11.

References

[1]   A. Balog. "An elementary Tauberian theorem and the prime number theorem". In: *Acta Math. Acad. Sci. Hungar.* 37.1-3 (1981), pp. 285–299. https://doi.org/10.1007/BF01904891 (cit. on p. 243).

[2]   E. Bombieri. "The asymptotic sieve". In: *Rend., Accad. Naz. XL, V. Ser. 1-2* (1976), pp. 243–269 (cit. on p. 242).

[3]   E. Bombieri. "On twin almost primes". In: *Acta Arith.* 28.2 (1975/76), pp. 177–193, 457–461 (cit. on p. 242).

[4]   K. Chandrasekharan. *Arithmetical functions.* Die Grundlehren der mathematischen Wissenschaften, Band 167. Springer-Verlag, New York-Berlin, 1970, pp. xi+231 (cit. on p. 243).

[5]   N. Debouzy. "Nombres presque premiers jumeaux sous une conjecture d'Elliott-Halberstam". PhD thesis. École doctorale 184 Aix-Marseille, 2018 (cit. on p. 243).

[6]   H.G. Diamond and J. Steinig. "An Elementary Proof of the Prime Number Theorem with a Remainder Term". In: *Inventiones math.* 11 (1970), pp. 199–258 (cit. on p. 241).

[7]   J.B. Friedlander and H. Iwaniec. "On Bombieri's asymptotic sieve". In: *Ann. Sc. Norm. Sup. (Pisa)* 5 (1978), pp. 719–756 (cit. on p. 242).

[8]   J.B. Friedlander and H. Iwaniec. "Bombieri's sieve". In: *Analytic number theory. Vol. 1. Proceedings of a conference in honor of Heini Halberstam, May 16-20, 1995, Urbana, IL, USA. Boston, MA.* Ed. by Bruce C. (ed.) et al. Berndt. Vol. 138. Birkhäuser. Prog. Math. 1996, pp. 411–430 (cit. on p. 242).

[9]   J.B. Friedlander and H. Iwaniec. *Opera de cribro.* Vol. 57. American Mathematical Society Colloquium Publications. American Mathematical Society, Providence, RI, 2010, pp. xx+527 (cit. on p. 243).

[10]   O. Ramaré. "On Bombieri's asymptotic sieve". In: *J. Number Theory* 130.5 (2010), pp. 1155–1189 (cit. on p. 242).

[11]   O. Ramaré. "From explicit estimates for the primes to explicit estimates for the Moebius function". In: *Acta Arith.* 157.4 (2013), pp. 365–379 (cit. on p. 241).

[12]   A. Selberg. "On an elementary method in the theory of primes". In: *Norske Vid. Selsk. Forh., Trondhjem* 19.18 (1947), pp. 64–67 (cit. on p. 239).

[13]   A. Selberg. "On elementary problems in prime number-theory and their limitations". In: *C.R. Onzième Congrès Math. Scandinaves, Trondheim, Johan Grundt Tanums Forlag* (1949), pp. 13–22 (cit. on p. 239).

[14]   T. Tatuzawa and K. Iseki. "On Selberg's elementary proof of the prime-number theorem". In: *Proc. Japan Acad.* 27 (1951), pp. 340–342 (cit. on p. 239).

Part V
Higher Ground: Applications/Extensions

Chapter 25
Rankin's Trick and Brun's Sieve

In 1938, R.A. Rankin extended considerably our knowledge of the possible large gaps between primes in the five pages long paper [8]. This result impressed the community and one of the ideas used by Rankin got tagged as "Rankin's trick". In fact, the same trick had been used way before, see*, for instance, the first page of [6] by H. Heilbronn and E. Landau, but the name "Rankin" sticked for this elegant process.

25.1 Rankin's Trick in its Simplest Form

In its simplest installment, what we do is the following. Suppose we need to evaluate the tail $\sum_{n \geq x} \mu^2(n)/\varphi(n)^2$. Rankin's trick is to use the upper bound

$$\sum_{n \geq x} \frac{\mu^2(n)}{\varphi(n)^2} \leq \sum_{n \geq x} \frac{\mu^2(n)}{\varphi(n)^2} \left(\frac{n}{x}\right)^a,$$

which is valid for any real parameter $a \geq 0$. The next step is the key: we extend the summation to every integers and use multiplicativity to write (on assuming that $a < 1$ to get convergent sums)

$$\sum_{n \geq x} \frac{\mu^2(n)}{\varphi(n)^2} \leq x^{-a} \prod_{p \geq 2} \left(1 + \frac{p^a}{(p-1)^2}\right).$$

There are several ways to continue. It may be enough to just select a value of a, say $a = 2/3$. One can also optimize this choice. For the above, we note that

$$\prod_{p \geq 2} \left(1 + \frac{p^a}{(p-1)^2}\right) \leq C \prod_{p \geq 2} \left(1 - \frac{p^a}{p^2}\right)^{-1} = \zeta(2-a),$$

---

* Let D.S. Ramana be thanked for this remark.

© The Author(s), under exclusive license to Springer Nature Switzerland AG 2022   251
O. Ramaré, *Excursions in Multiplicative Number Theory*, Birkhäuser Advanced Texts Basler Lehrbücher, https://doi.org/10.1007/978-3-030-73169-4_25

where

$$C = \prod_{p \geq 2}\left(1 + \frac{p^a}{(p-1)^2}\right)\left(1 - \frac{p^a}{p^2}\right) \leq \prod_{p \geq 2}\left(1 + p^a\left(\frac{1}{(p-1)^2} - \frac{1}{p^2}\right)\right)$$

$$\leq \prod_{p \geq 2}\left(1 + p\left(\frac{1}{(p-1)^2} - \frac{1}{p^2}\right)\right) \leq 5.$$

We furthermore have $\zeta(2-a) \leq (2-a)/(1-a)$ as proved in Exer. 3-4 for instance. This reduces the above bound to

$$\sum_{n \geq x} \frac{\mu^2(n)}{\varphi(n)^2} \leq \frac{10}{x}\frac{x^{1-a}}{1-a}.$$

A standard optimization process leads to the choice

$$1 - a = \frac{1}{\log x}$$

resulting in

$$\sum_{n \geq x} \frac{\mu^2(n)}{\varphi(n)^2} \leq \frac{10e \log x}{x}.$$

A more precise analysis of the initial sum we studied shows that it is equivalent to C'/x for some positive constant C', so that our very elementary procedure has gained considerably in flexibility, at the expense of loosing only one logarithm.

Exercise 25-1. (Additive Rankin's Trick). By using $x^n/n! \leq \exp x$ valid for any non-negative real number x, show that $n! \geq (n/e)^n$. Deduce that $\binom{n}{m} \leq (en/m)^m$.

Here is another idea we borrow from P. Letendre's PhD memoir.

Exercise 25-2. The aim of this exercise is to show that

$$d(n) \leq \exp\left(\frac{5}{2}\sqrt{\omega(n)\log n}\right).$$

1◇ Use Rankin's trick to prove that $d(n) \leq n^\sigma \prod_{p|n}\left(1 + \frac{1}{p^\sigma - 1}\right)$ for any non-negative σ.

2◇ Show that $p^\sigma - 1 \geq \sigma \log p$.

3◇ Set $a(n) = \sum_{p|n} \frac{1}{\log p}$. Prove that $d(n) \leq \exp\left(2\sqrt{a(n)\log n}\right)$ and conclude.

25.2 Rankin's Trick and Brun's Sieve

Rankin's trick can be used to simplify neatly the usually very combinatorial presentation of Brun's sieve; this has been done by R. Murty and N. Saradha in [7]. We do not discuss much of sieve theory in this book, but one of the problems is to find ways to manipulate numbers that are free of small prime factors, say up to some bound z. We illustrate the technique by a result of H. Daboussi and J. Rivat in [2, Lemma 4], which is inspired by [3, around page 81]. A notation and a warning are required. We define

$$P(z) = \prod_{p<z} p . \tag{25.1}$$

This definition is harmless enough but we draw the attention of the readers on the inequality $p < z$ which is *not* $p \leq z$. As a matter of fact, any definition would work for us, but if the readers roam around combinatorial sieve theory, several formulae may become false with the wrong choice.

Proposition 25.1

Let f be a non-negative multiplicative function, and $z \geq 2$ be some real parameter. We define

$$S = \sum_{p<z} \frac{f(p)}{1+f(p)} \log p .$$

We assume $S > 0$ and write $K(t) = \log t - 1 + 1/t$ for $t \geq 1$. Then for any y such that $\log y \geq S$, we have

$$\sum_{\substack{d \geq y, \\ d|P(z)}} \mu^2(d) f(d) \leq \prod_{p<z} (1 + f(p)) \exp\left(-\frac{\log y}{\log z} K\left(\frac{\log y}{S}\right)\right) .$$

When $\log y \geq 7S$, we have $K((\log y)/S) \geq 1$.

Proof Let $\eta \geq 0$ be some real parameter to choose. We start with

$$\sum_{\substack{d \geq y, \\ d|P(z)}} \mu^2(d) f(d) \leq \sum_{\substack{d \geq y, \\ d|P(z)}} \mu^2(d) f(d) \left(\frac{d}{y}\right)^\eta \leq y^{-\eta} \prod_{p<z} (1 + f(p) p^\eta) .$$

We next use the following identity:

$$1 + f(p) p^\eta = (1 + f(p)) \left(1 + \frac{f(p)}{1 + f(p)} (p^\eta - 1)\right) .$$

We further notice that the function $(e^x - 1)/x$ is non-decreasing when $x \geq 0$ (its series expansion has only non-negative coefficients), so that, when p is not more than z,

$$p^{\eta} - 1 \le \log p \, \frac{z^{\eta} - 1}{\log z} \, .$$

Therefore

$$\sum_{\substack{d \ge y, \\ d \mid P(z)}} \mu^2(d) f(d) \le y^{-\eta} \prod_{p<z} (1 + f(p)) \exp \sum_{p<z} \frac{f(p) \log p}{1 + f(p)} \, \frac{z^{\eta} - 1}{\log z}$$

$$\le \prod_{p<z} (1 + f(p)) \exp\left(\frac{(z^{\eta} - 1)S}{\log z} - \eta \log y \right).$$

We change parameter and set $v = \eta \log z$. The argument of the exponential reads

$$\frac{S}{\log z} \left(e^v - 1 - v \frac{\log y}{S} \right).$$

The parameter v can be chosen as we want provided it remains non-negative. We select

$$v = \log \frac{\log y}{S}$$

and this gives our result. We next check that the function K is non-decreasing. The last script explains the value 7.

```
K(t) = log(t) - 1 + 1/t;
solve( t = 2, 7, K(t)-1)
```

Here is an application.

Theorem 25.1 (Partial Fundamental Lemma)

Let $z_0 \ge 2$ and $z \ge$ be two real parameters, z being the largest. Let x be another real parameter, this time larger than z. We have

$$\sum_{\substack{n \le x, \\ (n, P(z)/P(z_0))=1}} 1 \ll x \frac{\log z_0}{\log z}$$

provided that $\log z \le \dfrac{\log x}{2 \log \log x}$.

A more general *Fundamental Lemma* exists in this area, and since we want the readers to be able to follow different sources, we have felt obliged to add the modifier *Partial* to the above name.

What we are counting here is the number of integers that do not have any prime factors in the interval $[z_0, z)$. The range we can afford is typical of the Brun sieve: if we were able to take $\log z$ as large as $\frac{1}{2} \log x$, then the only surviving n's would be primes. We cannot do that; however, we can already remove a lot of possible prime factors. We shall see in next chapter how to reach primes from there onwards.

Proof Let us first note that we can assume that z and x are large enough. We detect the coprimality condition with the Möbius function as in (29.3), i.e. on setting $Q = P(z)/P(z_0)$, we use

$$\mathbb{1}_{(m,Q)=1} = \sum_{\substack{d|m, \\ d|Q}} \mu(d) .$$

This holds simply because these divisors d divide the gcd of m and Q. The second idea, partially due to V. Brun, is to treat differently the large d's and the small ones. In the classical treatment of Brun, this notion of size is according to the number of prime factors of d but we shall only truncate according to the natural size. We introduce a truncation parameter D and write

$$\sum_{\substack{m \le x, \\ (m,Q)=1}} 1 = S(d \le D) + S(d > D),$$

where

$$S(d \le D) = \sum_{\substack{d \le D, \\ d|P(z)}} \mu(d) \sum_{d|m \le x} 1$$

and $S(d > D)$ is the complementary sum. Let us treat this one first. We find by using Proposition 25.1 with f being the multiplicative function that takes the value $1/d$ when d is prime to $P(z_0)$ and 0 otherwise that:

$$|S(d > D)| \le \sum_{\substack{d > D, \\ d|Q}} \mu^2(d)\frac{x}{d} \le x \prod_{z_0 \le p < z} \left(1 + \frac{1}{p}\right) \exp\left(-\frac{\log D}{\log z}\right)$$

provided that

$$\log D \ge 7 \sum_{z_0 \le p < z} \frac{\log p}{p + 1} . \tag{25.2}$$

Note that Mertens' third theorem (i.e. Theorem 12.4) ensures us that $\prod_{p \le z}(1 + p^{-1}) \ll \log z$. Concerning the sum $S(d \le D)$, we use

$$S(d \le D) = \sum_{\substack{d \le D, \\ d|Q}} \mu(d)\frac{x}{d} + O^*(D)$$

$$= \sum_{d|Q} \mu(d)\frac{x}{d} + O\left(\sum_{\substack{d > D, \\ d|Q}} \mu^2(d)\frac{x}{d}\right) + O^*(D) .$$

On combining both, we get

$$S(d \leq D) + S(d > D) = x \prod_{z_0 \leq p < z}\left(1 - \frac{1}{p}\right) + O\left(x(\log z)D^{-1/\log z} + D\right).$$

Mertens' theorem again tells that $\prod_{z_0 \leq p < z}\left(1 - \frac{1}{p}\right) \ll \frac{\log z_0}{\log z}$. We choose $D = \exp(2\log z \log\log z)$. With this choice, condition (25.2) is satisfied if z is large enough. Furthermore

$$D \leq \exp\left(2\frac{\log x}{2\log\log x}(\log\log x - \log\log\log x)\right)$$

$$\leq x\exp\left(-\frac{\log x}{\log\log x}\log\log\log x\right),$$

and the last factor is surely $\leq 1/\log x$ when x is large enough; this bound is itself not more than $1/\log z$. □

Exercise 25-3. Let $x \geq 1$ and let q be an integer all whose prime factors are below x. We want to show that there exists a constant C independent of x and q such that $\sum_{\substack{n \leq x, \\ (n,q)=1}} 1 \leq C\frac{\varphi(q)}{q}x$.

1 ⋄ Show that there exists a constant c_1 such that the product, say q_1, of the primes that divide q and that are not more than $z = \sqrt{\log q}$ is less than $\exp(c_1\sqrt{\log x})$.

2 ⋄ Let $D = \sqrt{x}$. Show that when x is large enough, we have $S = \sum_{p|q_1}\frac{\log p}{p+1} \leq (\log D)/7$ and that $\sum_{\substack{d>D, \\ d|q_1}} \mu^2(d)/d \ll 1/\log x$.

3 ⋄ Adapt the proof of Theorem 25.1 to show that $\sum_{\substack{n \leq x, \\ (n,q_1)=1}} 1 \ll \frac{\varphi(q_1)}{q_1}x$.

4 ⋄ Prove finally that $\frac{\varphi(q_1)}{q_1} \ll \frac{\varphi(q)}{q}$ and conclude.

Exercise 25-4. Let $x \geq 1$ and let q be an integer all whose prime factors are below x. We consider

$$C(x) = \max_{y \leq x}\ \max_{\substack{q \geq 1, \\ p|q \implies p \leq y}}\ \frac{q}{y\varphi(q)}\sum_{\substack{n \leq y, \\ (n,q)=1}} 1.$$

1 ⋄ Show that $C(x) = C([x])$ and that it is enough to consider integer values of y

2 ⋄ Show that $C(1) = C(2) = 1$ and that $C(3) = 4/3$.

3◇ Use the `forsubset` loop of Pari/GP to write a script that computes $C(x)$ for $x \in \{1, 2, \ldots, 40\}$.

4◇ Can you formulate a conjecture about $C(x)$?

V. Brun became famous when, at age 29, he published in [1] the first non-trivial upper bound concerning the number of prime twins. The readers have now enough material to achieve part of what Brun did.

Exercise 25-5. Let $z \geq 2$ and $D \geq z$ be two parameters to be chosen in due course We define

$$S = \sum_{\substack{m \leq x, \\ (m(m+2), P(z))=1}} 1 \, .$$

1◇ Show that we have $S = S(d \leq D) + S(d > D)$ where

$$S(d \leq D) = \sum_{\substack{d \leq D, \\ d \mid P(z)}} \mu(d) \sum_{\substack{m \leq x, \\ d \mid m(m+2)}} 1$$

and where $S(d > D)$ is defined similarly, simply by replacing the condition $d \leq D$ by $d > D$.

2◇ Define the multiplicative function ρ on prime powers by $\rho(2) = 1$, $\rho(p) = 2$ when $p \geq 3$ and $\rho(p^k) = 0$ when p is a prime and $k \geq 2$. Show that

$$\sum_{\substack{m \leq x, \\ d \mid m(m+2)}} 1 = \frac{\rho(d)}{d} x + O^*(\rho(d)) \, .$$

3◇ Show that $\prod_{p \leq z}(1 + 2/p) \ll (\log z)^2$.

4◇ Show that $\sum_{d \leq D} \rho(d) \ll D \log D$.

5◇ Show that, when $\log D \geq 7 \log z$, we have $\displaystyle\sum_{\substack{d > D, \\ d \mid P(z)}} \rho(d)/d \ll (\log z)^2 D^{-1/\log z}$.

6◇ Select $D = \sqrt{x}$ and choose z by $\log z = (\log x)/(14 \log \log x)$ and conclude that

$$\sum_{\substack{p \leq x, \\ p+2 \text{prime}}} 1 \ll \frac{x(\log \log x)^2}{(\log x)^2} \, .$$

7◇ Show that the series $\sum_p 1/p$ where p ranges the prime twins is convergent.

Chap. 29 proposes a very different proof of a similar result. Both methods have advantages and drawbacks.

References

[1]   V. Brun. "Sur les nombres premiers de la forme $ap + b$." In: *Arch. Math. Naturvid.* 34.14 (1914), p. 9 (cit. on p. 251).

[2]   H. Daboussi and J. Rivat. "Explicit upper bounds for exponential sums over primes". In: *Math. Comp.* 70.233 (2001), pp. 431–447 (cit. on p. 247).

[3]   P.D.T.A. Elliott. *Probabilistic number theory. I.* Vol. 239. Grundlehren der Mathematischen Wissenschaften. Mean-value theorems. Springer-Verlag, New York-Berlin, 1979, xxii+359+xxxiii pp. (2 plates) (cit. on p. 247).

[4]   J.B. Friedlander and H. Iwaniec. *Opera de cribro.* Vol. 57. American Mathematical Society Colloquium Publications. American Mathematical Society, Providence, RI, 2010, pp. xx+527 (cit. on p. 251).

[5]   H. Halberstam and H.-E. Richert. "Sieve methods". In: *Academic Press (London)* (1974), 364pp (cit. on p. 251).

[6]   H. Heilbronn and E. Landau. "Bemerkungen zur vorstehenden Arbeit von Herrn Bochner". German. In: *Math. Z.* 37 (1933), pp. 10–16 (cit. on p. 245).

[7]   M.R.P.M. Murty and N. Saradha. "On the sieve of Eratosthenes". In: *Can. J. Math.* 39.5 (1987), pp. 1107–1122 (cit. on p. 247).

[8]   R.A. Rankin. "The difference between consecutive prime numbers". In: *J. Lond. Math. Soc.* 13 (1938), pp. 242–247 (cit. on p. 245).

Chapter 26
Three Arithmetical Exponential Sums

In this chapter, we study three (linear) exponential sums, i.e. for us, sums of the shape

$$\sum_{n \le x} b(n) \, e(n\alpha)$$

for some arithmetical coefficients $b(n)$. The first case is the one of $b(n)$ being the characteristic function of the numbers that have no prime factors below some bound z. The case when $b(n)$ is the characteristic function of the primes will follow from the first case. The third situation will be when $b(n)$ equals $\mu(n)$. In each of these three cases, some rational approximation of the real number α comes into play, but we do not want to dig into this kind of problems. So rather than considering a general α, we restrict our attention to α being the golden ratio $\rho = (1 + \sqrt{5})/2$, which we know very well how to approximate. Theorem 26.3, Cor. 26.1 and Theorem 26.5 are highpoints of this chapter.

26.1 Rational Approximations of the Golden Ratio and Other Lemmas

Very good rational approximations of ρ are obtained by considering the *Fibonacci numbers* defined by the recursion

$$F_{k+2} = F_{k+1} + F_k , \quad F_1 = F_2 = 1 . \tag{26.1}$$

Lemma 26.1

We have, for any positive integer $k \ge 2$, $\left| \rho - \dfrac{F_{k+1}}{F_k} \right| \le \dfrac{1}{2F_k^2}$.

Furthermore, any interval $(x, 2x]$ contains a F_k, where x is any real number larger than 1, and finally F_{k+1} is prime to F_k.

© The Author(s), under exclusive license to Springer Nature Switzerland AG 2022
O. Ramaré, *Excursions in Multiplicative Number Theory*, Birkhäuser Advanced Texts
Basler Lehrbücher, https://doi.org/10.1007/978-3-030-73169-4_26

Proof Binet's Formula* tells us that

$$F_k = \frac{\rho^k - (-\rho)^{-k}}{\sqrt{5}} \tag{26.2}$$

and we compute that $5(\rho F_k^2 - F_{k+1} F_k) = -(-1)^k(\rho + \rho^{-1}) + (\rho - \rho^{-1})\rho^{-2k}$. We check numerically that this quantity is not more than $5/2$ when $2 \le k \le 10$, and readily extend this inequality to every $k \ge 2$. Therefore

$$\left| \rho - \frac{F_{k+1}}{F_k} \right| \le \frac{1}{2F_k^2} \;.$$

The quotient F_{k+1}/F_k tends to $\rho < 2$, and, by checking by hand the first few cases, it is not difficult to prove that $F_{k+1}/F_k \le 2$ from which we deduce that any interval $(x, 2x]$ contains a F_k, when x is any real number larger than 1. We take this opportunity to mention that the Fibonacci numbers can be computed in Pari/GP by the function `fibonacci`. This implies that they are accessible in Sage via the function `pari.fibonacci`. Concerning the coprimality, we see that any divisor of F_{k+1} and F_k is also a divisor of F_{k-1} and of F_k, and by recursion, of $F_1 = 1$.   □

Exercise 26-1. For any integer k, we define $N_k = \dfrac{(1 + \sqrt{2})^k - (1 - \sqrt{2})^k}{2\sqrt{2}}$.

1 ⋄ Show that $N_1 = 1$, $N_2 = 2$ and that $N_{k+2} - 2N_{k+1} - N_k = 0$ when $k \ge 1$. Deduce that N_k is an integer.

2 ⋄ Show that $(N_{k+1} - N_k)/N_k$ tends to $\sqrt{2}$ and that $N_k \ge (1 + \sqrt{2})^k /2$.

3 ⋄ Show that $\left| \sqrt{2} - \dfrac{N_{k+1} - N_k}{N_k} \right| \le \dfrac{1}{2N_k^2}$.

We shall also require the next technical lemma. Here we denote by $\|\theta\|$ the *distance to the nearest integer* which can be defined by

$$\|\theta\| = \min(\{\theta\}, 1 - \{\theta\}) \;. \tag{26.3}$$

Lemma 26.2

Let a, q be two coprime positive integer and let β be a real number smaller in absolute value than 1. Define $\alpha = a/q + \beta/(2q^2)$. We have

---

* Though A. de Moivre already knew this formula at least in 1718, and Euler gave a rigorous proof in 1765, this formula is attributed to J.P.M. Binet who published it in 1843 in [2, p. 563]. It seems that the date 1834 that one sees often on the web is due to some slippery fingers who wanted to participate to this joyous historical confusion.

$$\sum_{n\leq y} \min\left(\frac{x}{n}, \frac{1}{\|\alpha n\|}\right) \leq \frac{3x}{q}(1 + \log y) + 4(y + q)(1 + \log q),$$

where x and $y \geq 1$ are two arbitrary positive real parameters.

Proof We split the integer interval $[1, y]$ in at most $1 + yq^{-1}$ intervals of length at most q. In a typical interval $[1 + kq, q + kq]$, we define $\gamma = kq\alpha$ and look at $h \mapsto h\alpha + \gamma$, where h varies from 1 to q. The points $h\alpha$ are distant from one another by at least $1/(2q)$. Hence, at most two of them can be such that $\|h\alpha + \gamma\| \leq 1/(4q)$. The total contribution is thus at most

$$\frac{2x}{1 + kq} + 2\sum_{1\leq r\leq q/2} \frac{1}{r/(2q)} \leq \frac{2x}{1 + kq} + 4q(\log(q/2) + 1).$$

For the initial interval (i.e. $k = 0$), the bad case can only happen at the end of the interval (i.e. when $n = q$), so the bound is $xq^{-1} + 4q(\log q + 1)$ in that case. In total, this gives us at most

$$\frac{3x}{q}\left(\log\frac{y}{q} + 1\right) + \left(1 + \frac{y}{q}\right)4q(\log q + 1) \leq \frac{3x}{q}(1 + \log y) + 4(y + q)(1 + \log q).$$

The lemma is now proved. □

In what follows, the next elementary lemma will be of crucial importance.

Lemma 26.3

Let α be a real number. We have $\left|\sum_{n\in I} e(n\alpha)\right| \leq \min\left(N, \frac{1}{2\|\alpha\|}\right)$, where I is any interval containing at most N integers.

Proof By using an additive shift, we see that it is enough to study the sum

$$\sum_{0\leq n\leq N-1} e(n\alpha).$$

Its absolute value is bounded by N, which is optimal and reached for $\alpha = 0$. Let us assume α to be non-zero. We can further assume that α belongs to $(0, 1)$, and, since the sum for α and $1 - \alpha$ is conjugate, that α lies in $(0, 1/2]$. The quantity we study can be written as

$$\frac{e(N\alpha) - 1}{e(\alpha) - 1} = e((N - 1)\alpha/2)\frac{\sin \pi N\alpha}{\sin \pi\alpha}.$$

We recall that, over the interval $(0, \pi/2]$, the function $(\sin x)/x$ decreases, and is thus at least $2/\pi$. This gives us

$$\left|\sum_{0\leq n\leq N-1} e(n\alpha)\right| \leq \frac{\pi}{2\pi\alpha} = \frac{1}{2\|\alpha\|}$$

as required.   □

Among the easy lemmas, here is one that incorporates what we call a *dyadic decomposition*.

Lemma 26.4 (Dyadic Decomposition)

Let f be some function and A and B be two positive real numbers. We have

$$\left| \sum_{A < n \le B} f(n) \right| \le \frac{\log(2B/A)}{\log 2} \max_{\substack{A \le Q < Q' \le B, \\ Q' \le 2Q}} \left| \sum_{Q < n \le Q'} f(n) \right|,$$

where the maximum is over every pairs of real numbers (Q, Q') satisfying the stated conditions.

In short, provided we are ready to loose a logarithmic factor, we can restrict our variable to an interval $(Q, Q']$ with $Q' \le 2Q$. We can often avoid the loss, but most of the time, we do not need to be *so* precise (in Theorem 26.4, we will need this added precision). Some authors restrict to intervals $(Q, 2Q]$, and though it is usually of the same difficulty, it is nonetheless formally false.

Proof We start from $Q' = B$ and $Q = Q'/2$, and continue by descending in this manner. The last interval may need to be shortened. The final $Q = B/2^k$ has $2Q > A$ and therefore $2^k \le 2B/A$, which gives us a bound for the number of required intervals.   □

26.2 An Exponential Sum over Integers Free of Small Primes

Theorem 26.1

When $\log z \le \sqrt{\log x}/2$, we have $\displaystyle\sum_{\substack{m \le x, \\ (m, P(z)) = 1}} \exp(2i\pi\rho m) \ll x/z$, where $P(z)$ is defined in (25.1).

We stated the result with the lengthy notation $\exp(2i\pi\rho m)$, but we now shorten this to $e(\rho m)$. The same result holds with m being restricted to $(x/2, x]$.

Proof We detect the coprimality condition by using the Möbius function, i.e. we write

$$\mathbb{1}_{(m, P(z)) = 1} = \sum_{\substack{d \mid m, \\ d \mid P(z)}} \mu(d)$$

as in the proof of Theorem 25.1. Again similarly, we introduce a truncation parameter D and write

$$\sum_{\substack{m \le x, \\ (m, P(z))=1}} \exp(2i\pi\rho m) = S(d \le D) + S(d > D),$$

where

$$S(d \le D) = \sum_{\substack{d \le D, \\ d \mid P(z)}} \mu(d) \sum_{d \mid m \le x} e(\rho m),$$

while $S(d > D)$ is the complementary sum. Let us treat this one first. We find that, by Proposition 25.1, we have

$$|S(d > D)| \le \sum_{\substack{d > D, \\ d \mid P(z)}} \mu^2(d)\frac{x}{d} \le x \prod_{p < z}\left(1 + \frac{1}{p}\right)\exp\left(-\frac{\log D}{\log z}\right),$$

provided that $\log D \ge 7 \sum_{p < z} \frac{\log p}{p + 1}$. Concerning the sum $S(d \le D)$, we forget about the condition $d \mid P(z)$ and invoke Lemma 26.2 with any approximation F_{k+1}/F_k given by Lemma 26.1. This leads to (by Lemma 26.3)

$$|S(d \le D)| \le \sum_{d \le D} \min\left(\frac{x}{d}, \frac{1}{\|\rho d\|}\right)$$

$$\le \frac{3x}{F_k}(1 + \log D) + 4(D + F_k)(1 + \log F_k).$$

We select the integer k so that $F_k \in (D, 2D]$, as we may, so that the previous bound reduces to

$$|S(d \le D)| \le \left(\frac{3x}{D} + 24D\right)(1 + \log D).$$

We select D with $\log D = 2(\log z)^2$. By our assumption on z, we have $D^2 \le x$. Furthermore $(1 + \log D)/D \ll 1/z$ and $(\log z)D^{-1/\log z} \ll 1/z$. This concludes the proof. $\square$

The best upper bound for the trigonometric polynomial $S(a/q) = \sum_\ell c_\ell \, e(\ell a/q)$ is $\sum_\ell |c_\ell|$ as there may happen that $c_\ell = e(-\ell a/q)$. I.M. Vinogradov discovered that, by assuming some structure of the coefficients c_ℓ, this bound can be improved drastically. The next exercise exemplifies the situation.

Exercise 26-2. (The Toy Lemma). Let $(u_m)_{m \le M}$ and $(v_n)_{n \le N}$ be two complex sequences with $|u_m|, |v_n| \le 1$. Let a/q be a rational and assume $(a, q) = 1$ and $M, N \ge q$. Prove that $\left|\sum_{m,n} u_m v_n \, e(mna/q)\right| \le 2MN/\sqrt{q}$.

26.3 An Exponential Sum over the Primes

In 1937, I.M. Vinogradov introduced in [28] a very innovative technique to handle sums over primes (his book [29] is an excellent source). Since Vinogradov's investigations started with the Erathostenes sieve, it seems natural to believe that there should be an ancestor of Vinogradov's method that relates to it. We recall that $P(z)$ is defined in (25.1).

Theorem 26.2

Let z and x be two real parameters such that $4 \leq z^2 \leq x$. Let $r(n)$ denote the number of prime factors of n which lie in the interval $(z, \sqrt{x}]$. We finally define

$$\tilde{r}(n) = \begin{cases} 1/(1 + r(n)) & \text{when } \gcd(n, P(z)) = 1, \\ 0 & \text{otherwise.} \end{cases}$$

For any arithmetical function g such that $|g(n)| \leq 1$ for all n, we have

$$\sum_{x/2 < p \leq x} g(p) = \sum_{\substack{x/2 < \ell \leq x, \\ \gcd(\ell, P(z)) = 1}} g(\ell) - \sum_{\substack{z \leq p \leq \sqrt{x}, \\ x/(2p) < m \leq x/p}} \tilde{r}(m) g(mp) + O^*(x/z) .$$

Therefore, we can express a sum over the primes as a sum we know how to handle (this is the sum over integers coprime with $P(z)$), and a rather mysterious one that has a *bilinear* structure. As written before Exer. 26-2, this bilinearity was the main idea of Vinogradov. We take the version above from [23, Chap. 5]. The readers may also have a look at the presentation contained in [16, Section 13.2] by H. Iwaniec and E. Kowalski and at the methodological paper [1] of A. Balog.

Proof We need to detect the prime numbers from the interval $(x/2, x]$. We start from those integers ℓ in the said intervals that are coprime to Q. We remove from those the integers n that have a prime factor $p \leq \sqrt{x}$, and forcibly larger than or equal to z. Such integers are of the shape pm, though such a representation is not unique. Hence

$$\sum_{x/2 < p \leq x} g(p) = \sum_{\substack{x/2 < \ell \leq x, \\ \gcd(\ell, P(z)) = 1}} g(\ell) - \sum_{\substack{z \leq p \leq \sqrt{x}, \\ x/(2p) < m \leq x/p, \\ (m, P(z)) = 1}} \frac{g(mp)}{r(mp)} .$$

The next step consists in separating the variables in the weight $r(pm)$. We note that $r(pm) = 1 + r(m)$ whenever m is prime to p. If not, say $m = ps$, we have

$$\frac{1}{r(p^2 s)} = \frac{1}{1 + r(ps)} + \frac{1}{r(ps)(1 + r(ps))} .$$

We thus find that

$$\sum_{x/2<p\leq x} g(p) = \sum_{\substack{x/2<\ell\leq x, \\ \gcd(\ell,P(z))=1}} g(\ell) - \sum_{\substack{z\leq p\leq \sqrt{x}, \\ x/(2p)<m\leq x/p, \\ (m,Q)=1}} \frac{g(mp)}{1+r(m)}$$

$$- \sum_{\substack{z\leq p\leq \sqrt{x}, \\ x/(2p^2)<s\leq x/p^2, \\ (s,P(z))=1}} \frac{g(sp^2)}{r(ps)(1+r(ps))} \ .$$

On bounding the (absolute value of the) last term trivially, we obtain that

$$\sum_{\substack{z\leq p\leq \sqrt{x}, \\ x/(2p^2)<s\leq x/p^2, \\ (s,P(z))=1}} \frac{1}{r(ps)(1+r(ps))} \leq \sum_{p\geq z} \frac{x}{2p^2} \leq \frac{x}{2(z-1)} \leq \frac{x}{z} \ .$$

The theorem readily follows.   □

One can use Theorem 26.2 to prove that, for any irrational α, as $x \to \infty$ we have

$$\frac{\log x}{x} \sum_{p\leq x} \exp(2i\pi\alpha p) \longrightarrow 0 ,$$

which is one of the essential conclusions of Vinogradov's method. As previously, we restrict ourselves to the case when α is the golden ratio ρ.

Theorem 26.3

Let ρ be the golden ratio. As x goes to infinity and with $\pi(x) = \sum_{p\leq x} 1$, we have

$$\frac{1}{\pi(x)} \sum_{p\leq x} \exp(2i\pi\rho p) \longrightarrow 0 .$$

Note that we can replace $\pi(x)$ by $x/\log x$ in the above statement (a task for which a Chebyshev lower bound is enough).

Proof The proof is somewhat lengthy, but otherwise straightforward. We start from Theorem 26.2 with some parameter z such that

$$\log z \leq \sqrt{\log x}/2 ,$$

so that we can use Theorem 26.1 (but with the summation over m restricted to the interval $(x/2, x]$) to bound the first sum by x/z. We are left with

$$S = \sum_{\substack{z\leq p\leq \sqrt{x}, \\ x/(2p)<m\leq x/p}} \tilde{r}(m)\, e(\rho mp) \ . \tag{26.4}$$

We split dyadically the first sum via Lemma 26.4. Let thus Q and Q' be two real numbers such that $\sqrt{x}/2 \leq Q < Q' \leq \min(x/z, 2Q)$. We now study

$$S(Q, Q') = \sum_{Q < m \leq Q'} \tilde{r}(m) \sum_{\max(z, x/(2m)) \leq p \leq \min(x/m, \sqrt{x})} e(\rho m p) . \qquad (26.5)$$

We apply Cauchy's inequality and obtain, since $\tilde{r}(m) \leq 1$,

$$S(Q, Q')^2 \leq (Q' - Q + 1) \sum_{Q < m \leq Q'} \left| \sum_{\max(z, \frac{x}{2m}) < p \leq \min(\frac{x}{m}, \sqrt{x})} e(\rho m p) \right|^2 .$$

We open the square and invert the summation, getting

$$S(Q, Q')^2 \ll Q \sum_{\max(z, x/(2Q')) \leq p_1, p_2 \leq \min(x/Q, \sqrt{x})} \sum_{m \in I(p_1, p_2)} e(\rho m (p_1 - p_2)) ,$$

where $I(p_1, p_2)$ is a subinterval of $(Q, Q']$. Let n be the variable $|p_1 - p_2|$. It has at most $2x/Q$ representations in this form and it is non-zero if we consider only p_1 and p_2 distinct. We just consider the case $p_1 = p_2$ (the so-called *diagonal case*) independently and obtain

$$S(Q, Q')^2 \ll Q\left(\frac{x}{Q}Q + \frac{x}{Q} \sum_{1 \leq n \leq x/Q} \min\left(Q, \frac{1}{\|\rho n\|}\right)\right) .$$

This does not exactly have the required format to use Lemma 26.2. However noticing that $Q \leq X/n$ solves this issue. Therefore

$$S(Q, Q')^2 \ll xQ + Q\frac{x}{Q}\left(\frac{x}{F_k}\log x + \frac{x}{Q} + F_k\right) ,$$

by using Lemma 26.2 and by approximating the golden ratio ρ by some F_{k+1}/F_k for some positive integer k. We select k so that $F_k \in (\sqrt{x}, 2\sqrt{x}]$, whence

$$S(Q, Q')^2 \ll xQ + Q\frac{x}{Q}\left(\sqrt{x}\log x + x/Q\right) \ll \frac{x^2}{z} .$$

This gives us $S(Q, Q') \ll x/\sqrt{z}$ and by summing over Q, we get $S \ll x(\log x)/\sqrt{z}$. On gathering our estimates, we have reached

$$\sum_{x/2 < p \leq x} \exp(2i\pi\rho p) \ll \frac{x}{z} + \frac{x \log x}{\sqrt{z}} \ll x\exp(-\tfrac{1}{4}\sqrt{\log x} + \log\log x) ,$$

which is more than needed.   □

To get an idea of the actual size of the trigonometric sum $z_N = \sum_{p \leq N} \exp(2i\pi\rho p)$, F.M. Dekking and M. Mendès-France got the idea in [9] to plot the complex numbers z_N, as N increases, joining z_N to z_{N+1} by a straight line. On Fig. 26.1a is what we obtain for the primes ≤ 20, while on Fig. 26.1b is the path when we take all the primes up to 10000.

These sums seem to be much smaller than what we know how to prove! One sees also regions of accumulation followed by some longer leaps: the whole behaviour is to be clarified.

Exercise 26-3. Write a Sage script to produce similar figures for $z_N^* = \sum_{p \leq N} \exp(2i\pi\sqrt{2}p)$ and for $\tilde{z}_N = \sum_{n \leq N} \mu(n) \exp(2i\pi\rho n)$. Write a more efficient code that allows to go up to $N = 10^7$.

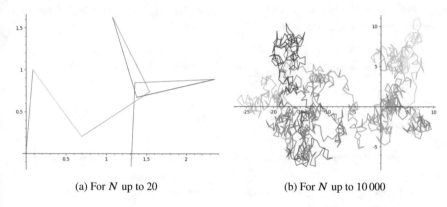

(a) For N up to 20   (b) For N up to 10 000

Fig. 26.1: Exponential sum path for (z_N)

Exercise 26-4. (From [13] by A. Granville) Prove that, when n has no prime factors below z, we have

$$\Lambda(n) = \mathbb{1}_{(n, P(z))=1} \log n - \sum_{\substack{\ell m = n, \\ (\ell m, P(z))=1, \\ \ell, m > 1}} \Lambda(m).$$

Use this identity to prove Theorem 26.3.

Here is a typical consequence of the previous theorem.

Corollary 26.1

Let ρ be the golden ratio and $[a, b] \subset [0, 1]$ be an interval such that $b - a > 1/2$. There are infinitely many primes p such that the fractional part $\{\rho p\}$ belongs to $[a, b]$.

Proof Let $J = (b, 1 + a)$ be the complementary interval of $I = [a, b]$ modulo 1. This means that, when x is any real number, either there exists an integer k such that $x + k$ belongs to I, or there exists an integer k such that $x + k$ belongs to J. The readers may certainly grasp more easily this notion by considering only the case when $a = 0$. Let $\theta = (1 + a + b)/2$ be the mid point of J, so that $J = (\theta - \xi, \theta + \xi)$ with $\xi = (1 + a - b)/2 \in [0, 1/4)$. For any $y \in J$, we have $\cos(2\pi(y - \theta)) \geq \cos(2\pi\xi) > 0$. Once these preparations are over, we define a sum S by $S = \sum_{p \leq x} \cos(2\pi(\{\rho p\} - \theta))$ and find that

$$S = \cos(2\pi\theta) \sum_{p \leq x} \cos(2\pi\rho p) - \sin(2\pi\theta) \sum_{p \leq x} \sin(2\pi\rho p),$$

which is $o(\pi(x))$ by Theorem 26.3. However, if only finitely primes p where such that $\{\rho p\} \notin [a, b]$, then we would have $S \geq \cos(2\pi\xi)\pi(x) - O(1)$, which leads to a contradiction and ends the proof. □

Exercise 26-5. Show that $\dfrac{1}{\pi(x)} \sum_{p \leq x} e(\sqrt{2}p) \longrightarrow 0$ as x goes to infinity.

Exercise 26-6. Let ρ be the golden ratio.

1 ◇ Show that $\dfrac{1}{\pi(x)} \sum_{p \leq x} \exp(2i\pi h\rho p) \longrightarrow 0$, for any non-zero fixed integer h.

2 ◇ By using Weyl's Equidistribution Theorem from [30], show that the sequence $(\cos 2\pi\rho p)_p$ is dense in $[-1, 1]$.

Concerning equidistribution, we refer the readers to the classical book [18] by L. Kuipers and H. Niederreiter and to [21, Chap. 1] by Montgomery.

Exercise 26-7. (Vaughan's Identity, see [26]) Let g be an arbitrary function and let y, z be two real positive parameters and $P > 4z$. Show that

$$\sum_{n} \Lambda(n)g(n) = \sum_{n \leq P} (u_n - v_n)g(n) + \sum_{\substack{yz < \ell \leq P/z, \\ yz < m \leq P/(yz)}} w_\ell g(\ell m)$$

with $u_n = \sum_{\substack{b|n, \\ b \leq y}} \mu(b) \log(n/b)$, $v_n = \sum_{\substack{bc|n, \\ b \leq y, c \leq z}} \mu(b)\Lambda(c)$ and $w_\ell = \sum_{\substack{bc = \ell, \\ b > y, c > z}} \mu(b)\Lambda(c)$.

Concerning this identity, the readers may also read [12] by P.X. Gallagher; Lemma 1 of [25] contains a modified version of it. The paper [10] by S. Drappeau and B. Topacogullari contains in Lemma 3.4 an identity to handle exponential sums on integers free of large prime factor.

Exercise 26-8. Let $k \geq 1$. For $n \in \mathbb{N}$, recall that we have defined $\Lambda_k(n) = \sum_{d|n} \mu(d) \log^k \dfrac{n}{d}$ in (24.3). Deduce a Vaughan-type identity for $\Lambda_k(n)$.

Exercise 26-9. (From Linnik in [19], Eq. (0.6.13)) Show that for an even integer K and arbitrary function $g \geq 0$, we have

$$\sum_n \frac{\Lambda(n)}{\log n} g(n) \geq \sum_{1 \leq k \leq K} \frac{(-1)^{k+1}}{k} \sum_n d_k^*(n) g(n),$$

where $d_k^*(n)$ is the number of k-uples $(a_1, a_2, \ldots, a_k)$ of integers *strictly* greater than 1, such that $d_1 d_2 \cdots d_k = n$.
Show that for odd K, the reverse inequality holds true.

An identity similar to the one of Exer. 26-9 but for the Moebius function may be found in the book [11, Section 17.2] by Friedlander & Iwaniec. The readers will also find related material in [15] by D.R. Heath-Brown.

26.4 An Exponential Sum with the Möbius Function

In [22, Theorem 1], H.L. Montgomery and R.C. Vaughan study exponential sums with multiplicative coefficients and obtain an optimal result. Interestingly, they rely on the Levin–Faĭnleĭb technique described in Chap. 13. We present an approach to this question similar to the one used in Theorem 26.2 and that has been employed by Kaisa Matomäki and M. Radziwiłł in their outstanding paper [20]. The first ingredient of their argument is a construction that leads to the following lemma.

Lemma 26.5

Let N be a large real parameter and $z_0 \leq z$ be two real parameters, larger than 2. For every $G : [1, N] \to \mathbb{C}$ with $|G| \leq 1$, we have

$$\sum_{n \leq N} G(n) = \sum_{\substack{pm \leq N, \\ z_0 < p \leq z}} \frac{G(pm)}{1 + \#\{p' \in (z_0, z] : p'|m\}} + O\left(N \frac{\log z_0}{\log z} + \frac{N}{z_0}\right).$$

Proof Let $\mathcal{N}$ be the set of integers from $[1, N]$ that have (at least) one prime factor in $(z_0, z]$. The Partial Fundamental Lemma (this is our Theorem 25.1) yields the bound $N - |\mathcal{N}| \ll (N \log z_0)/\log z$. Next we find that

$$\sum_{n \in \mathcal{N}} G(n) = \sum_{\substack{pm \leq N, \\ z_0 < p \leq z}} \frac{G(pm)}{\#\{p' \in (z_0, z] : p'|pm\}}.$$

We proceed as in the proof of Theorem 26.2:

$$\sum_{n\in N} G(n) = \sum_{\substack{pm\leq N, \\ z_0<p\leq z}} \frac{G(pm)}{1 + \#\{p' \in (z_0, z] : p'|m\}}$$

$$+ \sum_{\substack{pm\leq N, \\ z_0<p\leq z, \\ p|m}} \frac{G(pm)}{(1 + \#\{p' \in (z_0, z] : p'|m\})\#\{p' \in (z_0, z] : p'|m\}}$$

and this last quantity is at most $O(N/z_0)$, as required. $\qquad\qquad\square$

Theorem 26.4

Let $F : \mathbb{N} \to \mathbb{C}$ with $|F| \leq 1$ and let v be a multiplicative function with $|v| \leq 1$. Let $\tau > 0$ be some (small) parameter and assume that for all primes $p_1, p_2 \leq e^{1/\tau}$, $p_1 \neq p_2$, we have that for $M \geq M_0(\tau)$

$$\left|\sum_{m\leq M} F(p_1 m)\overline{F(p_2 m)}\right| \leq \tau M .$$

Then for $N \geq N_0(\tau)$, we have $(1/N)\left|\sum_{n\leq N} v(n)F(n)\right| \ll \sqrt{\tau} \log(1/\tau)$.

This theorem is due to J. Bourgain, P. Sarnak and Tamar Ziegler in [3, Theorem 2], though our proof is simpler. It sparked a flurry of activity in ergodic theory around the *Sarnak conjecture*, see [24]. In [4], M. Cafferata, A. Perelli and A. Zaccagnini improve on the above by removing the $\log(1/\tau)$.

Proof We start by Lemma 26.5 and set $z = e^{1/\tau}$ together with $z_0 = \exp(1/\sqrt{\tau})$ and $r(m) = 1 + \#\{p \in (z_0, z] : p|m\}$. We notice that

$$\sum_{\substack{pm\leq N, \\ z_0<p\leq z}} \frac{v(pm)}{r(m)}F(pm) = \sum_{\substack{pm\leq N, \\ z_0<p\leq z}} F(pm)\frac{v(p)v(m)}{r(m)}$$

$$+ \sum_{\substack{p^2\ell\leq N, \\ z_0<p\leq z}} F(p^2\ell)\frac{v(p^2\ell) - v(p)v(p\ell)}{r(p\ell)} .$$

The last sum is $O^*\left(\frac{N}{2}\sum_{p\geq z_0} 1/p^2\right)$ which in turn is at most N/z_0. Furthermore, for $z_0 \leq Q \leq Q' \leq \min(2Q, z)$, we find that

$$\left| \sum_{m \le N/Q} \frac{v(m)}{r(m)} \sum_{\substack{Q < p \le Q', \\ p \le N/m}} F(pm)v(p) \right|^2 \le \frac{N}{Q} \sum_{m \le N/Q} \left| \sum_{\substack{Q < p \le Q', \\ p \le N/m}} F(pm)v(p) \right|^2$$

$$\le \frac{N}{Q} \sum_{Q < p_1, p_2 \le Q'} v(p_1)\overline{v(p_2)} \sum_{m \le N/\max(p_1, p_2)} F(p_1 m)\overline{F(p_2 m)}$$

$$\ll \frac{N}{Q}\left(\frac{N}{Q} \frac{Q}{\log Q} + \tau \frac{N}{Q} \frac{Q^2}{(\log Q)^2} \right)$$

$$\ll N^2 \left(\frac{1}{Q \log Q} + \frac{\tau}{(\log Q)^2} \right).$$

We want to use a dyadic decomposition but Lemma 26.4 is too loose. We use the method though, meaning that we sum over $Q = 2^k z_0$ for k from 0 to $\log(z/z_0)/\log 2$, and use $\sqrt{a^2 + b^2} \le |a| + |b|$ to infer that

$$\left| \sum_{m \le N/Q} v(m) \sum_{z_0 < p \le z, p \le N/m} F(pm)v(p) \right| \ll \frac{N}{\sqrt{z_0 \log z_0}} + N\sqrt{\tau} \log \frac{2 \log z}{\log z_0}.$$

We should explain the last term: we have to evaluate $\displaystyle\sum_{0 \le k \le \log(z/z_0)/\log 2} \frac{1}{k + \log z_0}$. We simply consider this quantity as the difference of two harmonic sums. The theorem follows readily. □

A weaker version of the previous theorem had been obtained long before by H. Daboussi in [5] (see also [6] and [7]) and used by I. Kátai, for instance, in [17]. This method is very flexible. The next exercise is inspired by it. See also Exer. 28-5 for another proof of the same result.

Exercise 26-10. In this exercise, we prove the following equality, valid for any complex-valued function G of modulus bounded by 1. Let $z_0 < z$ be two positive real parameters. We set $D = \sum_{z_0 < p \le z} 1/p$ which we assume to be positive. We have

$$\sum_{n \le N} G(n) = \frac{1}{D} \sum_{\substack{z_0 < p \le z, \\ pm \le N}} G(pm) + O(N/\sqrt{D}).$$

1 ⋄ Using Exer. 12-1 prove that $\displaystyle\sum_{n \le N} \left| \sum_{\substack{p | n, \\ z_0 < p \le z}} 1 - D \right|^2 \le ND.$

2 ⋄ Conclude.

3 ⋄ Use the decomposition just obtained to prove a result similar to Theorem 26.4: start with the same assumptions and see the bound you can get instead of $\sqrt{\tau} \log(1/\tau)$. See [14] by A.J. Harper.

Theorem 26.5

We have $\dfrac{1}{x}\displaystyle\sum_{n\le x}\mu(n)\,e(n\rho)\longrightarrow 0$ as x goes to infinity, where $\rho=(1+\sqrt{5})/2$ is the Golden ratio.

H. Davenport was the first to obtain a result of this form in [8], but he relied on I.M. Vinogradov's result. The proof we follow is direct.

Proof We use Theorem 26.4 with $\nu=\mu$ and $F(n)=e(n\rho)$. Let then p_1 and p_2 be two distinct primes below some $e^{1/\tau}$. We have

$$\left|\sum_{m\le M}e\big((p_1-p_2)m\rho\big)\right|\ll\frac{1}{\|(p_1-p_2)\rho\|}\,.$$

Choose an integer k so that $F_k\in(e^{1/\tau},2e^{1/\tau}]$. Then $|p_1-p_2|/(5F_k^2)\le 1/(2F_k)$ which means, by using the approximation given by Lemma 26.1, that

$$(p_1-p_2)\rho=\frac{(p_1-p_2)F_{k+1}}{F_k}+O^*\left(\frac{1}{2F_k}\right).$$

Since $|p_1-p_2|<F_k$, it cannot be divisible by F_k and thus

$$\|(p_1-p_2)\rho\|\ge\frac{1}{2F_k}\ge\frac{1}{4e^{1/\tau}}\,.$$

We conclude from these estimates that $\left|\displaystyle\sum_{m\le M}e\big((p_1-p_2)m\rho\big)\right|\le\tau\cdot M$, provided that $M\ge Ce^{1/\tau}/\tau$ for some (possibly large) constant C. Theorem 26.4 thus applies. We then let τ go to 0 to obtain our result. $\qquad\square$

Exercise 26-11.

1 ⋄ Show that $\dfrac{1}{x}\displaystyle\sum_{n\le x}\mu^2(n)\,e(n\rho)\longrightarrow 0$ as x goes to infinity.

2 ⋄ Let $[\alpha,\beta]\subset[0,1]$ be an interval such that $\beta-\alpha>1/2$. Show that there are infinitely many square-free integers n such that the fractional part $\{\rho n\}$ belongs to $[\alpha,\beta]$.

Exercise 26-12.

1 ⋄ Show that $\dfrac{1}{x}\displaystyle\sum_{n\le x}\lambda(n)\,e(n\rho)\longrightarrow 0$ as x goes to infinity, where λ is the Liouville function.

2 ⋄ Let $[\alpha,\beta]\subset[0,1]$ be an interval such that $\beta-\alpha>1/2$. Show that there are infinitely many *irregular numbers* n, i.e. integers having an odd number of prime factors, such that the fractional part $\{\rho n\}$ belongs to $[\alpha,\beta]$.

3 ◇ By using Weyl's Equidistribution Theorem, show that the sequence $(\cos 2\pi \rho n)$ where n ranges the irregular integers dense in $[-1, 1]$.

Exercise 26-13. Let χ_3 be the only non-principal character modulo 3. Show that
$$\frac{1}{x} \sum_{n \leq x} \mu(n) \chi_3(n) \, e(n\rho) \longrightarrow 0 \text{ as } x \text{ goes to infinity.}$$

Further Reading

The usage of exponential sums with arithmetical coefficients is very developed. A first course can be found in the book [29] by I.M. Vinogradov, and should be furthered with a study of the *Circle Method* that can be found in [27] by R.C. Vaughan.

References

[1] A. Balog. "On sums over primes". In: *Elementary and analytic theory of numbers (Warsaw, 1982)*. Vol. 17. Banach Center Publ. PWN, Warsaw, 1985, pp. 9–19 (cit. on p. 258).

[2] M.J. Binet. "Mémoire sur l'intégration des équations linéaires aux différences finies, d'un ordre quelconque, à coefficients variables". In: *Comptes Rendus Acad. Sciences Paris* 17 (1843), pp. 559–567 (cit. on p. 254).

[3] J. Bourgain, P.C. Sarnak, and T. Ziegler. "Distjointness of Moebius from horocycle flows". In: Dev. Math. 28 (2013), pp. 67–83. https://doi.org/10.1007/978-1-4614-4075-8_5 (cit. on p. 263).

[4] M. Cafferata, A. Perelli, and A. Zaccagnini. "An Extension of the Bourgain.Sarnak.Ziegler Theorem with Modular Applications". In: *Q. J. Math.* 71.1 (2020), pp. 359–377. https://doi.org/10.1093/qmathj/haz048 (cit. on p. 264).

[5] H. Daboussi. "Fonctions multiplicatives presque périodiques B". In: (1975). D'après un travail commun avec Hubert Delange, 321–324. Astérisque, No. 24–25 (cit. on p. 264).

[6] H. Daboussi and H. Delange. "Quelques propriétés des fonctions multiplicatives de module au plus égal à 1". In: *C. R. Acad. Sci. Paris Sér. A* 278 (1974), pp. 657–660 (cit. on p. 264).

[7] H. Daboussi and H. Delange. "On multiplicative arithmetic functions whose modulus does not exceed one". In: *J. London Math. Soc. (2)* 26.2 (1982), pp. 245–264. https://doi.org/10.1112/jlms/s2-26.2.245 (cit. on p. 264).

[8] H. Davenport. "On some infinite series involving arithmetical functions. II". In: *Quart. J. Math., Oxf. Ser.* 8 (1937), pp. 313–320 (cit. on p. 265).

[9] F. M. Dekking and M. Mendès France. "Uniform distribution modulo one: a geometrical viewpoint". In: *J. Reine Angew. Math.* 329 (1981), pp. 143–153. https://doi.org/10.1515/crll.1981.329.143 (cit. on p. 260).

[10] S. Drappeau and B. Topacogullari. "Combinatorial identities and Titchmarsh's divisor problem for multiplicative functions". In: *Algebra and Number Theory* 13.10 (2020), pp. 2383–2425. https://doi.org/10.2140/ant.2019.13.2383 (cit. on p. 262).

[11] J.B. Friedlander and H. Iwaniec. *Opera de cribro*. Vol. 57. American Mathematical Society Colloquium Publications. American Mathematical Society, Providence, RI, 2010, pp. xx+527 (cit. on p. 262).

[12] P.X. Gallagher. "Bombieri's mean value theorem". In: *Mathematika* 15 (1968), pp. 1–6 (cit. on p. 262).

[13] A. Granville. "An alternative to Vaughan's identity". In: *Proceedings of the 'Second Symposium on Analytic Number Theory'*. Ed. by Rivista di Matematica della Universita di Parma. Cetraro (Italy), 8-12 July 2019, 2000, 3pp (cit. on p. 261).

[14]   A.J. Harper. "A different proof of a finite version of Vinogradov's bilinear sum inequal-
       ity (NOTES)". In: *Preprint* (2011). http://warwick.ac.uk/fac/sci/maths/people/staff/harper/
       finitebilinearnotes.pdf (cit. on p. 265).

[15]   D.R. Heath-Brown. "Sieve identities and gaps between primes". English. In: *Journees arith-
       métiques, Metz, 21-25 september 1981*. 1982. http://www.numdam.org/item/AST_1982_
       94__61_0 (cit. on p. 262).

[16]   H. Iwaniec and E. Kowalski. *Analytic number theory*. American Mathematical Society
       Colloquium Publications. xii+615 pp. American Mathematical Society, Providence, RI,
       2004 (cit. on p. 258).

[17]   I. Kátai. "A remark on a theorem of H. Daboussi". In: *Acta Math. Hungar.* 47.1-2 (1986),
       pp. 223–225. https://doi.org/10.1007/BF01949145 (cit. on p. 264).

[18]   L. Kuipers and H. Niederreiter. *Uniform distribution of sequences*. Pure and Applied Math-
       ematics. Wiley-Interscience [John Wiley & Sons], New York-London-Sydney, 1974, pp.
       xiv+390 (cit. on p. 262).

[19]   Y.V. Linnik. "The dispersion method in binary additive problems". In: *Leningrad* (1961),
       208pp (cit. on p. 262).

[20]   K. Matomäki and M. Radziwiłł. "Multiplicative functions in short intervals". In: *Ann. of
       Math. (2)* 183.3 (2016), pp. 1015–1056. https://doi.org/10.4007/annals.2016.183.3.6 (cit. on
       p. 263).

[21]   H.L. Montgomery. *Ten lectures on the interface between analytic number theory and har-
       monic analysis*. Vol. 84. CBMS Regional Conference Series in Mathematics. Published for
       the Conference Board of the Mathematical Sciences,Washington, DC, 1994, pp. xiv+220
       (cit. on p. 262).

[22]   H.L. Montgomery and R.C. Vaughan. "Exponential sums with multiplicative coefficients".
       In: *Invent. Math.* 43.1 (1977), pp. 69–82. https://doi.org/10.1007/BF01390204 (cit. on p.
       263).

[23]   O. Ramaré. *Un parcours explicite en théorie multiplicative*. vii+100 pp. Éditions universi-
       taires europénnes, 2010 (cit. on p. 258).

[24]   P. Sarnak. *Three Lectures on the Mobius Function Randomness and Dynamics*. Tech.
       rep. Institute for Advanced Study, 2011. http://publications.ias.edu/sites/default/files/
       MobiusFunctionsLectures(2)_0.pdf (cit. on p. 264).

[25]   A. Sedunova. "A logarithmic improvement in the Bombieri-Vinogradov theorem". In: *J.
       Theor. Nombres Bordx.* 31.3 (2019), pp. 635–651 (cit. on p. 262).

[26]   R.C. Vaughan. "Mean value theorems in prime number theory". In: *J. London Math Soc. (2)*
       10 (1975), pp. 153–162 (cit. on p. 262).

[27]   R.C. Vaughan. *The Hardy-Littlewood method*. Vol. 80. Cambridge Tracts in Mathematics.
       Cambridge University Press, Cambridge-New York, 1981, pp. xi+172 (cit. on p. 266).

[28]   I.M. Vinogradov. "Representation of an odd number as a sum of three primes". In: *Dokl.
       Akad. Nauk SSSR* 15 (1937), pp. 291–294 (cit. on p. 257).

[29]   I.M. Vinogradov. *The method of trigonometrical sums in the theory of numbers*. Translated
       from the Russian, revised and annotated by K.F. Roth and Anne Davenport, Reprint of the
       1954 translation. Mineola, NY: Dover Publications Inc., 2004, pp. x+180 (cit. on pp. 257,
       266).

[30]   H. Weyl. "Über die Gleichverteilung von Zahlen mod. Eins". German. In: *Math. Ann.* 77
       (1916), pp. 313–352 (cit. on p. 262).

Chapter 27
Convolution Method and Non-Positive Functions

In our installment of the convolution method, the estimates in Lemma 8.1 and 10.2 were valid for $t > 0$ and not only for $t \geq 1$; this fact played a leading role in getting simple expressions. We cannot rely on such a trick when dealing with functions whose prototype is $\mu(n)/n$. We show in this chapter how Rankin's trick offers a way around. Most of the material is taken from the paper [2] by Nathalie Debouzy.

We need an input that we take from [4, Theorem 1.2].

Lemma 27.1

For $x \geq 3\,500$, we have $\left| \sum_{n \leq x} \frac{\mu(n)}{n} \right| \leq \frac{1}{25 \log x}$.

Better estimates are available, though only up to now for larger values of x. A still unimproved explicit bound concerning the Möbius function may be found in [1] by E. Cohen and F. Dress. It is possible to infer Lemma 27.1 by summation by parts from Theorem 23.4 for large values of x. One could hope to complete the argument by some finite computation, and this is theoretically possible. A closer look at the process discloses that we should check by hand the above inequality for every integer values of x below $\exp(10^{21})$, a bound that is astronomically large!!

Here is our main theorem.

Theorem 27.1

For $x > 1$, we have $\left| \sum_{n \leq x} \frac{\mu(n)}{\varphi(n)} \right| \leq \frac{4}{\log x}$.

In [3], G.H. Hardy shows inter alia that $\sum_{n \geq 1} \mu(n)/\varphi(n) = 0$. The above theorem quantifies explicitly the rhythm of convergence. Please notice that our proof is far from optimal! The truncation at $\sqrt{x}$ below and Rankin's exponent $2/3$ can be better tuned, variants of Lemma 27.1 in other ranges are available and the numerical

© The Author(s), under exclusive license to Springer Nature Switzerland AG 2022
O. Ramaré, *Excursions in Multiplicative Number Theory*, Birkhäuser Advanced Texts
Basler Lehrbücher, https://doi.org/10.1007/978-3-030-73169-4_27

verification can be extended. We encourage the readers to get a better constant than the poor 4 above. Since in essence we compare $\mu(n)/\varphi(n)$ to $\mu(n)/n$ and we have the constant $1/25$ for the latter function, the present proof loses a factor one hundred!

We start by verifying the theorem for $x \geq 2 \cdot 10^6$ with Pari/GP.

```
{Check( upperlimit) = my( somme = 1.0 );
   for(k = 2, upperlimit, somme += moebius(k)/eulerphi(k);
   if( abs(somme) > 4/log(k+1),
      print("Problem at ", k)));
}
```

Here again, the readers should be careful when reducing the verification from every real number x in $(1, 2 \cdot 10^6]$ to every *integer* in the same interval: we have to compare our sum to $4/\log(k + 1)$ and not to $4/\log k$.

Proof Let us set $f_6(n) = \mu(n)/\varphi(n)$ and define $g(n) = \mu(n)/n$. We look for a multiplicative function h_6 such that $f_6 = g \star h_6$. The Euler factor of the Dirichlet series of f_6 at the prime p reads

$$1 - \frac{1}{(p-1)p^s} \,,$$

while we have $D(g, s) = 1/\zeta(s + 1)$. The Euler factor of the Dirichlet series of h_6 at the prime p is thus given by

$$D_p(h_6, s) = \left(1 - \frac{1}{(p-1)p^s}\right)\left(1 - \frac{1}{p^{s+1}}\right)^{-1}$$

$$= 1 - \frac{1}{(p-1)p^{s+1}}\frac{p^{s+1}}{p^{s+1}-1} = 1 - \frac{1}{p-1}\sum_{k \geq 1}\frac{1}{p^{ks+k}} \,,$$

so that the function h_6 we are looking for is given on prime powers by

$$h_6(p^k) = \frac{-1}{(p-1)p^k} \,. \tag{27.1}$$

A compact expression for h_6 is also given by

$$h_6(n) = \frac{(-1)^{\omega(n)}}{\varphi(n)k(n)} \,,$$

where $k(n) = \prod_{p^\alpha \mid n} p$ is the *square-free kernel* of n. Notice further that, at $p = 2$, the Euler factor of $D(h_6, 0)$ reads

$$1 - \sum_{k \geq 1}\frac{1}{2^k} = 0 \,,$$

whence $D(h_6, 0) = 0$. The product $D(|h_6|, s)$ is absolutely convergent in the sense of Godement when $\Re s > -1$. A Pari/GP script gives us that (see also Exer. 27-2)

$$D(|h_6|, 0) = \prod_{p \geq 2} \left(1 + \frac{1}{(p-1)^2}\right) \leq 2.827 \ .$$

We are ready for the core of the proof. The convolution identity gives us

$$\sum_{n \leq x} f_6(n) = \sum_{\ell \leq x} h_6(\ell) \sum_{m \leq x/\ell} g(m)$$

$$= \sum_{\ell \leq \sqrt{x}} h_6(\ell) \sum_{m \leq x/\ell} g(m) + \sum_{\sqrt{x} \leq \ell \leq x} h_6(\ell) \sum_{m \leq x/\ell} g(m) \ .$$

Let us assume that $x \geq 2 \cdot 10^6$ (as we may). In the first term of the previous equality, we use Lemma 27.1, since $x/\ell \geq \sqrt{x} \geq 3\,500$ and get

$$\left| \sum_{m \leq x/\ell} g(m) \right| \leq \frac{1}{25 \log(x/\ell)} \leq \frac{1}{25 \log \sqrt{x}} \ .$$

Therefore

$$\left| \sum_{\ell \leq \sqrt{x}} h_6(\ell) \sum_{m \leq x/\ell} g(m) \right| \leq \frac{1}{25 \log \sqrt{x}} \sum_{\ell \leq \sqrt{x}} |h_6(\ell)| \leq \frac{2}{25 \log x} D(|h_6|, 0) \ .$$

For the second term, we recall that $\left|\sum_{m \leq x/\ell} g(m)\right|$ is bounded by 1 by Gram's inequality proved in Exer. 6-5. On using Rankin's trick, we get

$$\sum_{\sqrt{x} \leq \ell \leq x} |h_6(\ell)| \leq \sum_{\sqrt{x} \leq \ell} |h_6(\ell)| \left(\frac{\ell}{\sqrt{x}}\right)^{2/3}$$

$$\leq \sum_{\ell \geq 1} \frac{|h_6(\ell)|}{\ell^{-2/3}} x^{-1/3} = D(|h_6|, -2/3)/x^{1/3} \ .$$

We readily compute that

$$D(|h_6|, -2/3) = \prod_{p \geq 2} \left(1 + \frac{1}{(p-1)(p^{1/3}-1)}\right) \leq 27 \ . \tag{27.2}$$

Indeed, this value is exactly computed in Exer. 9-12, but let us provide another argument. We estimate the product over the primes not more than 10^5 with Pari/GP and get the approximate value 26.0527.... We bound the remainder term by using an integral. Its logarithm is equal to

$$\sum_{p \geq 10^5} \log\left(1 + \frac{1}{(p-1)(p^{1/3}-1)}\right) \leq \sum_{n \geq 10^5+1} \frac{1}{(n-1)(n^{1/3}-1)}$$

$$\leq \sum_{n \geq 10^5+1} \frac{1.1}{n^{4/3}} \leq 1.1 \int_{10^5}^{\infty} \frac{dt}{t^{4/3}} \ .$$

The last integral is equal to $3/10^{5/3} = 0.0646\ldots$ and so, finally, the infinite product is indeed bounded by 27. We have used the slightly simplifying trick that a prime p larger than 10^5 is... larger than $10^5 + 1$.

We finally note that $\frac{2}{25} + \frac{27 \log x}{x^{1/3}} \le 4$ when $x \ge 2 \cdot 10^6$. $\qquad\square$

Exercise 27-1. Show that, for $x > 1$, we have $\left| \sum_{n \ge x} \frac{\mu(n)}{\varphi(n)} \right| \le \frac{4}{\log x}$.

Exercise 27-2. Let $F(X) = 1 - 2X + 2X^2$. With the notation of Theorem 9.5 and 9.6, show that $s_F(1) = 2$, $s_F(2) = 0$ and $s_F(k) = 2s_F(k-1) - 2s_F(k-2)$ Prove that

$$\prod_{p \ge 2} \left(1 + \frac{1}{(p-1)^2} \right) = 2.82641\ 99970\ 67591\ 57554\ 63917\ 47236\ 95374\ldots$$

Exercise 27-3. (Nathalie Debouzy). Let f be such that the function h defined by $f = g \star h$ where $g(n) = \mu(n)/n$ is such that $D(|h|, s)$ converges absolutely for $\Re s > -1$. Show that, for every $x \ge 2 \cdot 10^6$, we have

$$\left| \sum_{n \le x} f(n) \right| \le \frac{C}{\log x} + \frac{C'}{x^{1/3}},$$

where $C = \frac{2}{25} D(|h|, 0)$ and $C' = D(|h|, -2/3) + D(|h|, 0)$.

The readers should also notice that, in Exer. 23-7, we have already used the convolution method with the Möbius function, though in a less precise manner.

Further Reading

Explicit estimates in multiplicative number theory are being more and more developed. A good though difficult read is the forthcoming book of H. Helfgott.

References

[1] H. Cohen and F. Dress. "Estimations numériques du reste de la fonction sommatoire relative aux entiers sans facteur carré". In: *Prépublications mathématiques d'Orsay : Colloque de théorie analytique des nombres, Marseille* (1988), pp. 73–76 (cit. on p. 269).

[2] N. Debouzy. "Convolution Method". In: *Integers* (), 15 pp. (Cit. on p. 269).

[3] G.H. Hardy. "Note on Ramanujan's trigonometrical function c_q (n) and certain series of arithmetical functions". English. In: *Proc. Camb. Philos. Soc.* 20 (1921), pp. 263–271 (cit. on p. 269).

[4] O. Ramaré. "Explicit estimates on several summatory functions involving the Moebius function". In: *Math. Comp.* 84.293 (2015), pp. 1359–1387 (cit. on p. 269).

Chapter 28
The Large Sieve Inequality

We have given some glimpse of the Brun sieve in the preceding chapters. We have also shown how the idea of Vinogradov to get a *bilinear structure* for some sums was enough to deal with them. This idea of bilinear structure has been very fecund in analytic number theory and is also found in another form of the sieve which is called the *large sieve*. and stems from the work of Y. Linnik in [8].

28.1 An Approximate Parseval Equation

Let $\mathcal{H}$ be a complex vector space endowed with a Hermitian form $\langle f|g\rangle$, left linear and right sesquilinear, i.e. such that

$$\langle \lambda f + \mu h|g\rangle = \lambda\langle f|g\rangle + \mu\langle h|g\rangle, \quad \langle f|g\rangle = \overline{\langle g|f\rangle} \quad \text{and} \quad \langle f|f\rangle \geq 0,$$

for every $\lambda, \mu \in \mathbb{C}$ and $f, g, h \in \mathcal{H}$.

We consider a finite family $(\varphi_i^*)_{i\in I}$ of points in $\mathcal{H}$ and a similarly indexed family $(M_i)_{i\in I}$ of positive real numbers such that

$$\forall(\xi_i)_i \in \mathbb{C}^I, \quad \left\|\sum_i \xi_i\varphi_i^*\right\|^2 \leq \sum_i M_i|\xi_i|^2 . \tag{28.1}$$

The next two lemmas form the core of our approach.

> **Lemma 28.1**
>
> We may take $M_i = \sum_j |\langle \varphi_i^*|\varphi_j^*\rangle|$.

Here is an enlightening reading of this lemma due to P.D.T.A. Elliott in [3]: the Hermitian form that appears has a matrix whose diagonal terms are the $\|\varphi_i^*\|^2$'s. A theorem of S. Geršhgorin says that all the eigenvalues of this matrix lie in the

O. Ramaré, *Excursions in Multiplicative Number Theory*, Birkhäuser Advanced Texts Basler Lehrbücher, https://doi.org/10.1007/978-3-030-73169-4_28

union of the so-called Geršhgorin's *discs* centred at the points $\|\varphi_i^*\|^2$, with radius $\sum_{j \neq i} |[\varphi_i^* | \varphi_j^*]|$. Notice, however, we do not know that each Geršhgorin disc does indeed contain an eigenvalue, a flaw that is somehow repaired in the above lemma.

Proof We first notice that

$$\left\| \sum_i \xi_i \varphi_i^* \right\|^2 = \left\langle \sum_i \xi_i \varphi_i^* \,\middle|\, \sum_i \xi_i \varphi_i^* \right\rangle = \sum_{i,j} \xi_i \overline{\xi_j} \langle \varphi_i^* | \varphi_j^* \rangle \,.$$

We then separate ξ_i and ξ_j by appealing to the inequality $2|\xi_i \overline{\xi_j}| \leq |\xi_i|^2 + |\xi_j|^2$. On rearranging the terms, we reach

$$\left\| \sum_i \xi_i \varphi_i^* \right\|^2 \leq \sum_i |\xi_i|^2 \sum_j |\langle \varphi_i^* | \varphi_j^* \rangle| \,,$$

as required. $\qquad\square$

Lemma 28.2

For every $f \in \mathcal{H}$ and with M_i as in (28.1), we have $\sum_i M_i^{-1} |\langle f | \varphi_i^* \rangle|^2 \leq \|f\|^2$.

This lemma is due to A. Selberg according to E. Bombieri in [2, section 2] and in [1].

Proof For the proof, we simply write

$$\left\| f - \sum_i \xi_i \varphi_i^* \right\|^2 \geq 0$$

and expand the square. We take care of $\| \sum_i \xi_i \varphi_i^* \|^2$ by using (28.1), getting

$$\|f\|_2^2 - 2\Re \sum_i \overline{\xi_i} \langle f | \varphi_i^* \rangle + \sum_i M_i |\xi_i|^2 \geq 0 \,.$$

The optimal choice is $\xi_i = \langle f | \varphi_i^* \rangle / M_i$, the lemma readily follows. $\qquad\square$

Combining Lemma 28.2 together with Lemma 28.1 yields what is usually known as "Selberg's lemma" in this context.

Exercise 28-1. (Halász's Inequality). Under the hypotheses of Lemma 28.2, prove the inequality $\left(\sum_i |\langle f | \varphi_i^* \rangle| \right)^2 \leq \|f\|^2 \sum_{i,j} |\langle \varphi_i^*, \varphi_j^* \rangle|$.

Exercise 28-2. (Jutila's Inequality).

$1 \diamond$ Let $(a_n)_n$ and $(x_{m,n})$ be two sequences of complex numbers such that $|a_n| \leq A_n$ for some sequence (A_n). Prove the inequality

$$\sum_{r,s} \left| \sum_n a_n x_{r,n} \overline{x_{s,n}} \right|^2 \leq \sum_{r,s} \left| \sum_n A_n x_{r,n} \overline{x_{s,n}} \right|^2 .$$

2 ◇ Let $(a_n)_n$ be complex numbers whose absolute value is not more than A and let $(t_r)_r$ be real numbers. Show that we have

$$\sum_{r,s} \left| \sum_{n \leq N} a_n n^{-\frac{1}{2} + i(t_r - t_s)} \right|^2 \leq A^2 \sum_{r,s} \left| \sum_{n \leq N} n^{-\frac{1}{2} + i(t_r - t_s)} \right|^2 .$$

The second inequality comes from Lemma 2 of [7] by M. Jutila while the first one is Theorem 4, Chap. 7 of [11] by H.L. Montgomery.

Exercise 28-3. Prove the inequality $\left| \sum_{i,j} \xi_i \overline{\xi_j} f(i,j) \right| \leq \sum_i |\xi_i|^2 \sum_j |f(i,j)|.$

28.2 The Large Sieve Inequality

We recall the following classical notation:

$$e(\alpha) = \exp(2i\pi\alpha) . \tag{15.6}$$

We also use $\|x\|$ to denote not only the norm associated to the Hermitian form, but the distance to the nearest integer as well.

Theorem 28.1 (Large Sieve Inequality)

Let $\mathcal{X}$ be a finite set of points of $\mathbb{R}/\mathbb{Z}$. Set

$$\delta = \min \left\{ \|x - x'\|, x \neq x' \in \mathcal{X} \right\} .$$

For any sequence of complex numbers $(u_n)_{1 \leq n \leq N}$, we have

$$\sum_{x \in \mathcal{X}} \left| \sum_n u_n e(nx) \right|^2 \leq \sum_n |u_n|^2 (N - 1 + \delta^{-1}) .$$

The left-hand side can be thought as a Riemann sum over the points in $\mathcal{X}$; at least when the set $\mathcal{X}$ is dense enough. The spacing between two consecutive points being at least δ, this left-hand side multiplied by δ can be thought as approximating

$$\int_0^1 \left| \sum_n u_n e(n\alpha) \right|^2 d\alpha = \sum_n |u_n|^2 .$$

This is essentially so when δ^{-1} is much greater than N, but it turns out that the case of interest in number theory is the opposite one. In this latter case, we can look at $\sum_n u_n \, e(nx)$ as being a linear form in $(u_n)_n$. The spacing condition implies that $\mathcal{X}$ has less than δ^{-1} elements, so that the number of linear forms implied is indeed less than the dimension of the ambient space (which is N). In that case, these linear forms are independent (the corresponding determinant is a Vandermonde determinant). So what is really at stake here is more almost orthogonality than approximation, which is how we have chosen the method of proof.

Theorem 28.1 in this version is due to Selberg (the proof can be found in [10]). In the same year and by a different method, a marginally weaker version (without the -1 on the right-hand side) was proved by Montgomery and Vaughan in [12]. We prove an even less precise result, namely, with $N + 1 + 2\delta^{-1}$ instead of $N - 1 + \delta^{-1}$. First, we recall what is the Fourier transform of the de la Vallée-Poussin kernel.

28.2.1 A Fourier Transform

Let N' and L be two given positive integers. Consider the function $F(n)$ whose graph is given in Fig. 28.1.

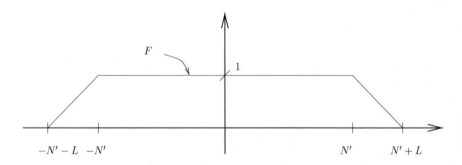

Fig. 28.1: The function F

Since computing its discrete Fourier transform can be cumbersome, we consider two functions G and H, drawn in Fig. 28.2, satisfying $F = (G - H)/L$. This gives us

$$
\begin{aligned}
L \sum_{n \in \mathbb{Z}} F(n) \, e(ny) &= \sum_{n \in \mathbb{Z}} G(n) \, e(ny) - \sum_{n \in \mathbb{Z}} H(n) \, e(ny) \\
&= \sum_{0 \le |n| \le N'+L} (N' + L - |n|) \, e(ny) - \sum_{0 \le |n| \le N'} (N' - |n|) \, e(ny) \\
&= \left| \sum_{0 \le m \le N'+L} e(my) \right|^2 - \left| \sum_{0 \le m \le N'} e(my) \right|^2
\end{aligned}
$$

and so

$$\sum_{n\in\mathbb{Z}} F(n)\,e(ny) = \frac{1}{L}\left|\frac{\sin \pi(N'+L)y}{\sin \pi y}\right|^2 - \frac{1}{L}\left|\frac{\sin \pi N'y}{\sin \pi y}\right|^2. \tag{28.2}$$

The value at $y = 0$ is $\sum_{n\in\mathbb{Z}} F(n) = 2N' + L$.

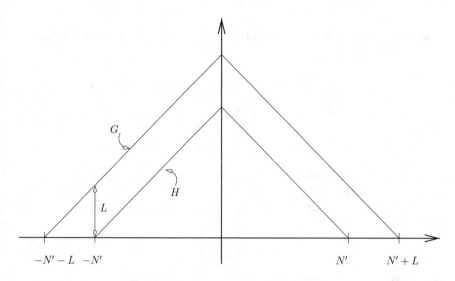

Fig. 28.2: The functions H and G

28.2.2 Proof of a weak form of Theorem 28.1

In this section, we prove Theorem 28.1 with a factor $N + 1 + 2\delta^{-1}$ rather than the optimal $N - 1 + \delta^{-1}$. Such a change is immaterial for most applications.

We apply Lemma 28.2 together with Lemma 28.1. First notice that we may assume N to be an integer. Next set $N' = [N/2]$ the integer part of $N/2$ and $f(n) = u_{N'+1+n}$ (with $u_{N+1} = 0$ if N is even) so that f is supported on $[-N', N']$. The Hilbert space we take is $\ell^2(\mathbb{Z})$ with its standard scalar product so that f belongs to it when extended by setting $f(n) = 0$ for any integer n out of the above interval. Notice also that

$$\|f\|^2 = \sum_n |u_n|^2.$$

We need to define our almost orthogonal system. We select

$$\forall x \in X, \quad \varphi_x^*(n) = e(nx)\sqrt{F(n)}, \tag{28.3}$$

where F is as defined in Sect. 28.2.1. We then find that

$$\langle f|\varphi_x^*\rangle = \mathrm{e}(-(N'+1)x)\sum_{1\leq n\leq N}\sqrt{F(n)}u_n\,\mathrm{e}(nx)\,. \tag{28.4}$$

The computations of the preceding section show that

$$\|\varphi_x^*\|_2^2 = 2N' + L, \quad |\langle\varphi_x^*|\varphi_{x'}^*\rangle| \leq \frac{1}{4L\|x-x'\|^2} \quad \text{if } x \neq x'$$

by using the classical inequality $|\sin x| \leq 2\|x\|/\pi$ used in Lemma 26.3. When x is fixed, we find that

$$\sum_{\substack{x'\in X\\ x'\neq x}} |\langle\varphi_x^*|\varphi_{x'}^*\rangle| \leq \sum_{\substack{x'\in X\\ x'\neq x}} \frac{1}{4L\|x-x'\|^2}$$

$$\leq 2\sum_{k\geq 1}\frac{1}{4L(k\delta)^2} \leq \frac{\pi^2}{12L\delta^2}\,,$$

since the definition of δ implies that the worst case that could happen for the sequence $(\|x-x'\|)_{x'}$ would be if all x''s were located at $x + \ell\delta$ with ℓ an integer taking the values $\pm 1, \pm 2, \pm 3, \ldots$. Next we choose L to be an integer so as to nearly minimize $2N' + L + \pi^2/(12L\delta^2)$, i.e.

$$L = \left\lceil\frac{\pi}{2\sqrt{3}\delta}\right\rceil, \tag{28.5}$$

which gives $2N' + L + \pi^2/(12L\delta^2) \leq N + 1 + \frac{\pi}{\sqrt{3}}\delta^{-1}$. We conclude by noting that $\pi/\sqrt{3} \leq 1.82 \leq 2$.

Let us end this section by a remark: Montgomery proved in an appendix of [9] the inequality $(\sin\pi x)^{-2} \leq (\pi\|x\|)^{-2} + 1$ for $0 \leq x \leq 1/2$. With it we obtain a better bound for $[\varphi_x^*|\varphi_{x'}^*]$, and consequently improve the $N+1+\pi\delta^{-1}/\sqrt{3}$ to $N+3+2\delta^{-1}/\sqrt{3}$.

28.3 Introducing Farey Points

In most arithmetical applications, the set X is a truncation of the Farey series, that is,

$$X = \{a/q,\ q \leq Q,\ a \bmod^* q\},$$

where Q is a parameter to be chosen and $a \bmod^* q$ means a ranges over the invertible residue classes modulo q. When a/q and a'/q' are two distinct points of X, we have

$$\left|\frac{a}{q} - \frac{a'}{q'}\right| = \frac{|aq' - a'q|}{qq'} \geq \frac{1}{qq'} \geq Q^{-2}$$

since $aq' - a'q$ is a non-zero integer.* We set classically

---

* By discussing whether $q = q'$ or not, one can enlarge this bound to $1/(Q^2 - Q)$.

$$S(a/q) = \sum_{1 \le n \le N} u_n \, e(na/q)$$

and get

$$\sum_{q \le Q} \sum_{a \bmod^* q} |S(a/q)|^2 \le \sum_n |u_n|^2 (N + Q^2), \qquad (28.6)$$

which is essentially what is referred to as *the large sieve inequality*.

Further material on the large sieve inequality may be found in [10] by Montgomery and in [6, Chap. 7] by Iwaniec and Kowalski.

Exercise 28-4. Let (u_n) be a sequence of complex numbers. Define $S(\alpha) = \sum_n u_n \, e(n\alpha)$ and, when $q \ge 1$ is a modulus, $S(q; b) = \sum_{n \equiv b[q]} u_n$.

1 ⋄ Define $W^*(q) = \sum_{a \bmod^* q} |S(a/q)|^2$ and $V(q) = q \sum_{b \bmod q} |S(q; b)|^2$. Show that
$W^*(d) = q \sum_{\delta|q} \mu(q/\delta) V(\delta)$.

2 ⋄ Let δ be a divisor of q. Establish that $\displaystyle\sum_{\substack{d_1|q, d_2|q, \\ (d_1,d_2)=\delta}} \mu(q/d_1)\mu(q/d_2) = \mu(q/\delta)$.

3 ⋄ Define further $V^*(q) = \displaystyle\sum_{b \bmod q} \left| \sum_{d|q} \mu(q/d) d S(d; b) \right|^2$. Prove first that we have
$V^*(d) = q^2 \sum_{\delta|q} \mu(q/\delta) V(\delta)$ and conclude that $V^*(q) = q W^*(q)$.

4 ⋄ Show that

$$\sum_{p \le Q} p \sum_{1 \le b \le p} \left| \sum_{n \equiv b[p]} u_n - \frac{\sum_n u_n}{p} \right|^2 \le (N - 1 + Q^2) \sum_n |u_n|^2, \qquad (28.7)$$

where the summation is restricted to prime moduli p.

5 ⋄ Assume the sequence satisfies $|u_n| \le 1$ and Q is not more than $\sqrt{N}$. Prove that

$$\left| \left\{ (b,p)/p \le Q, 1 \le b \le p, \left| \sum_{n \equiv b[p]} u_n - \frac{\sum_n u_n}{p} \right| > \frac{N \log^2 Q}{Q \sqrt{p}} \right\} \right| \ll \frac{Q^2}{(\log Q)^2}$$

hence, for almost every pairs (b, p), we have $S(p; b) \sim \sum_n u_n / p$.

A general version of inequality (28.7) may be found in Theorem 2.1 of the book [13] (see also the proof of Theorem 5.1 therein). We proceed with an exercise inspired by the papers [4] and [5] of Hildebrand.

Exercise 28-5. Let G be any complex-valued function of modulus bounded by 1 Show by using (28.7) above that, for any $z \le \sqrt{N}$, we have

$$\sum_{n \leq N} G(n) = \frac{1}{D} \sum_{\substack{mp \leq N, \\ p \leq z}} G(mp) + O(N / \sqrt{\log \log z}),$$

where $D = \sum_{p \leq z} 1/p$, leading to another proof of the result of Exer. 26-10.

We now propose a different approach to the Turán-Kubilius inequality proved in Exer. 12-11.

Exercise 28-6. (A partial Turán-Kubilius Inequality). Let f be a complex-valued function on the primes and $X \geq 1$ be some real parameter. Define

$$M(f) = \sum_{p \leq \sqrt{X}} \frac{f(p)}{p}, \quad D(f) = \sum_{p \leq \sqrt{X}} \frac{|f(p)|^2}{p}.$$

We want to prove that $(\star)$ $\displaystyle\sum_{n \leq X} \left| \sum_{p | n, p \leq \sqrt{X}} f(p) - M(f) \right|^2 \leq 2XD(f).$

$1 \diamond$ Define $S(n) = \displaystyle\sum_{p | n, p \leq \sqrt{X}} f(p) - M(f)$ and $\kappa(p, n) = \begin{cases} 1 - 1/p & \text{when } p | n, \\ -1/p & \text{when } p \nmid n. \end{cases}$ Write

the left-hand side of $(\star)$ as $\displaystyle\sum_p f(p) \sum_n \overline{S(n)}\kappa(p, n)$ and conclude by using (28.7)

$2 \diamond$ Let $\omega(n)$ be number of prime factors of n counted without multiplicity. Show that, when $X \geq 3$, we have $\displaystyle\sum_{n \leq X} |\omega(n) - \log \log X|^2 \ll X \log \log X.$

References

[1]   E. Bombieri. "A note on the large sieve". In: *Acta Arith.* 18 (1971), pp. 401–404 (cit. on p. 274).

[2]   E. Bombieri. "Le grand crible dans la théorie analytique des nombres". In: *Astérisque* 18 (1987/1974), 103pp (cit. on p. 274).

[3]   P.D.T.A. Elliott. "On inequalities of large sieve type". In: *Acta Arith.* 18 (1971), pp. 405-422 (cit. on p. 273).

[4]   A. Hildebrand. "The prime number theorem via the large sieve". In: *Mathematika* 33.1 (1986), pp. 23–30. https://doi.org/10.1112/S002557930001384X (cit. on p. 279).

[5]   A. Hildebrand. "Some new applications of the large sieve". In: *Number theory (New York, 1985/1988)*. Vol. 1383. Lecture Notes in Math. Springer, Berlin, 1989, pp. 76–88. https://doi.org/10.1007/BFb0083571 (cit. on p. 279).

[6]   H. Iwaniec and E. Kowalski. *Analytic number theory*. American Mathematical Society Colloquium Publications. xii+615 pp. American Mathematical Society, Providence, RI, 2004 (cit. on p. 279).

[7]   M. Jutila. "Zero-density estimates for L-functions". In: *Acta Arith.* 32 (1977), pp. 55–62 (cit. on p. 275).

[8]   Y.V. Linnik. "The large sieve". In: *Doklady Akad. Nauk SSSR* 30 (1941), pp. 292–294 (cit. on p. 273).

[9]   H.L. Montgomery. "Topics in Multiplicative Number Theory". In: *Lecture Notes in Mathematics (Berlin)* 227 (1971), 178pp (cit. on p. 278).

[10]   H.L. Montgomery. "The analytic principle of the large sieve". In: *Bull. Amer. Math. Soc.* 84.4 (1978), pp. 547–567 (cit. on pp. 276, 279).

[11]   H.L. Montgomery. *Ten lectures on the interface between analytic number theory and harmonic analysis.* Vol. 84. CBMS Regional Conference Series in Mathematics. Published for the Conference Board of the Mathematical Sciences,Washington, DC, 1994, pp. xiv+220 (cit. on p. 275).

[12]   H.L. Montgomery and R.C. Vaughan. "The large sieve". In: *Mathematika* 20.2 (1973), pp. 119–133 (cit. on p. 276).

[13]   O. Ramaré. *Arithmetical aspects of the large sieve inequality.* Vol. 1. Harish-Chandra Research Institute Lecture Notes. With the collaboration of D.S. Ramana. New Delhi: Hindustan Book Agency, 2009, pp. x+201 (cit. on p. 279).

Chapter 29
Montgomery's Sieve

Sieving is all about extracting information from a sequence that is imbedded in a *host sequence*. The leading example is one of the primes: we start from the integers of the interval $[1, x]$ and remove all the ones that are divisible by a prime not more than $\sqrt{x}$. What remains are the primes from $(\sqrt{x}, x]$ together with the integer 1. The target sequence, the one we aim at understanding, is thus this latter. Let us formalize the process. Denote by $\mathcal{P}$ the set of prime numbers $p \le Q$. For every p in $\mathcal{P}$, denote by Ω_p a subset of $\mathbb{Z}/p\mathbb{Z}$ which cardinality we denote by $\omega(p)^*$. These are the classes we want to remove. In the case of the primes, we remove $\Omega_p = \{0\}$. More examples are given below. A feature of sieves is to be noticed: we aim generally at getting an *upper* or a *lower* bound for the cardinality of our target set. Which means that our sieving process can be somewhat imprecise. Upper estimates are easier to get.

29.1 The Statement

> **Theorem 29.1**
>
> Let $Q \ge 1$ be a real number and $\mathcal{P}$ be a set of primes $p \le Q$. For every p in $\mathcal{P}$, choose a subset $\Omega_p \subset \mathbb{Z}/p\mathbb{Z}$ of cardinality $\omega(p)$. Let $\mathcal{Q}$ be the set of square-free integers less than Q composed of prime factors from $\mathcal{P}$. Finally, let Z be the number of integers from $[M + 1, M + N]$ that, on reduction modulo p, do not belong to any Ω_p. Then
>
> $$Z \le (N + Q^2)/L,$$
>
> where

---

* The notations Ω_p and $\omega(p)$ are standard. They bear no relation with the functions Ω and ω that count the number of prime factors with and, respectively, without multiplicity.

© The Author(s), under exclusive license to Springer Nature Switzerland AG 2022
O. Ramaré, *Excursions in Multiplicative Number Theory*, Birkhäuser Advanced Texts Basler Lehrbücher, https://doi.org/10.1007/978-3-030-73169-4_29

$$L = \sum_{q \in Q} \mu^2(q) \prod_{p|q} \frac{\omega(p)}{p - \omega(p)} .$$

This theorem results from combining the modern version of the large sieve inequality contained in Theorem 28.1 together with the proof of Montgomery in [11]. The quantity L is usually handled through Theorem 13.3. Some examples:

1. Choose $\Omega_p = \{0\}$ for every $p \le Q$. The integers that are in no Ω_p for any $p \le Q$, are the integers not divisible by any $p \le Q$. This set contains, in particular, the primes $p > Q$. We thus obtain

$$\#\{p \in (\max(Q, M + 1), M + N]\} \le (N + Q^2) \Big/ \sum_{q \le Q} \frac{\mu^2(q)}{\varphi(q)} ,$$

and, by using Exer. 10-4, we deduce from this that

$$\#\{p \in (\max(Q, M + 1), M + N]\} \le (N + Q^2)/ \log Q .$$

On choosing $Q = \sqrt{N}/ \log N$, we obtain the case $q = 1$ of the *Brun–Titchmarsh inequality*.

2. Select $\Omega_p = \{0, -2\}$ for every $p \le Q$. The integers that are in no Ω_p for any $p \le Q$ are the integers n such that neither n nor $n + 2$ is divisible by a prime $p \le Q$. This set contains, in particular, the prime numbers $p > Q$ such that $p + 2$ is also prime. We thus conclude that

$$\#\{p \in (\min(Q, M + 1), M + N], p + 2\text{prime}\}$$

$$\le (N + Q^2) \Big/ \sum_{q \le Q} \mu^2(q) \prod_{\substack{p|q, \\ p \ne 2}} \frac{2}{p - 2} .$$

We leave it to the readers to proceed and use Theorem 13.3, cf. Theorem 29.2.

***Proof (of Theorem* 29.1)** Let N be the sequence of integers in $[M + 1, M + N]$ that are not in Ω_p for any p in $\mathcal{P}$. Let $(u_n)_{M+1 \le n \le M+N}$ be an arbitrary sequence of complex numbers satisfying

$$n \notin N \implies u_n = 0 .$$

Consider the trigonometric polynomial $S(\alpha) = \sum_{n \in N} u_n \, e(n\alpha)$. The central point of the argument (aside from the large sieve inequality) is that, for every $q \in Q$, we have

$$\sum_{\substack{1 \le a \le q, \\ \gcd(a,q)=1}} |S(a/q)|^2 \ge |S(0)|^2 J(q) , \quad J(q) = \prod_{p|q} \frac{\omega(p)}{p - \omega(p)} . \qquad (29.1)$$

Note that, if (29.1) is true for q and q', and if q and q' are coprime, then the inequality also holds true for qq'. Actually, the Chinese Remainder Theorem (see Exer. 18-4) allows us to write

$$\sum_{\substack{1 \le c \le qq', \\ \gcd(c,qq')=1}} \left| S\left(\frac{c}{qq'}\right) \right|^2 = \sum_{\substack{1 \le a \le q, \\ \gcd(a,q)=1}} \sum_{\substack{1 \le b \le q', \\ \gcd(b,q')=1}} \left| S\left(\frac{a}{q} + \frac{b}{q'}\right) \right|^2.$$

Let a be fixed. Note that $S\left(\frac{a}{q} + \frac{b}{q'}\right) = \sum_{n \in N} u_n\, e(na/q) e(nb/q')$. Since (29.1) is true for q' and changing u_n into $u_n\, e(na/q)$, we get

$$\sum_{\substack{1 \le b \le q', \\ \gcd(b,q')=1}} \left| S\left(\frac{a}{q} + \frac{b}{q'}\right) \right|^2 \ge J(q') \left| S\left(\frac{a}{q}\right) \right|^2.$$

We then apply (29.1) for q and deduce its validity for qq'. Hence, it suffices to show that (29.1) holds when q is a prime, say p. We directly check that in this case we have

$$\sum_{\substack{1 \le a \le p, \\ \gcd(a,p)=1}} |S(a/q)|^2 = p \sum_{b \bmod p} |S(b;p)|^2 - |S(0)|^2 \quad \text{where } S(b;p) = \sum_{n \equiv b[p]} u_n.$$

On the right-hand side we restrict b to $\mathbb{Z}/p\mathbb{Z} \setminus \Omega_p$. The Cauchy–Schwarz inequality guarantees us that

$$|S(0)|^2 = \left| \sum_{b \in \mathbb{Z}/p\mathbb{Z} \setminus \Omega_p} S(b;p) \right|^2 \le (p - \omega(p)) \sum_{b \bmod p} |S(b;p)|^2,$$

and hence

$$\sum_{\substack{1 \le a \le p, \\ \gcd(a,p)=1}} |S(a/q)|^2 \ge \left(\frac{p}{p - \omega(p)} - 1 \right) |S(0)|^2,$$

as required. Now we can invoke the large sieve inequality (28.6). This gives us

$$|S(0)|^2 \sum_{q \in Q} \mu^2(q) J(q) \le \sum_n |u_n|^2 (N + Q^2).$$

We choose $u_n = 1$ if $n \in N$ and $u_n = 0$ otherwise. The proof is over. $\qquad \square$

The next exercise shows how to get a "sieving effect" from the large sieve in a different manner. It is due to Selberg in the seventies (unpublished) and is used, for instance, in a simple manner by Elliott in [3].

Exercise 29-1.

1 ⋄ Show that, for any complex sequence $(u_n)_{M < n \le M+N}$, we have

$$\sum_{q \le Q} \frac{\mu^2(q)}{\varphi(q)} \left| \sum_{M < n \le M+N} u_n c_q(n) \right|^2 \le \sum_{M < n \le M+N} |u_n|^2 (N + Q^2),$$

where $c_q(n)$ is the Ramanujan sum.

2 ◊ Deduce from the above that the number of primes in the interval $(M, M + N]$ is at most $(2 + o(1))N / \log N$, getting another proof (of part) of the Brun–Titchmarsh theorem. Show also that, when $M \geq \sqrt{N}$, we have

$$\sum_{M < p \leq M+N} 1 \leq \frac{4N}{\log N} \tag{29.2}$$

for every $N \geq 2$, without any error term.

Here is now a consequence from (29.2) that we borrow from Lemma 2.6 of [7] by A.Granville, A.J. Harper and K. Soundararajan.

Exercise 29-2. For any real parameter $T \geq 100$, we define $K(u) = \max(0, 1 - |u|/T)$

1 ◊ Establish that $\hat{K}(t) = \int_{-\infty}^{\infty} K(u) \, e(tu) du = T \sin(\pi t/T)^2 / (\pi t)^2$.

2 ◊ Show that $|\pi^2 T \hat{K}(t)| \geq 1$ when $|t| \leq T$.

3 ◊ Show that $|\operatorname{sh} x| \leq \frac{5}{4}|x|$ when $|x| \leq 1$.

4 ◊ Show that, for any $q \geq T$, the number of integer points in the interval $[qe^{-2\pi/T}, qe^{2\pi/T}]$ is not more than $17q/T$.

5 ◊ Show that, when $q \geq T^2$, we have $4 \log(eq + 1) / \log(17q/T) \leq 8$.

6 ◊ By using Exer. 28-3 and (29.2), prove the inequality

$$\int_{-T}^{T} \left| \sum_{T^2 < p \leq x} \frac{a(p) \log p}{p^{it}} \right|^2 dt \leq 1500 \sum_{T^2 < p \leq x} p |a(p)|^2 \log p,$$

where $(a(p))$ is an arbitrary sequence of complex numbers.

The readers may want to compare this result with Corollary 3 of [13] by H.L. Montgomery and R.C. Vaughan, as well as with Theorem 1 of [6] by P.X. Gallagher.

29.2 Applications

Here are three typical consequences of Theorem 29.1. In these three cases, the density L is evaluated by invoking Theorem 13.3.

Theorem 29.2

Let $h > 1$ be an even integer. We have, as x tends to infinity,

$$\#\{p \leq x, p + h \text{ prime}\} \leq \frac{8(1 + o(1))x}{(\log x)^2} \prod_{p \geq 3} \left(1 - \frac{1}{(p-1)^2}\right) \prod_{2 < p | h} \frac{p-1}{p-2}.$$

Theorem 29.3

We have, as x tends to infinity, $\#\{p \le x, p + 2 \text{ and } p + 6 \text{ primes}\} \ll \dfrac{x}{(\log x)^3}$.

Exercise 29-3.

1 ◊ Show that, if p, $p + 2$ and $p + 4$ are simultaneously prime, then $p = 3$.

2 ◊ Enumerate the triples $(n, n + 2, n + 4)$ where $n \le 3^{100}$ is a prime power.

Theorem 29.4

We have, as x tends to infinity, $\#\{p \le x, p^2 - 2 \text{ prime}\} \ll \dfrac{x}{(\log x)^2}$.

For the last proof, the readers may use (5.8). Here is how to guess the right order of magnitude: we start from an integer n in the interval $[1, x]$, we assume that n is a prime, an event that happens with "probability" $1/\log x$, so we are left with about $x/\log x$ solutions, and then, we assume that $n + 2$ is also prime, an event which has also probability $1/\log x$. The general belief is that the events "n is prime" and "$n + 2$ is prime" are independent, so that the final set should have size of order $x/(\log x)^2$. The above theorems thus give upper bounds of the conjectured true size.

Exercise 29-4. Show that, as x tends to infinity,

$$\#\{p \le x, (p^2 + 1)/2 \text{prime}\} \ll \frac{x}{(\log x)^2} .$$

Exercise 29-5. We introduced *prime magic squares* in Exer. 17-8. Show that the number of 3×3 prime magic squares whose common sum is at most N is a bounded above by $CN^2/(\log N)^9$ for some constant C.

Exercise 29-6. Let N be an even integer and let $r(N)$ be the number of representations of N as a sum of two prime numbers. Use Montgomery's sieve to prove that $r(N) \le (8 + o(1)) \, \mathfrak{S}(N)\dfrac{N}{\log^2 N}$, where $\mathfrak{S}(N) = \mathfrak{S}_\infty \displaystyle\prod_{\substack{p \mid N \\ p \ne 2}} \dfrac{p - 1}{p - 2}$ and $\mathfrak{S}_\infty = 2 \displaystyle\prod_{p \ge 3} \left(1 - \dfrac{1}{(p - 1)^2}\right)$.

The quantitative version of the Goldbach's conjecture of Hardy and Littlewood in [10] asserts that $r(N) \sim \mathfrak{S}(N)N/(\log N)^2$, so that the above looses only a constant! Lots of work has been put to obtain a better constant than the above 8 (which roughly dates from the fifties). The latest record at the time of writing is 3.911 due to Jie

Wu in [16]. Though the 1742-conjecture of Goldbach* is the most renowned, his 1752-conjecture is also very intriguing: *it asserts that every odd integer, save 1, 5777 and 5993, can be written as the sum of twice a square plus a prime.*

Exercise 29-7. Find an arithmetic progression none of whose member can be represented as a sum of two primes.

Exercise 29-8 (From [4] by Erd).

1 ◇ Show by using Exer. 5-4 that any power of 2 belongs to at least one of the sets $1 + 2\mathbb{Z}$, $1 + 7\mathbb{Z}$, $2 + 5\mathbb{Z}$, $2^3 + 17\mathbb{Z}$, $2^7 + 13\mathbb{Z}$ and $2^{23} + 241\mathbb{Z}$.

2 ◇ Show that there exists an arithmetic progression consisting only of odd numbers, no term of which is of the form $2^k + p$.

L. Habsieger and X.-F. Roblot proved more on this issue in [8].

Exercise 29-9. Let $\mathcal{A}$ be the sequence of even integers that can be written as a sum of two primes and let $A(x) = \#\{a \in \mathcal{A}, \ a \leq x\}$ be the number of elements of $\mathcal{A}$ not exceeding x. Show that $A(x) \gg x$ when $x \geq 6$, i.e. the set of integers that are a sum of two primes has a strictly positive lower density.

The process described in the above exercise is due to L. Schnirelmann in [15]. It may be deduced from this result by using *additive combinatorics* (partly invented by Schnirelmann for this very purpose) that there exists a constant C such every integer larger than 2 can are a sum of at most C prime numbers. The minimal value of C is known as *Schnirelmann's constant*. It is now known that $C \leq 4$ and the Goldbach Conjecture is equivalent to showing that $C = 3$.

Further Reading

On sieves, we refer the readers to [9] by Halberstam and Richert, on which many have learned the trade, and to [5] by Friedlander and Iwaniec, two masters of the discipline. Two classical references for the large sieve and its applications are [12] by Montgomery and [1] by Bombieri. The book [14] is dedicated to more arithmetical aspects. To sieve polynomial values and not only quadratic ones, a substitute to Lemma 18.1 is needed. Pertaining classical material is exposed in a very readable way in [2, Section 10.1] by Diamond, Halberstam and Galway.

---

* We take this opportunity to tell the readers that there are no known visual representation of Christian Goldbach, either drawing or painting, and that the ones that float on the web are all wrong (it is often Euler or H. Grassmann with specs, and sometimes it is B. Riemann).

References

[1] E. Bombieri. "Le grand crible dans la théorie analytique des nombres". In: *Astérisque* 18 (1987/1974), 103pp (cit. on p. 288).

[2] H.G. Diamond, H. Halberstam, and W.F. Galway. *A higher-dimensional sieve method*. Vol. 177. Cambridge tracts in mathematics. Cambridge University Press, 2008 (cit. on p. 288).

[3] P.D.T.A. Elliott. "On maximal variants of the Large Sieve. II". In: *J. Fac. Sci. Univ. Tokyo, Sect. IA* 39.2 (1992), pp. 379–383 (cit. on p. 285).

[4] P. Erdös. "On integers of the form 2^k p and some related problems". In: *Summa Brasil. Math.* 2 (1950), pp. 113–123 (cit. on p. 288).

[5] J.B. Friedlander and H. Iwaniec. *Opera de cribro*. Vol. 57. American Mathematical Society Colloquium Publications. American Mathematical Society, Providence, RI, 2010, pp. xx+527 (cit. on p. 288).

[6] P.X. Gallagher, "A large sieve density estimate near $\sigma = 1$". In: *Invent. Math.* 11 (1970), pp. 329–339. https://doi.org/10.1007/BF01403187 (cit. on p. 286).

[7] A. Granville, A.J. Harper, K. Soundarararajan, "'A new proof of Halász's theorem, and its consequences". In: *Compos. Math.* 155.1 (2019), pp. 126–163. https://doi.org/10.1112/s0010437x18007522 (cit. on p. 286).

[8] L. Habsieger, X.-F. Roblot, "On integers of the form $p\, 2^k$". In: *Acta Arith.* 122.1 (2006), pp. 45–50. https://doi.org/10.4064/aa122-1-4 (cit. on p. 288).

[9] H. Halberstam and H.-E. Richert. "Sieve methods". In: *Academic Press (London)* (1974), 364pp (cit. on p. 288).

[10] G.H. Hardy and J.E. Littlewood. "Some problems of "Partitio Numerorum" III. On the expression of a number as a sum of primes". In: *Acta Math.* 44 (1922), pp. 1–70 (cit. on p. 287).

[11] H.L. Montgomery. "A note on the large sieve". In: *J. London Math. Soc.* 43 (1968), pp. 93–98 (cit. on p. 284).

[12] H.L. Montgomery. "Topics in Multiplicative Number Theory". In: *Lecture Notes in Mathematics (Berlin)* 227 (1971), 178pp (cit. on p. 288).

[13] R.C. Montgomery H.L. nd Vaughan. "Hilbert's inequality". In: *J. Lond. Math. Soc., II Ser.* 8 (1974), pp. 73–82 (cit. on p. 286).

[14] 14 O. Ramaré. *Arithmetical aspects of the large sieve inequality*. Vol. 1. Harish-Chandra Research Institute Lecture Notes. With the collaboration of D.S. Ramana. New Delhi: Hindustan Book Agency, 2009, pp. x+201 (cit. on p. 288).

[15] L.G. Schnirelmann. "Über additive Eigenschaften von Zahlen". In: *Math. Ann.* 107 (1933), pp. 649–690 (cit. on p. 288).

[16] Jie Wu. "Chen's double sieve, Goldbach's conjecture and the twin prime problem". In: *Acta Arith.* 114.3 (2004), pp. 215–273 (cit. on p. 287).

Hints and Solutions for Selected Exercises

This chapter contains some hints and solutions to the 296 exercises of this book. The numbering is *Chapter–Exercise–Question*, so that 17-5-ii refers to the second question of Exercise 5 of Chapter 17. The question number is simply omitted when the exercise is not split in several questions.

———————————— §*Chapter* 1 ————————————

Solution to 1-1. Suppose we are given two coprime integers m and n. Then we have

$$f_1(mn) = \prod_{p \mid mn} \frac{p+1}{p+2} .$$

By Euclid's lemma, the prime factors p of mn are divided into two groups: the ones that divide m and the ones that divide n. These two groups have no intersection since any prime that belongs to both of these groups would divide both m and n, hence their greatest common divisor (m, n), which is 1. Thus,

$$f_1(mn) = \prod_{p \mid m} \frac{p+1}{p+2} \prod_{p \mid n} \frac{p+1}{p+2} = f_1(m) f_1(n) ,$$

which is what we wanted to show. What about the condition $f_1(1) = 1$? In the expression $\prod_{p \mid 1} (p + 1)/(p + 2)$, the set of primes in the product is empty, and, by convention, the product over the empty set is equal to 1.

Hint for 1-2. Use the decomposition into prime factors.

Hint for 1-4. Use the decomposition into prime factors.

Solution to 1-6. It is enough to sum directly: the divisors form a geometric progression with common ratio p.

Solution to 1-7-i. By contradiction: let m be the smallest integer > 1 with the property that such an n exists. The largest prime factor p of m is coprime to $\varphi(m)$ and, hence, divides the denominator of $\varphi(n)/n$. Thus, p divides n as well and we can

O. Ramaré, *Excursions in Multiplicative Number Theory*, Birkhäuser Advanced Texts Basler Lehrbücher, https://doi.org/10.1007/978-3-030-73169-4

divide both m and n by this prime (consider the exact power v of p that divides m). This contradicts the minimality of m.

Solution to 1-7-iii. Compute the function at 2 and at 15, where the value is 3.

Solution to 1-7-iv. Compute the function at 3 and at 5×13, where the value is $5/4$.

Hint for 1-8. One can establish the claim with the constant $\sqrt{3}$ and it is optimal.

Hint for 1-9. The best constant is $18^{1/3}$.

Hint for 1-10-ii. One may consider the multiplicative function $n \mapsto d(n)/n^{\varepsilon}$.

Hint for 1-17-iv. The readers may start by showing that f takes the value $f(1)$ on each prime.

Hint for 1-18. One can calculate this sum directly by using the fact that the characteristic function of integers m coprime to n can be written as $\displaystyle\sum_{\substack{d \mid n \\ d \mid m}} \mu(d)$, a fact the readers should first establish.

Solution to 1-22. By multiplicativity, it is enough to show the property for prime powers $n = p^a$, and this is obvious.

Hint for 1-25. Start by assuming that $n \le m$. Find an integer $k_1 \le n$ such that $n + k_1$ is a power of 2. Then so is $m + k_1$. Consider then $n + (n + 2k_1)$ and $m + (n + 2k_1)$ and deduce that it is impossible to have $n < m$.

---------------------------- §*Chapter 2* ----------------------------

Solution to 2-3-i. We use multiplicativity and write the power series as

$$\sum_{a \ge 0} d(p^{2a}) z^a = \sum_{a \ge 0} (2a + 1) z^a = 2 \sum_{a \ge 0} (a + 1) z^a - \sum_{a \ge 0} z^a$$

$$= \frac{2}{(1 - z)^2} - \frac{1}{1 - z} = \frac{1 + z}{(1 - z)^2} = \frac{1 - z^2}{(1 - z)^3}.$$

The above implies that $\displaystyle\sum_{n \ge 1} d(n^2)/n^s = D(\mathbb{1}, s)^3 / D(\mathbb{1}, 2s)$.

Solution to 2-3-ii. We proceed again by multiplicativity and write

$$\sum_{a \ge 0} d(p^a)^2 z^a = \sum_{a \ge 0} (a + 1)^2 z^a = \sum_{a \ge 0} (a + 1)(a + 2) z^a - \sum_{a \ge 0} (a + 1) z^a$$

$$= \frac{2}{(1 - z)^3} - \frac{1}{(1 - z)^2} = \frac{1 + z}{(1 - z)^3} = \frac{1 - z^2}{(1 - z)^4}.$$

Therefore $\displaystyle\sum_{n \ge 1} d(n)^2/n^s = \mathcal{D}(\mathbb{1}, s)^4 / \mathcal{D}(\mathbb{1}, 2s)$.

Solution to 2-3-iii. It is enough to compare the Dirichlet series.

Solution to 2-6-ii. The function that associates $\sigma(n)^2$ to n is multiplicative. On prime powers $n = p^a$, we have $\sigma(p^a)^2 = \left(\dfrac{p^{a+1} - 1}{p - 1} \right)^2$, which is also valid for $a = 0$ giving $\sigma(1) = 1$. We now compute $g \star \sigma^2$ at the prime power p^a and find that

$$g \star \sigma^2(p^a) = \left(\frac{p^{a+1} - 1}{p - 1}\right)^2 - p\left(\frac{p^a - 1}{p - 1}\right)^2$$

$$= \frac{p^{2a+2} - 2p^{a+1} + 1 - p^{2a+1} + 2p^{a+1} - p}{(p - 1)^2}$$

$$= \frac{p^{2a+1} - 1}{p - 1} = \sigma(p^{2a}),$$

as required.

Solution to 2-11-ii.

```
d2 = direuler(p = 2, 1000, 1/(1-X^2), 1000);
d3 = direuler(p = 2, 1000, 1/(1-X^3), 1000);
dm6 = direuler(p = 2, 10000, 1-X^6, 1000);
dpowerful = dirmul(d2, dirmul(d3, dm6));
beg = 1;
for(n = 2, 1000,
  if(dpowerful[n] == 1,
     print("gap at ", n, ": ", n-beg);
     beg = n,));
```

Hint to 2-11-iii. Compute the gaps up to 500 000. The first gap of size 14 appears between $n = 30\,459\,375$ and $n + 14$. I have no example of a gap of size 34.

———————————— §*Chapter* 3 ————————————

Hint for 3-3. Consider $f(n) = 1/\log(n + 1)^2$.

Hint for 3-4. Compare with an integral.

Hint for 3-5. The readers may want to write, for any $n \geq 1$,

$$\frac{1}{n^s} = s \int_n^\infty \frac{dt}{t^{s+1}} \, .$$

(A technique known as *summation by parts*)

Hint for 3-6. Compare to an integral.

Hint for 3-8-i. The series $L(1)$ and $L'(1)$ are *alternating*.

Hint for 3-8-ii. The readers may first prove the identity $\zeta(s) = L(s)/(2^{1-s} - 1)$.

Hint for 3-9-ii. Differentiate the previous expression.

Hint for 3-9-iii. We have

$$\int_1^\infty \frac{\{t\} - 1/2}{t^{s+1}} dt = \sum_{n \geq 1} \int_n^{n+1} \frac{t - n - 1/2}{t^{s+1}} dt = \sum_{n \geq 1} \int_0^1 \frac{u - 1/2}{(u + n)^{s+1}} dt$$

$$= \sum_{n \geq 1} \frac{s + 1}{2} \int_0^1 \frac{u^2 - u}{(u + n)^{s+2}} dt < 0 \, .$$

Hint for 3-9-iv. Show first that $\dfrac{1}{s-1} + \dfrac{\zeta'(s)}{\zeta(s)} > \dfrac{1}{2s(s-1)\zeta(s)}$.

Solution to 3-13. $\displaystyle\sum_{n\geq 1} 2^{\omega(n)}\lambda(n)/n^s = \zeta(2s)/\zeta^2(s)$.

Hint for 3-14-ii. Since these two functions are multiplicative, it is enough to compute the Euler factors of their Dirichlet series.

Hint for 3-15-i. The characteristic function of the irregular numbers is given by $(1 - \lambda(n))/2$, where $\lambda(n)$ is the Liouville function.

Solution to 3-20-i. The function $\mathbb{1}_{n=1}$ is the only such function.

Solution to 3-20-ii. It is $\mu \cdot f$. We can prove this by using Dirichlet series. The local factor of the Dirichlet series for f is $1 + \dfrac{f(p)}{p^s - f(p)} = \dfrac{p^s}{p^s - f(p)}$, hence its inverse is $1 - f(p)/p^s$.

─────────────── §*Chapter 4* ───────────────

Solution to 4-3. For $n = 48$, we have that $d(48) = 10$ and for $n = 60$ we have $d(60) = 12$, so the smallest n such that $d(n) = 11$ is 1024.

Hint for 4-5. We have $2^{\omega(n)} \leq d(n)$.

Hint for 4-7. For the first part, simply note that $\sigma(n)/n = \sum_{d\mid n} 1/d$. For the second one, recall that $\sigma(n)/n$ is multiplicative, find an explicit expression and consider $\log(\sigma(n)/n)$ as in the argument of Theorem 4.1.

Hint for 4-11-iii. Proceed by recursion over q.

Hint for 4-11-iv. For any integer n, introduce the greatest divisor of n in $\mathcal{D}$, say $k_D(n)$.

Solution to 4-12-i. Use the Dirichlet series $\zeta(s)^r$ and inspect its local factors. We have $d_r(p^a) = \dbinom{r-1+a}{r-1}$. The readers may also look at the generalization of the Binomial theorem recalled in Exer. 19-4.

Solution to 4-12-ii. In order to show "complete sub-multiplicativity", it is enough to consider prime powers, i.e. we aim at proving that $d_r(p^{u+v}) \leq d_r(p^u)d_r(p^v)$. Let $A(w)$ be a set of r-tuples of integers $a_i \geq 0$ such that $a_1 + a_2 + \cdots + a_r = w$. We wish to construct an injection from $A(u+v)$ to $A(u) \times A(v)$. Given $(a_1, \ldots, a_r) \in A(u+v)$ we define

$$a_1' = \min(a_1, u),\ a_2' = \min(a_2, u - a_1'),\ a_3' = \min(a_3, u - a_1' - a_2'),\ \ldots$$
$$\ldots,\ a_r' = \min(a_r, u - a_1' - a_2' - \cdots - a_{r-1}').$$

We have $a_1' + a_2' + \cdots + a_r' \leq u$. If the inequality is strict, then $a_1' < u$, hence $a_1 = a_1'$; $a_2' < u - a_1$ and hence $a_2' = a_2$. In general, we obtain $a_r' = a_r$. The sum of $(a_i')_{1 \leq i \leq r}$ is equal to $u + v \geq u$, which leads to contradiction. Thus $(a_1', a_2', \ldots, a_r') \in A(u)$ and $(a_1'', a_2'', \ldots, a_r'') \in A(v)$, where $a_i'' = a_i - a_i'$. The application that associates $(a_1', a_2', \ldots, a_r')$ and $(a_1'', a_2'', \ldots, a_r'')$ to $(a_1, \ldots, a_r)$ is injective since $a_i = a_i' + a_i''$.

The remainder of the proof is classical:

$$\sum_{n \leq N} \frac{d_r(n)^2}{n} = \sum_{n \leq N} \frac{d_r(n)}{n} \sum_{n_1 n_2 \cdots n_r = n} 1$$

$$\leq \sum_{n_1 n_2 \dots n_r \leq N} \frac{d_r(n_1) d_r(n_2) \cdots d_r(n_r)}{n_1 n_2 \cdots n_r}$$

$$\leq \left(\sum_{n \leq N} d_r(n)/n \right)^r \leq \left(\sum_{n \leq N} 1/n \right)^{r^2}.$$

Solution to 4-12-iii. We prove the stated inequality for $n = p^a$. We set $k = r^\ell$. We need to find the maximum of

$$x(a, r, \ell) = \frac{(a + r - 1)!^\ell a! (k - 1)!}{(r - 1)!^\ell a!^\ell (a + k - 1)!} = \frac{(k - 1)!}{(r - 1)!^\ell} \frac{(a + r - 1)!^\ell}{a!^{\ell - 1}(a + k - 1)!}$$

Note $x(1, r, \ell) = 1$. We further check that

$$\frac{x(a + 1, r, \ell)}{x(a, r, \ell)} = \left(1 + \frac{r - 1}{a + 1} \right)^\ell \frac{a + 1}{a + k} = \frac{(1 + u)^\ell}{1 + \frac{k-1}{r-1} u} = q(u)$$

with $u = (r - 1)/(a + 1)$. The logarithmic derivative with respect to u of the latter expression is

$$\frac{(\ell - 1)(k - 1)u + \ell(r - 1) - k + 1}{(1 + u)(r - 1 + (k - 1)u)}.$$

Since $\ell \geq 1$, the sign of this expression is non-positive at $u = 0$, and it may become non-negative at $u = (r - 1)/2$, but in any case, $q(u)$ takes its maximum on the side of the considered interval. At $u = 0$, we have $q(u) = 1$ while at $a = 1$, its value is

$$\left(\frac{r + 1}{2} \right)^\ell \frac{2}{r^\ell + 1}.$$

This expression is non-increasing in r (compute its derivative). When $r = 2$, it is equal to 1 when $\ell = 1$ and is ≤ 1 for larger ℓ's. As a conclusion, $q(u) \leq 1$, which implies that the $x(a, r, \ell)$ is non-increasing in a, as required.

Hint for 4-13-i. First show that $\sum_{n \leq x} d_r(n) \leq x \sum_{n \leq x} d_{r-1}(n)/n$. Continue with an integration by parts to reduce all of that to an upper bound for $\sum_{n \leq x} d_{r-1}(n)$.

Hint for 4-13-iii. The readers may start by proving that

$$\sum_{n \leq x} d_r(n)^2 \leq x \sum_{m_1, m_2 \leq x} \frac{d_{r-1}(m_1) d_{r-1}(m_2)}{[m_1, m_2]},$$

then use the preceding question and set $m_2 = dn_2$.

Hint for 4-13-iv. The readers may first want to prove that

$$\sum_{m_1 \leq x} d_{r-1}(m_1) d(m_1) \leq \sum_{n_1 \leq x} d_{r-1}(n_1) \sum_{n_2 \leq x/n_1} d_{r-1}(n_2).$$

Hint for 4-14. Define the auxiliary function $G(n) = \displaystyle\sum_{\substack{\delta \mid n \\ \delta \leq n^{1/k}}} d^{ks}(\delta)$ and show that this

function is *sub-multiplicative*, i.e. that $G(n_1 n_2) \geq G(n_1)G(n_2)$ for $\gcd(n_1, n_2) = 1$. Then we prove that $d^s(p^\nu) \leq d^{ks}(p^{\lfloor \nu/k \rfloor}) \leq G(p^\nu)$ when $\nu \geq k$ and that $d^s((p_1 \cdots p_r)^\nu) \leq k^{ks} G((p_1 \cdots p_r)^\nu)$ when $\nu < k$, since for $r < k$ we have $d^s(n) \leq k^{ks}$. Similarly, for $r \geq k$ we have $d^s(n) \leq k^{ks} d^{ks}((p_1 \cdots p_{[r/k]})^\nu)$ (on arranging $p_1 < p_2 < \ldots < p_r$ we have $(p_1 \cdots p_{[r/k]})^\nu \leq n^{1/k}$).

Solution to 4-15-iii. We can take

$$r_n(1,1) = \frac{1}{1 + g(p)}, \quad r_n(1,p) = \frac{g(p)}{1 + g(p)} - \frac{f(p)}{1 + f(p)}, \quad r_n(p,p) = \frac{f(p)}{1 + f(p)} \ .$$

Hint for 4-15-v. The readers may note that $\dfrac{f(p)}{1 + f(p)} = 1 - \dfrac{1}{1 + f(p)}$.

———————————— §*Chapter 5* ————————————

Hint for 5-2. Consider the product of all the elements in the image set of h.

Hint for 5-3. Let us write $b - 1 = 3q + r$ with q and r integers and $r \in \{0, 1, 2\}$. The remark that $n^r \equiv 1 \pmod{b}$ should lead to a proof!

Hint for 5-5-i. Pair an integer $m \leq b - 1$ with the integer $m^* \leq b - 1$ that is such that $mm^* \equiv 1 \pmod{b}$ (hence m^* is a representant of the class of the inverse of m). Prove that there are only two cases when $m^* = m$ and conclude.

Hint for 5-5-iii. Pair integers m and n between 1 and $b - 1$ if we have $mn \equiv a \pmod{b}$ and proceed otherwise as above.

Hint for 5-8-i. Show that the R.H.S. can be brought modulo b, to the form $(-1)^x \left(\frac{b-1}{2}\right)!$ for some integer x.

Hint for 5-9-ii. The readers may want to write $k(k + 1) \equiv k(k + k\overline{k})[b]$.

Hint for 5-9-iii. The integer k from $\{1, \ldots, p - 2\}$ is to be counted in N if and only if $\left(1 + \left(\dfrac{k}{b}\right)\right)\left(1 + \left(\dfrac{k+1}{b}\right)\right) = 4$, while this quantity takes the value 0 otherwise.

Hint for 5-10-iv. Distinguish according to whether $d \leq y$ or not, with y a suitable parameter. When $d > y$, set $m = [x/d]$ and sum over m.

Hint for 5-12. We know by Eq. (5.7) that -1 is a square modulo p, i.e. $-1 \equiv a^2 \pmod{p}$. Use Thue's lemma for a and bound $u^2 + v^2$ from above to conclude.

Hint for 5-14. Let δ be a square root in $\mathbb{C}$ of $-D$. Notice that $a^2 + Db^2 = (a + \delta b)(a - \delta b)$ and that $(a + \delta b)(a' + \delta b') = aa' - Dbb' + \delta(ab' + a'b)$.

———————————— §*Chapter 6* ————————————

Hint for 6-1. Use the inverse of $\mathbb{1}$ for the unitary convolution found in Exer. 1-21.

Hint for 6-3-ii. The readers may use the decomposition $\dfrac{1}{X + 1} = \dfrac{1}{X - 1} - \dfrac{2}{X^2 - 1}$.

Hint for 6-5. Let d' be the product of the primes $\leq x$ that are coprime to d. Study the quantity $\sum_{n \leq x, (n, d')=1} 1$ by using

$$\sum_{\substack{\ell \mid n \\ \ell \mid d'}} \mu(\ell) = \begin{cases} 1 & \text{when } \gcd(n, d') = 1, \\ 0 & \text{otherwise.} \end{cases}$$

Notice that we may assume that x is an integer.

Hint for 6-7. The readers may first show the auxiliary identity:

$$\int_1^x \left[\frac{x}{t}\right] \frac{M(t)dt}{t} = \log x .$$

Hint for 6-8-i. The easiest way is surely to use Dirichlet series.

Hint for 6-8-iii. The initial function f may then be recovered by the formula $f(x) = \sum_{n \le x} g^{(-\star)}(n)F(x/n)$ where $g^{(-\star)}$ is the convolution inverse of g. This inverse exists thanks to the condition $g(1) \ne 0$, as per Exer. 1-15.

Hint for 6-9-i. This product is simply $2\delta_{n=2} - \delta_{n=1}$. The convolution sum can be directly handled; the readers may also prove this identity first computing the Dirichlet series of g and applying Theorem 2.1.

Solution to 6-10. Select $H(t) = 1 - 1/t$, $g(n) = 1/n$ and $k(u) = uh(u)$ in Theorem 6.3.

Hint for 6-11-i. The readers may want to select $f(n) = \mu(n)1_{(n,7)=1}$, $g(n) = 1_{(n,7)=1}$, then $h(t) = 1/t$ and $H(t) = \frac{6}{7}t \log t - a \log t - bt$ for some properly chosen parameters a and b. Inspiration can also be obtained by looking at Exer. 7-4 of the next chapter!

Hint for 6-11-ii. The readers may want to select $f(n) = \mu(n)1_{(n,7)=1}$, $g(n) = 1_{(n,7)=1}$, then $h(t) = (\log t)/t$ and $H(t) = at \log^2 t + bt \log t + ct + d \log^2 t + f \log t$ for some properly chosen parameters a, b, c, d and f.

———————————— §*Chapter 7* ————————————

Hint for 7-1. The readers may start from $\dfrac{1}{n} = \displaystyle\int_n^{n+1} \dfrac{dt}{[t]}$ and $\log \dfrac{n+1}{n} = \displaystyle\int_n^{n+1} \dfrac{dt}{t}$.

Hint for 7-4-i. The readers may notice that the sum under examination is the difference between $\sum_{n \le x} 1/n$ and $\sum_{m \le x/2} 1/(2m)$.

Hint for 7-4-ii. The readers may handle the coprimality condition via the identity

$$1_{(n,q)=1} = \sum_{\substack{d \mid n, \\ d \mid q}} \mu(d) \tag{29.3}$$

that they should first establish. When using this identity, people from the trade say "we handle the coprimality condition by the Möbius function".

Hint for 7-5-ii. We have $2 \sum_{n \le x} n^2 = B_2([x])$ and $|B_2(x) - B_2([x])|$ may be bounded above that the Mean Value Theorem. Note also that the proposed approximation trivially holds when $x \in [0, 2]$.

Hint for 7-6-i. Use $2(x - n) = B_2(x - n + 1) - B_2(x - n)$.

Hint for 7-6-iv. One inequality is obvious while for the other one, notice that $(2\{t\} - 1)t + \{t\} - \{t\}^2 - t = (\{t\} - 1)(2t - \{t\})$.

Hint for 7-7. The case $N < d$ is easy. Let $N \geq d$ and b be the integer from $[1, d]$ which is congruent to N modulo d. We find that d times the sum we consider equals

$$\sum_{1 \leq a \leq b} \left(\frac{N-a}{d} - \frac{b-a}{d} + 1\right)^2 + \sum_{b+1 \leq a \leq d} \left(\frac{N-a}{d} - \frac{b-a}{d}\right)^2.$$

——————————————— §*Chapter 8* ———————————————

Hint for 8-1. The function $\mathbb{1}_{(n,6)=1} f_0(n)$ is multiplicative and close to Id.

Hint for 8-4-i. Use the convolution method for $f(n) = \dfrac{2^{\omega(n)} n}{\varphi(n)} = (d \star g)(n)$.

Hint for 8-4-ii. Use the convolution method for $f(n) = \mu^2(n) \dfrac{2^{\omega(n)} n}{\varphi(n)} = (d \star g)(n)$.

——————————————— §*Chapter 9* ———————————————

Solution to 9-1-i. We first find that

$$1 + \frac{4}{\sqrt{p}(p-1)} = \left(1 - \frac{1}{p^{3/2}}\right)^{-4} \left(1 - \frac{1}{p^{3/2}}\right)^4 \left(1 + \frac{4}{\sqrt{p}(p-1)}\right).$$

The computations being cumbersome, it is better to use some software help. We asked Pari/GP to expand (here q is a symbol for $\sqrt{p}$)

```
( (1-1/q^3)^4*(1+4/q/(q^2-1)) -1 )*q^13*(q^2-1)
```

and deduce from the answer that

$$\left(1 - \frac{1}{p^{3/2}}\right)^4 \left(1 + \frac{4}{\sqrt{p}(p-1)}\right)$$
$$= 1 + \frac{4p^5 - 10p^{9/2} - 6p^{7/2} + 20p^3 + 4p^2 - 15p^{3/2} - \sqrt{p} + 4}{p^{13/2}(p-1)}.$$

Hint for 9-1-iii. The readers may use the script

```
{g(p) = (4*p^5 - 10*p^(9/2) - 6*p^(7/2) + 20*p^3 + 4*p^2
         - 15*p^(3/2) - p^(1/2) + 4) / (p^(13/2)*(p-1);}
bigP = 1000000;
Aux = prodeuler( p = 2, bigP, 1 + g(p))*exp(8/3/bigP^(3/2));
zeta(3/2)^4 * Aux
```

Pari/GP computes $\zeta(3/2)$ with a many decimal accuracy.

Hint for 9-2. The readers may use logarithmic differentiation and Möbius inversion from Sect. 6.1 to establish this fact.

Hint for 9-3. Recall (1.11).

Hint for 9-4. The coefficients s_F, when $F = 1 - z - z^2$, can be expressed in terms of the Fibonacci numbers. The readers may then notice that the corresponding sequence b_F is made of integers.

Hint for 9-6. The readers may first notice that

$$(1/\rho)^\delta = -a_1(1/\rho)^{\delta-1} - a_2(1/\rho)^{\delta-2} - \ldots - a_\delta .$$

Hint for 9-10. The readers may want to adapt the above script.

Hint for 9-11. The readers may follow the argument of Theorem 9.4, though with $1/p^s$ rather than $1/p$ and let J go to infinity.

Hint for 9-17-i. The readers may compare the first value with $\zeta'(2)/\zeta(2)$ and proceed similarly with the second one.

Hint for 9-18-i. Concerning the last part, the readers may show that c_k is an integer that satisfies $|c_k| \le 2/\sqrt{3}$ by expressing c_k in terms of the roots of $1 - X + X^2$.

Hint for 9-18-ii. The only missing ingredient we need is a way to compute $\zeta'(j)$, and here is how to compute $\zeta'(2)$:

```
R(zetaderiv(1,2))
```

The readers may improve this script by keeping the values of the sequence $(b(j))$ in a list, or, more elaborately, to use the *decorator* `'#@cached_function'`, see a Sage manual for further explanations. With $\beta = 2$, $P = 20000$, $J = 70$ and R being the real interval field with 1000 bits, the readers should be able to get more than 250 decimal digits in precision and get the value

$$0.60838\ 17178\ 63324\ 72268\ 38345 \ldots$$

———————————— §*Chapter* 10 ————————————

Hint for 10-1. Let $F(X) = 1 - X^3 - X^4 - X^5$ and $G(X) = 1 - X^3$. The polynomial G is already in Witt expansion. For F, prove that $s_F(k) = s_F(k-3) - s_F(k-4) - s_F(k-5)$ while $s_F(1) = s_F(2) = 0$, $s_F(3) = 3$, $s_F(4) = -4$ and $s_F(5) = -5$.

Hint for 10-2-ii. Apply Lemma 10.4 to $\varphi(n)/n^2$.

Hint for 10-4-i. When n is a prime number, say p, we have $\dfrac{1}{p-1} = \dfrac{1}{p} \dfrac{1}{(1-1/p)}$. In the general case, n is a product of the form $p_1 p_2 \cdots p_k$ and we find that $\dfrac{1}{\varphi(n)} = \dfrac{1}{p_1 - 1} \dfrac{1}{p_2 - 1} \cdots \dfrac{1}{p_k - 1}$.

Hint for 10-8. A quick look at Exer. 9-17 discloses that $C = \dfrac{6}{\pi^2}(\gamma - 2\zeta'(2)/\zeta(2))$.

Hint for 10-10-iii. Indeed $\ell \le x/D^2$. Once ℓ is fixed, show that there is at most one integer d in the required interval.

$$\text{──────────────── } \S Chapter\ 11 \text{ ────────────────}$$

Hint for 11-3-ii. We recall that $\int_0^\infty t(\log t)e^{-t}\,dt = \Gamma'(2) = \Gamma(2)\psi(2)$ and that $\psi(2) = 1 - \gamma$ where exceptionally we have denoted by ψ the *Digamma function*.

Hint for 11-5. A summation by parts gives us that our sum, say S, is given by

$$
\begin{aligned}
S &= \int_1^x \sum_{n\le t} d(n)\frac{dt}{t^2} + \frac{\sum_{n\le x} d(n)}{x} \\
&= \int_1^x \sum_{n\le t}(t\log t + (2\gamma - 1)t)\frac{dt}{t^2} + \int_1^\infty \left(\sum_{n\le t} d(n) - t\log t - (2\gamma - 1)t\right)\frac{dt}{t^2} \\
&\quad - \int_x^\infty O^*(\sqrt{t} + \tfrac{1}{2})\,\frac{dt}{t^2} + \frac{x\log x + (2\gamma - 1)x}{x} + O^*\left(\frac{1}{\sqrt{x}} + \frac{1}{2x}\right)
\end{aligned}
$$

and the result follows readily.

$$\text{──────────────── } \S Chapter\ 12 \text{ ────────────────}$$

Hint for 12-3. The readers may start by proving that $[x/a_i] = [[x]/a_i]$.

Hint for 12-4-iii. One may proceed by contradiction.

Solution to 12-5. We see that

$$
\begin{aligned}
\sum_{n\le x} \Lambda(n) &= \sum_{d\le x} \mu(d)\left(\frac{x}{d}\log\frac{x}{d} - \frac{x}{d} + O(\log(2x/d))\right) \\
&= x\sum_{d\le x} \frac{\mu(d)}{d}\log\frac{x}{d} - x\sum_{d\le x} \frac{\mu(d)}{d} + O(x)
\end{aligned}
$$

and the estimate follows.

Hint for 12-8-iii. The Dirichlet Hyperbola Principle (see Chap. 11) gives us

$$
\begin{aligned}
S &= \sum_{d\le x/\lambda} f(d)[x/d] = \sum_{d\le D} f(d) \sum_{\ell \le x/d} 1 \\
&\quad + \sum_{\ell \le L} \sum_{d\le \min(x/\lambda, x/\ell)} f(d) - \sum_{\ell \le L} 1 \sum_{d\le D} f(d)\,.
\end{aligned}
$$

This leads us to

$$
S = x(M(f)\log x + C(f)) - M(f)x\left(1 - \gamma + \sum_{\ell \le \lambda}\left(\frac{1}{\ell} - \frac{1}{\lambda}\right)\right) + O(R)\,,
$$

where R is the error term.

Hint for 12-8-vi. See Sect. 10.6.

Hint for 12-8-vii. The readers may have a look at Eq. (10.5).

Hint for 12-8-ix. We remark that

$$\sum_{n \le x} \Lambda(n) = \sum_{n \le x} \left(\sum_{d \mid n} \mu(d) \left(\log \frac{x}{d} + \gamma \right) \right) - \log x - \gamma \ .$$

Hint for 12-9. The readers may restrict these series to prime summands.

Hint for 12-11-i. The readers may simply expand the square and compute each contribution.

Hint for 12-12-ii. The readers may first notice that

$$\sum_{n \le x} (f_{p_1} f_{p_2} f_{p_3} f_{p_4})(n) = \sum_{\delta \mid p_1 p_2 p_3 p_4} (f_{p_1} f_{p_2} f_{p_3} f_{p_4})(\delta) \sum_{\substack{n \le x, \\ (n, p_1 p_2 p_3 p_4) = \delta}} 1$$

and then use the previous question.

Hint for 12-14-i. Proceed as in the proof of Theorem 4.1 and first show that the number of the prime divisors of k that are larger than some parameter y is at most $(\log k)/\log y$.

Hint for 12-15-i. The readers may use

$$\sum_{2 < p \le x} \frac{\log p}{p - 2} \le \sum_{n \le x} \frac{\Lambda(n)}{n} - \frac{\log 2}{2} - \frac{\log 2}{4} + \sum_{2 < p \le x} \frac{2 \log p}{p(p - 2)} \ .$$

─────────────── §*Chapter* 13 ───────────────

Hint for 13-2. The readers may use the inequality $\sum_{p \le D} (\log p)/p \le \log D$, proved in (12.6) and which is valid when $D \ge 1$. We thus have

$$- \sum_{d \le D} \frac{\mu^2(d) g(d)}{d} + \prod_{p \le D} \left(1 + \frac{g(p)}{p} \right) = \sum_{\substack{n > D, \\ P^+(n) \le D}} \frac{\mu^2(n) g(n)}{n} \ ,$$

where $P^+(n)$ is the largest prime factor of n. It is then possible to bound above the right-hand side by $\displaystyle\sum_{P^+(n) \le D} \frac{\mu^2(n) g(n)}{n} \frac{\log n}{\log D}$.

Hint for 13-3-iii. The readers may remark (see Exer. 20-11) that $\sum_{n \ge 1} n^{-2} = \pi^2/6$ and remove the contribution of the first non-prime integers n.

Hint for 13-4-i. Write $S = \sum_{n \le x} (\mathbb{1} \star g)(n)$ in the form

$$S = \sum_{m \le x} g(m) [x/m] = x \sum_{m \le x} \frac{g(m)}{m} - \sum_{m \le x} g(m) \{x/m\} \ .$$

Hint for 13-5-i. Notice that $x/n = [x/n] + \{x/n\}$ and $[x/n] = \sum_{m \le x/n} 1$.

Hint for 13-5-iv. We can get the upper bound for $\sum_{n \le x} 1/2^{\omega(n)}$ from the one for $\sum_{n \le x} 1/(n 2^{\omega(n)})$ and then apply the Levin–Faĭnleĭb theorem.

Hint for 13-8-i. Use multiplicativity.

Solution to 13-8-iii. A summation by parts gives us

$$\sum_{n \le N} \frac{1}{\varphi(n)} = \sum_{n \le N} \frac{n}{\varphi(n)} \left(\int_n^N \frac{dt}{t^2} + \frac{1}{N} \right) = \int_1^N \sum_{n \le t} \frac{n}{\varphi(n)} \frac{dt}{t^2} + O(1)$$

$$= C_1 \log t + \int_1^\infty \left(\sum_{n \le t} \frac{n}{\varphi(n)} - C_1 t \right) \frac{dt}{t^2} - \int_N^\infty \left(\sum_{n \le t} \frac{n}{\varphi(n)} - C_1 t \right) \frac{dt}{t^2}, + O(1)$$

where $C_1 = \zeta(2)\zeta(3)/\zeta(6)$, whence the result.

Solution to 13-8-iv. We note that $\log d = \sum_{m|d} \Lambda(m)$ from which we deduce that

$$\sum_{d \ge 1} \frac{\mu^2(d) \log d}{d\varphi(d)} = \sum_{p \ge 2} \frac{\log p}{p(p-1)} \sum_{\substack{\ell \ge 1, \\ (\ell, p) = 1}} \frac{\mu^2(\ell)}{\ell\varphi(\ell)}$$

$$= \sum_{p \ge 2} \frac{\log p}{p(p-1)} \frac{1}{1 + \frac{1}{p(p-1)}} \prod_{p' \ge 2} \left(1 + \frac{1}{p'(p'-1)} \right)$$

as claimed. On using the estimate $\sum_{n \le N} 1/n = \log N + \gamma + O(1/N^{1/3})$ that is valid for any real number $N > 0$, we conclude that

$$\sum_{n \le N} \frac{1}{\varphi(n)} = \sum_{d \ge 1} \frac{\mu^2(d)}{\varphi(d)} \sum_{d | n \le N} \frac{1}{n}$$

$$= \sum_{d \ge 1} \frac{\mu^2(d)}{d\varphi(d)} \left(\log(N/d) + \gamma - O\big((d/N)^{1/3}\big) \right)$$

$$= \frac{\zeta(2)\zeta(3)}{\zeta(6)} (\log N + \gamma) - \sum_{d \ge 1} \frac{\mu^2(d) \log d}{d\varphi(d)} + O(N^{1/3}) .$$

Solution to 13-8-v. Indeed, let us consider the case when n is a prime number and call, respectively, $S(N)$ and $S_0(N)$ our sum till N as well as its smooth approximation. We let $S(N) = S_0(N) + R(N)$. We get

$$\frac{1}{N-1} = S(N) - S(N-1)$$

$$= S_0(N) - S_0(N-1) + R(N) - R(N-1)$$

$$= \frac{\zeta(2)\zeta(3)}{\zeta(6)N} + R(N) - R(N-1) + O(1/N^2) .$$

Since $\zeta(2)\zeta(3) \ne \zeta(6)$, this gives us, for large N (and restricted to prime values): $|R(N) - R(N-1)| \gg 1/N$. Therefore $|R(N)|$ or $|R(N-1)|$ is $\gg 1/N$.

Hint for 13-9. The readers may want to compare this series with $\sum_{d \le D} 1/\varphi(d)$.

Hint for 13-11-i. The readers may imitate the start of the usual Levin–Faĭnleĭb argument and bound

$$\sum_{\substack{p^k \le D/m, \\ (p,m)=1}} g(p^k) \log(p)$$

from below by $\kappa(D/m) - C\log m$.

——————————— §*Chapter* 14 ———————————

Solution to 14-2. We can assume that $x \ge 7$ and verify the other cases by hands. Let $N \ge 7$ be the integer part of x. Write $x = N + \xi$ for a certain ξ in the interval $[0, 1)$ (well, ξ is the fractional part of x). The inequality $x/2 < p \le x$ is stronger than $N/2 < p \le N$. For even N, say, $N = 2\ell$, the inequality $N/2 < p \le N$ becomes $\ell < p \le 2\ell$. But since $\ell \ge 2$, then 2ℓ is not a prime number. The inequality transforms into $(\ell - 1) + 1 < p \le 2(\ell - 1) + 1$, hence the corresponding product is $\le 4^{\ell-1} \le 2^x$.

For odd N, say, $N = 2\ell + 1$, the inequality $N/2 < p \le N$ becomes $\ell < p \le 2\ell + 1$. Hence

$$\prod_{x/2<p\le x} p \le (\ell + 1) \prod_{\ell+1<p\le 2\ell+1} p \le (\ell + 1)4^\ell = \frac{N+1}{4}2^N \le (7/3)^N \le (7/3)^x .$$

Hint for 14-6-i. Write $\binom{3n/2}{n/2} = \dfrac{(n+1)\dots\frac{3n}{2}}{(n/2)!}$ and check that $2p$ divides this binomial coefficient when p is a prime from the interval $(n/2, 3n/4]$.

——————————— §*Chapter* 15 ———————————

Hint for 15-2-iii. Build a partition of $(\mathbb{Z}/q\mathbb{Z})^\times$ with several sets $H(x)$.

Solution to 15-6.

```
chi3(n)=[0,1,-1][1+n%3]
```

Hint for 15-12-i. Show that, if the parameter A is chosen large enough, we have, when x is large enough in terms of A,

$$\sum_{\substack{x/A<p\le x, \\ p\equiv 1[3]}} \frac{\log p}{x/A} \ge \sum_{\substack{x/A<p\le x, \\ p\equiv 1[3]}} \frac{\log p}{p} \ge \frac{1}{4}\log A .$$

This proves the lower bound. For the upper bound, start similarly and cover the interval $[1, x]$ by intervals of the shape $(y/A, y]$ (save maybe some initial integers).

Hint for 15-14-ii. Follow the argument that led to (12.9).

Hint for 15-15. The readers may notice that the integer $4 \cdot m! + 3$ has at least one prime factor $\equiv 3 \pmod 4$.

Hint for 15-16. The readers may remark that the integer $6 \cdot m! + 5$ has at least one prime factor $\equiv 5 \pmod 6$.

Hint for 15-17-i. The readers should first prove an analogue to Lemma 15.3 and then follow the proof of Lemma 15.3. The non-vanishing of $L(1, \chi)$ is a consequence of Theorem 5.5.

─────────────── §*Chapter* 16 ───────────────

Hint for 16-1. Recall that n is a sum of two squares if and only if the exponents of the primes congruent to 3 modulo 4 that divide n are even.

─────────────── §*Chapter* 17 ───────────────

Hint for 17-1-ii. A recursion over r may be an idea.

Hint for 17-2-i. Decompose $G(z)$ in products of linear polynomials and use, when $\beta \neq 0$, that we have $1/(z - \beta) = -(1/\beta) \sum_{\ell \geq 0}(z/\beta)^\ell$. Notice also that the product of the roots of G has absolute value $|G(0)|$ and that $B_0 \geq 1$.

Hint for 17-2-ii. The readers may start with the case $r = 1$.

Hint for 17-2-iv. Have a look at Exer. 4-12.

Hint for 17-3. Expand the log of both sides in power series.

Hint for 17-4. The readers may check that $M(1, 0) = M(0, 1) = 1$, that $M(0, m) = M(m, 0) = 0$ when $m \geq 2$, and then notice that $M(m_1, m_2) \leq 2^{m_1 + m_2}$.

Hint for 17-5. The readers may consider $\zeta(s)H(s)$ and apply (17.2) with $k = 3$.

Hint for 17-6-i. Use the Levin–Faĭnleĭb theorem (i.e. Theorem 13.2) together with Theorem 15.3.

Hint for 17-6-ii. The readers may notice that $1 - \chi_3(p) = 2$ when $p \equiv 2[3]$ and that $1 - \chi_3(p) = 0$ when $p \equiv 1[3]$.

Hint for 17-6-iii. The usual theory ensures us that the coefficients v_ℓ do exist (alternatively, Exer. 17-2 reproves of this claim). On multiplying both sides by $2 - z + 2z^2$ and using the unicity of the development in power series, show that we have $2v_{\ell+2} - v_{\ell+1} + 2v_\ell = 0$ when $\ell \geq 1$.

Hint for 17-6-iv. $\texttt{Ustar[}\ell\texttt{/d][k/d]}/2^{k/d-\ell/d-1}$ is the k-th coefficient of $f(z^d)^{\ell/d}$ when $d|k$.

Solution to 17-6-v. The choice $J = 30$ and $P = 1000$ yields more than fifty digits and here are the first thirty ones:

$$\prod_{p \geq 3} C_p = 1.01295\,77929\,52369\,23346\,20100\,51487\,\ldots$$

Hint for 17-6-vi. Note first that, with $F^*(z) = -(5-8z+4z^2)z^2$ and $G^*(z) = 4(1-z)^3$, we have $B_p^2 = 1 - F^*(1/p)/G^*(1/p)$. We set $f^* = F^*/G^*$ and take for χ the function $\mathbb{1}$. We further set $(f^*(z))^r = \sum_{\ell \geq 0} v_\ell^*(r)z^\ell$. The readers may first determine explicitly $(v_\ell^*(1))$ by developing $1/(1 - z)^3$.

Solution to 17-6-vi. Here are the first thirty digits, obtained with $P = 10000$ and $J = 40$:

$$\prod_{p \geq 3} B_p^2 = 0.89413\,25164\,25882\,19817\,36476\,36342\,\ldots$$

Solution to 17-6. The final answer concerning the value of A is

$$A = 2.29273\,36791\,24018\,81682\,07097\,77426\,09177\,63688\,14399\,21551\,\ldots$$

Solution to 17-8. Let us denote by $(p_{i,j})$ the primes according to their location in the sought array of numbers and by N the common sum. We have ten unknowns and eight linear equations: however the sum of the three lines is also the sum of the three columns, so only seven of these equations are independent. In practice, we ignore the condition restraining the sum of the third line. We express all the variables in terms of N, $p_{1,1} = p$ and $p_{1,2} = p'$. In order to do so, we consider the initial seven-dimensional vector V whose components are $(p_{1,3}, p_{2,1}, p_{2,2}, p_{2,3}, p_{3,1}, p_{3,2}, p_{3,3})$, the final seven-dimensional vector $F = (N - p - p', N, N - p, N - p', N, N - p, N)$ and the matrix M that expresses the conditions: the first line of the matrix to express the fact that $p_{1,2} + p_{1,3} = N - p$, the second line for $p_{2,1} + p_{2,2} + p_{2,3} = N$ and so on, the last two lines being for expressing the diagonal constraints. The next Pari/GP script is built on these premises.

```
{ind(i , j) =
   if(i == 1, if( j == 3, return(1),),);
   if(i == 2, if( j == 1, return(2),
      if( j == 2, return(3),
      if( j == 3, return(4),))),);
   if(i == 3, if( j == 1, return(5),
      if( j == 2, return(6),
      if( j == 3, return(7),))),);}

{isint(x) = (type(x) == "t_INT");}

{magicsquares(bound, Verbose = 2)=
   my(V, F = vector(7), M = matrix(7, 7), Minv, totalnb = 0);
   \\ Build the matrix:
   \\ p13 = N - p - pprime
   M[1, ind(1, 3)] = 1;
   \\ p21 + p22 + p23 = N and ignore p31 + p32 + p33 = N
   M[2, ind(2,1)] = 1; M[2, ind(2,2)] = 1; M[2, ind(2,3)] = 1;
   \\ p21 + p31 = N - p
   M[3, ind(2, 1)] = 1; M[3, ind(3, 1)] = 1;
   \\ p22 + p32 = N - p'
   M[4, ind(2, 2)] = 1; M[4, ind(3, 2)] = 1;
   \\ p13 + p23 + p33 = N
   M[5, ind(1,3)] = 1; M[5, ind(2,3)] = 1; M[5, ind(3,3)] = 1;
   \\ p22 + p33 = N-p
   M[6, ind(2, 2)] = 1; M[6, ind(3, 3)] = 1;
   \\ p13 + p22 + p31 = N
   M[7, ind(1,3)] = 1; M[7, ind(2,2)] = 1; M[7, ind(3,1)] = 1;
   \\ Invert this matrix:
   Minv = M^(-1);
   forstep(N = 15, bound, 2,
```

```
   forprime(p = 3, N,
     forprime(pprime = 3, N - p,
       for(i = 1, 7, F[i] = N);
       F[1] += -p; F[3] += -p; F[6] += -p;
       F[1] += -pprime; F[4] += -pprime;
       V = Minv*(F~);
       \\ They should be integers:
       if(isint(V[1]) &&isint(V[2]) &&isint(V[3]) &&isint(V[4])
           && isint(V[5]) && isint(V[6]) && isint(V[7]),
         \\ They should be prime:
         if(isprime(V[1]) && isprime(V[2]) && isprime(V[3])
             && isprime(V[4]) && isprime(V[5]) && isprime(V[6])
             && isprime(V[7]),
           \\ They also should be distinct:
           if(length(Set(concat([p, pprime], V~))) == 9,
             totalnb += 1;
             if(Verbose >= 2, print(concat([p], V)));
                             print("Common sum is ", N),),),),)
       )));
   if(Verbose >= 1,
     print("There are ", totalnb, " 3x3 prime magic squares");
     print("  whose common sum is <= ", bound),),);
}
```

The call `magicsquares(2000, 1)` gives the answer: there are 1744 prime magic squares of size 3×3 whose common sum is under 2000. It seems that all the possible common sums are divisible by 3, a fact we leave to the readers!

─────────────── §*Chapter* 18 ───────────────

Hint for 18-1-ii. There are only two solutions.

Hint for 18-4. The readers should first check that this function is injective. To show the surjectivity one can use a cardinality argument or Bezout's theorem: there are two integers u and v such that $uq + vq' = 1$. The point $b/(qq')$ of $\Xi_{qq'}$ is equal to $\Psi(bv/q \bmod 1, bu/q' \bmod 1)$.

Hint for 18-6-ii. See Exer. 15-12.

Hint for 18-8-iii. Let n be some integer and let p be a prime divisor of $f(n)$. The readers may first notice that $(3n)^3 \equiv 1 \pmod{p}$ and then use Exer. 5-3. The conclusion comes from combining the two facts that $\mathcal{P}(f)$ is infinite and that any prime from this set is congruent to 1 modulo 3.

─────────────── §*Chapter* 19 ───────────────

Hint for 19-4-iii. Use $(n - 1)^{1-s} - n^{1-s} = n^{1-s}((1 - 1/n)^{1-s} - 1)$.

Hint for 19-6-ii. The readers may apply the logarithmic differential operator to both sides of the formula proved in Exer. 3-18.

———————————— §*Chapter* 20 ————————————

Solution to 20-2. For the first integral, when $x < 1$, shift the line of integration to the right-hand side. We encounter only two polar contributions, amounting to $2x - x^2/2$. When $x \geq 1$, we shift the line of integration to the left-hand side and find the contribution $\log x + 3/2$. For the second integral, when $x < 1$, shift the line of integration to the right-hand side. We encounter only two polar contributions, amounting to $x - x^2/2$. When $x \geq 1$, we shift the line of integration to the left-hand side and find the contribution $1/2$.

Hint for 20-5. Use an integral representation of arctan.

Hint for 20-7. It is given by $\check{w}_0(s) = \dfrac{3^{s+1} - 2^{s+2} + 1}{s(s + 1)}$.

Hint for 20-9. We recall that, when y is positive and β is any real number, we have

$$\int_{-\infty}^{\infty} e^{-yt^2 + i\beta t}\, dt = \sqrt{\frac{\pi}{y}}\, e^{-\frac{\beta^2}{4y}}.$$

Hint for 20-10-iii. The readers may show that $|B_3(v)|$ is bounded by $1/20$ over $[0, 1]$.

———————————— §*Chapter* 21 ————————————

Solution to 21-3-ii. By definition we have:

$$D(f_1, s) = \prod_{p \geq 2}\left(1 + \sum_{k \geq 1}\frac{1}{p^{ks}}\prod_{\ell \mid p^k}\frac{\ell + 1}{\ell + 2}\right).$$

Let us have a closer look at the local factors. Since both ℓ and p are prime numbers, we have $\ell = p$. It remains to insert this value above. This yields

$$\sum_{k \geq 1}\left(\frac{p + 1}{p + 2}\right)p^{-ks} = \frac{p + 1}{(p + 2)(p^s - 1)},$$

giving the Euler product we wanted.

Solution to 21-3-iii. The product $D(f_1, s)$ *looks like* $\prod_{p \geq 2}\left(1 + 1/(p^s - 1)\right)$ which is, indeed, $\zeta(s)$. We wish to take out this factor from our product and see what remains. Therefore we write

$$D(f_1, s) = \prod_{p \geq 2}\left(1 + \frac{p + 1}{(p + 2)(p^s - 1)}\right)$$

$$= \prod_{p \geq 2}\left(1 + \frac{p + 1}{(p + 2)(p^s - 1)}\right)\frac{p^s - 1}{p^s}\left(\frac{1}{1 - 1/p^s}\right) = K(s)\zeta(s)$$

as announced. The product given by $K(s)$ converges absolutely in the sense of Godement for such s for which the series $\sum 1/(p^s(p + 2))$ converges absolutely. By extending the summation to every integers, we see that this holds as soon as $\Re s > 0$.

Solution to 21-3-iv. Under the above constraints for σ and t, the function $K(s)$ (here $s = \sigma + it$) can be bounded above in absolute value via

$$\prod_{p \geq 2}\left(1 + \frac{1}{(p+2)p^{\sigma}}\right) \leq \prod_{p \geq 2}\left(1 + \frac{1}{(p+2)p^{1/4}}\right) \leq 2.5,$$

thanks to Pari/GP, see the small script:

```
P0 = 300000;
prodeuler(p = 2, P0, 1.0 + 1/(p+2)/p^(1/4)*exp(4/P0^(1/4)).
```

In this script, we have bounded the tail by using

$$\sum_{p > P_0} \log\left(1 + \frac{1}{(p+2)p^{1/4}}\right) \leq \int_{P_0}^{\infty} \frac{dt}{(t+2)t^{3/4}} \leq \frac{4}{P_0^{1/4}}.$$

On using Lemma 19.1, we prove easily that $|D(f_1, s)| \leq 3 \cdot (2 + 1/4) \cdot (4 + 4/3)$ when $s = 1/4 + it$ and $|t| \leq 2$.

Hint for 21-3-v. Use Lemma 20.1 with $a = 2$.

Solution to 21-3-vi. It is enough to write

$$\left|\int_{1/4-i\infty}^{1/4+i\infty} D(f_1, s)\frac{x^s \, ds}{s(s+1)}\right| \leq x^{1/4} \int_{1/4-i\infty}^{1/4+i\infty} |D(f_1, s)|\frac{|ds|}{|s(s+1)|}.$$

The last integral is finite. A more precise estimate is obtained via Lemma 19.1:

$$\left|\int_{1/4-i\infty}^{1/4+i\infty} D(f_1, s)\frac{x^s \, ds}{s(s+1)}\right| \leq 2x^{1/4}\left(\int_0^2 36\frac{dt}{1/4 \cdot 5/4} + \int_2^{\infty} 3 \cdot 24 t^{3/4} \log t \, \frac{dt}{t^2}\right),$$

$$\leq 2734 \cdot x^{1/4}.$$

Hint for 21-3-vii. See Exer. 9-9 in Chap. 9 for a method of computing a very accurate approximate value of the constant $\mathscr{C}_2$. A quick manner to get an approximate value follows by writing

$$\mathscr{C}_2 = \prod_{p \geq 2}\left(1 - \frac{1}{p(p+2)}\right) = \frac{1}{\zeta(2)}\prod_{p \geq 2}\left(1 + \frac{2}{(p^2-1)(p+2)}\right),$$

where this time, the final Euler product converges in $1/p^2$.

───────────────── §*Chapter* 22 ─────────────────

Hint for 22-1-i. Notice first that, for any parameter $L > 0$ to be chosen to serve us the best, we have

$$(x + L)\left(1 - \frac{n}{x+L}\right) - x\left(1 - \frac{n}{x}\right) = L.$$

Solution to 22-1-ii. Since $0 \le f_1(n) \le 1$, we find that

$$\sum_{x < n \le x+L} f_1(n) \le L + 1 .$$

Using the result of Question 7 of Exer. 21-3 twice, for x and $x + L$, deduce that

$$\sum_{n \le x} f_1(n) = \mathscr{C}_1 \frac{(x+L)^2 - x^2}{2L} + O\big((x+L)^{5/4} L^{-1} + L\big) .$$

On assuming $L \le x$ the above brings us to the result.

Solution to 22-1-iii. What is the best choice for L? We can, for instance, determine the minimum (in L) of the function $L \mapsto x^{5/4} L^{-1} + L$, and it is straightforward in this example, but this can easily become rather challenging (just think about the simple case $L \mapsto x^{5/4} L^{-1} + L/\log L$). Here is a way to make a choice: the first term $(x^{5/4} L^{-1})$ is decreasing, while the second one (meaning L) is increasing. We take L such that two curves cross, namely,

$$x^{5/4}/L = L .$$

This leads to $L = x^{5/8}$. Further, $x^{5/4}/L = L = x^{5/8}$. Our theorem is proved.

Hint for 22-2-ii. Start with $d(n) \le 2 \sum_{d | n, d \le \sqrt{n}} 1.$

Hint for 22-2-iii. We recall that $\sum_{n \le x} \dfrac{1}{n} \le \log x + 1$ and that $\sum_{n \le \sqrt{x}} \dfrac{1}{\sqrt{n}} \le 2\sqrt{x}.$

Hint for 22-4-vii. The readers may show that a sequence in $\mathscr{S}(\alpha_1, \ldots, \alpha_k)$ can either be written as $(\alpha_1, \ldots, \alpha_k, \beta_1, \ldots, \beta_\ell)$ with $(\beta_1, \ldots, \beta_\ell) < (\alpha_1, \ldots, \alpha_k)$ or, when $\alpha_1 = \ldots = \alpha_k$, as m copies of $(\alpha_1, \ldots, \alpha_k)$ followed by a sequence $(\beta_1, \ldots, \beta_\ell) < (\alpha_1, \ldots, \alpha_k)$. The sub-multiplicativity of Φ is then enough to conclude.

Solution to 22-4-ix.

```
def Divisors(alist):
    DivList = [[]]
    for i in range(0, len(alist)):
        auxDivList = [dd + [j] for dd in DivList
                                for j in range(0, alist[i] + 1)]
        DivList = auxDivList
    return(DivList)

def mygp(alpha):
    if alpha == 0:
        return 1
    elif alpha == 1:
        return(5/2)      #47/20 #12/5
    else:
```

```
          return((5*alpha-1)/2)

def myg(alist):
    return(prod( [mygp(alpha) for alpha in alist]))

def quotient(alist, adivisor):
    return([alist[i]-adivisor[i] for i in range(0, len(alist))]])

def Gstar(alist):
    return(sum([min(myg(dd), myg(quotient(alist, dd)))
                for dd in Divisors(alist)]))

def d3(alist):
    return(prod([(alpha+1)*(alpha+2)/2 for alpha in alist]))

def Phi(alist):
    return(d3(alist)/Gstar(alist))

def sigma(alist):
    k = len(alist)-1
    while (k >= 0) and (alist[k] == alist[len(alist)-1]):
        k -= 1
    return([alist[i] for i in range(0, k+1)] + [alist[k+1]+1])

def calculC(borne):
    myC, k, aux = 0, 1, 2
    # Special part for [1,1,...,1]:
    while aux > 1:
        if k%2 == 0:
            valgstar = 2*sum([binomial(k,j)*mygp(1)^j
                              for j in range(0, (k/2) +1)])
            valgstar += -binomial(k,k/2)*mygp(1)^(k/2)
        else:
            valgstar = 2*sum([binomial(k,i)*mygp(1)^i
                              for i in range(0, (k+1)/2)])
        aux = 3^k/valgstar
        myC = max(myC, aux)
        k += 1
    print("We went till k = ", k-1)
    if borne == 1:
        return myC

    alist = [2]        # Now, part after [2]
    while alist[0] < borne:
        aux = Phi(alist)
```

```
            while aux > 1:
                myC = max(myC, aux)
                print([alist, myC])
                alist = alist + [1]
                aux = Phi(alist)
            alist = sigma(alist)
        return(aux)
```

This script is not very fast as we did not use the symmetry $\delta \mapsto n/\delta$. Notice that we have to consider lists of more than twenty entries when $g(p) = 5/2$, of more than forty entries (see next question) when $g(p) = 12/5$ and of more than seventy of them when $g(p) = 47/20$.

Solution to 22-5.

```
R = RealIntervalField(200)

def g(n):
    res = 1
    l = factor(n)
    for p in l:
        res *= p[0]-2
    return(R(res))

def model(z):
    return(0.29261985704515491401* R(z^2))

def modelrest(z):
    return(R(z^(3/2)*log(z)^2))

def getbounds (zmin, zmax):
    zmin = max (0, floor (zmin))
    zmax = ceil (zmax)
    res = R(0)
    output = []
    for n in range (1, zmin + 1):
        res += g(n)
    maxi = abs((res - model (zmin))/modelrest(zmin)).upper()
    maxiall = maxi
    for n in range (zmin + 1, zmax + 1):
        m = model (n)
        mr = modelrest (n)
        maxi = max (maxi, abs((res - m)/mr).upper())
        res += g(n)
```

```
        maxi = max (maxi, abs((res - m)/mr).upper())
        if n % 100000 == 0:
            print("Upto ", n, " : ", maxi, cputime())
            output.append(maxi)
            maxiall = max (maxiall, maxi)
            maxi = R(-1000).upper()
    aux = abs ((res - model (zmax))/modelrest(zmax)).upper()
    maxi = max (maxi, aux)
    maxiall = max (maxiall, maxi)
    print("When ", zmin, " <= z <= ", zmax)
    str = "|sum_{n <= z} f0(n) - C0 z^2| / (z^(3/2) log(z)^2)"
    print(str, " <= ", maxiall)
    return([maxiall, output])
```

The readers may put this script in the file "Myscript-1.sage" and load it in Sage:

```
sage: load("Myscript-1.sage")
sage: myres = getbounds( 2, 10000000)
sage: list_plot(myres[1], plotjoined = True)
```

Hint for 22-6. The readers may have a look at [1, Lemma 9.3]. The sum $S_1(x) = \sum_{n \leq x} \frac{1}{\varphi(n)}$ is studied in Exer. 13-8.

———————————————— §*Chapter* 23 ————————————————

Hint for 23-1-i. Use Cauchy's inequality on $4\left|\Re D(t)\right|^2 = \left|\sum_{n \geq 1} a_n\left(n^{it} + n^{-it}\right)\right|^2$ and notice that $|n^{it} + n^{-it}|^2 = 2(1 + \Re n^{2it})$.

Hint for 23-3. Use Lemma 19.2.

Hint for 23-6-i. The readers may want to shift the line of integration to the far right.

Hint for 23-6-ii. We may shift the line of integration in the definition of w to $\Re s = -1$. On this line, we have $|s - \rho| \geq 1$ and $|s - 1|/|s + 2| \leq 2$. On gathering these estimates, we get the stated result.

Hint for 23-6-iv. We recall that $\frac{1}{\zeta(s)} = s \int_1^\infty M(u)du/u^{s+1}$.

Hint for 23-7. The readers may want to consider $(1+\lambda(n))/2$, where λ is the Liouville function and, by noticing that $\lambda(n) = \sum_{a^2 b = n} \mu(b)$, show that, for any A, we have

$$\sum_{n \leq x} \lambda(n) = O_A(x/(\log x)^A) .$$

Hint for 23-8. For going from $\Lambda(n)$ to $\log p$, the reader may apply Lemma 12.2.1.

Hint for 23-9. Remember Theorem 13.4!

─────────────── §*Chapter* 24 ───────────────

Hint for 24-4-i. First establish that

$$\sum_{d|n} \mu(d)\left(\log \frac{x}{d}\right)^2 = \left(\log \frac{x}{n}\right)^2 \delta_{n=1} + 2\log\frac{x}{n}\Lambda(n) + \Lambda_2(n) .$$

Hint for 24-4-iv. The readers may start by establishing successively that

$$F(x)(\log x)^2 + 2\sum_{n\leq x}\Lambda(n)\log n\, F\left(\frac{x}{dn}\right) = 2H_1(x) - H_2(x) + \sum_{n\leq x}\Lambda_2(n)F\left(\frac{x}{dn}\right)$$

and then that

$$F(x)(\log x)^2 - 2\sum_{n\leq x}(\Lambda \star \Lambda)(n)\log n\, F\left(\frac{x}{dn}\right) = 2H_1(x) - H_2(x) - \sum_{n\leq x}\Lambda_2(n)F\left(\frac{x}{dn}\right) .$$

Solution to 24-6. The next script computes the required maximum for all *real* values of x in the interval [initialbound, finalbound + 1).

```
{Check(initialbound = 10, finalbound = 100, whentotell = 20)=
 my(r0 = 0.0, r1 = 0.0, r2 = 0.0, r3 = 0.0, aux1, aux2, aux3,
    mymax = 0.0, root1, root2, delta, val0, val1, val2, val3);
 \\ initialisation:
 forsquarefree(dd = 1, initialbound-1,
   aux1 = moebius(dd[1])/dd[1]; aux2 = log(dd[1]);
   r0 += aux1; r1 += aux1*aux2;
   r2 += aux1*aux2^2; r3 += aux1*aux2^3);
 \\ running:
 for(d = initialbound, finalbound,
   aux1 = moebius(d)/d; aux2 = log(d); aux3 = log(d+1);
   r0 += aux1; r1 += aux1*aux2;
   r2 += aux1*aux2^2; r3 += aux1*aux2^3;
   \\ Does the derivative vanish?
   delta = (2*3*(r1+1))^2 - 4*(-3*r0)*(-3*r2-6*Euler);
   root1 = aux2; root2 = aux2;
   if(delta < 0,, \\ We assume that r0 does not vanish
     root1 = (-2*3*(r1+1)+sqrt(delta))/2/(-3*r0);
     root2 = (-2*3*(r1+1)-sqrt(delta))/2/(-3*r0);
     if((root1 > aux2) && (root1 < aux3),, root1 = aux2);
     if((root2 > aux2) && (root2 < aux3),, root2 = aux2));
   [val0, val1, val2, val3] =
     apply(logx -> abs(r3 + logx*(-3*r2-6*Euler)
                         + 3*logx^2*(r1+1) - logx^3*r0),
           [aux2, root1, root2, aux3]);
```

```
    mymax = vecmax([mymax, val0, val1, val2, val3]);
    if(d % whentotell == 0,
        print("At d = ", d, ", max = ", mymax),););
  return(mymax);}
```

It seems the computed quantity converges towards a limit.

Hint for 24-7. The readers may use the first half of Exer. 24-5.

———————————————— §*Chapter* 25 ————————————————

Hint for 25-2-iii. The choice $\sigma = \sqrt{a(n)}/\log n$ leads to the stated inequality.

Hint for 25-3-iv. The readers may first prove that $\prod_{z<p\le\log q}(1 + 1/p) \ll 1$.

Hint for 25-4-iii.

```
{getC0(x) =
  my( res = 0, nb = primepi(x), listprimes = primes(nb));
  forsubset(nb, s,
    my(ss = vector(length(s), n, listprimes[s[n]]));
    my(q, rho, aux = 0);
    if(length(ss) == 0, q = 1; rho = 1,
        q = vecprod([p| p<- ss]);
        rho = vecprod([p/(p-1)| p <- ss]));
    for( n = 1, x, if(gcd(n,q) == 1, aux++,));
    res = max(res, aux/x*rho));
  return(res);}

{getListC(begx, endx) =
  my(accumulator = 1);
  for(y = 1, x,
    accumulator = max(accumulator, getC0(y));
    print("C(", y, ") = ", accumulator));}
```

Hint for 25-5-iii. It may be useful to note that $1 + 2/p \le (1 + 1/p)^2$.

Hint for 25-5-iv. The readers may use Theorem 13.1.

———————————————— §*Chapter* 26 ————————————————

Hint for 26-2. Use Cauchy's inequality.

Solution to 26-3.

```
C = ComplexField(200);

def getListPoints(beg, end):
    mysum = 0.0
    rho = sqrt(2) # (1+sqrt(5))/2
    for p in Primes():
```

```
        if p <= beg:
            mysum += C(exp(C(I*2*pi)*rho*p))
        else:
            break
    mylist = [(real(mysum), imag(mysum))]
    for p in Primes():
        if p<= beg:
            continue
        elif p <= end:
            mysum += C(exp(C(I*2*pi)*rho*p))
            mylist = mylist + [(real(mysum), imag(mysum))]
        else:
            break
    return(mylist)

def makeexpsumPrimesPlot(bound, blocksize):
    myPlot = list_plot(getListPoints(1, blocksize),
                        plotjoined=True,
                        color = hue(0))
    for j in range(2, ceil(bound/blocksize)+1):
        low = (j-1)*blocksize;
        up = min(j*blocksize, bound);
        auxPlot = list_plot(getListPoints(low, up),
                            plotjoined=True,
                            color = hue(j*blocksize/bound))
        myPlot = myPlot + auxPlot
    return(myPlot)

expsumPrimesPlot = makeexpsumPrimesPlot(10000,500)
expsumPrimesPlot.show()
# save(expsumPrimesPlot, "expsumPrimesUpTo10000.png")
```

Hint for 26-5. The readers may want to use Exer. 26-1.

Hint for 26-9. Begin with the following:

$$\Lambda(n) \ge T(n) = \Lambda(n) - \sum_{\ell a = n} \Lambda(\ell) d_K^*(a) \,.$$

Solution to 26-9. Equation (12.3) allows us to rewrite $T(n)$ as

$$T(n) = \sum_{bc=n} \mu(b) \log c - \sum_{bca=n} \mu(b)\tau_K^*(a) \log c$$
$$= \sum_{cf=n} \left(\mu(f) - \sum_{ba=f} \mu(b)\tau_K^*(a) \right) \log c \,.$$

In the expression inside the brackets we set $a = qa'$, giving

$$\sum_{ba=f} \mu(b)d_K^*(a) = \sum_{\substack{bqa'=f, \\ q>1}} \mu(b)d_{K-1}^*(a') \, .$$

We *glue* b and q into one variable b' and employ

$$\sum_{\substack{bq=b', \\ q>1}} \mu(b) = \begin{cases} 0 & \text{if } b' = 1, \\ -\mu(b') & \text{otherwise.} \end{cases}$$

Finally,

$$\sum_{ba=f} \mu(b)d_K^*(a) = d_{K-1}^*(f) - \sum_{b'a'=f} \mu(b')d_{K-1}^*(a') \, .$$

The sum on the right-hand side is of the same type as the one we study, but with $K - 1$ instead of K. Proceeding by recursion on K we find that

$$\mu(f) - \sum_{ba=f} \mu(b)d_K^*(a) = \sum_{k=0}^{K-1}(-1)^k d_k^*(f) \, ,$$

where $d_0^*(f)$ is 1 for $f = 1$ and is zero otherwise.

Further, writing $\log n = \log(d_1 \cdots d_{k+1})$ on the right-hand side, we establish that

$$\sum_{cf=n} d_k^*(f)\log c = d_{k+1}^*(n)\log(n)/(k + 1) \, ,$$

which concludes the proof.

The readers may also use the Taylor development of $\log(1 - (\zeta(s) - 1))$ and derive another proof.

Hint for 26-10-ii. Replace $G(n)$ by $(1/D)\sum_{pm=n, z_0<p\leq z} G(pm)$ via the previous question.

Hint for 26-11-i. The function μ^2 is multiplicative and bounded.

Hint for 26-11-ii. Equation (10.10) gives an approximation of the number of square-free integers below any bound. The readers may then mimic the proof of Cor. 26.3.

Hint for 26-12-ii. The readers may look at Exers. 23-7 and 3-15, and then mimic the proof of Cor. 26.3.

———————————— §*Chapter 28* ————————————

Hint for 28-1. The readers may write $\sum_i |\langle f|\varphi_i^*\rangle| = \sum_i c_i\langle f|\varphi_i^*\rangle = \langle f| \sum_i c_i\varphi_i^*\rangle$ where $c_i = \overline{\langle f|\varphi_i^*\rangle}/|\langle f|\varphi_i^*\rangle|$ when $\langle f|\varphi_i^*\rangle \neq 0$ and $c_i = 0$ otherwise.

Hint for 28-2-i. The LHS may also be written $\sum_{n,m} a_n\overline{a_m}\left|\sum_r x_{r,m}\overline{x_{r,n}}\right|^2$.

Hint for 28-4-ii. There are essentially two ways to prove what is required. The first one goes by setting $d_1' = d_1/\delta$, $d_2' = d_2/\delta$, $q' = q/\delta$ and to dispense of the condition $(d_1', d_2) = 1$ by using $\mathbb{1}_{(d_1', d_2')=1} = \displaystyle\sum_{\substack{\ell \mid d_1', \\ \ell \mid d_2'}} \mu(\ell)$. The second method is to invoke multiplicativity and to first prove that

$$\sum_{\substack{d_1 \mid q, \\ d_2 \mid q, \\ (d_1, d_2)=\delta}} \mu(q/d_1)\mu(q/d_2) = \prod_{p^\alpha \| q} \sum_{\substack{d_1 \mid p^\alpha, d_2 \mid p^\alpha, \\ (d_1, d_2)=(\delta, p^\alpha)}} \mu(p^\alpha/d_1)\mu(p^\alpha/d_2) . \qquad (29.4)$$

The sum with p^α is then easy to handle with a case-by-case analysis, according to the values of (δ, p^α) (either p^α or strictly smaller). Proving Eq. (29.4) is an interesting task. The easiest way is surely to convert the condition $d_1 \mid q$ into a function $\mathbb{1}_{d \mid q}$ and to use the fact that

$$\mathbb{1}_{d \mid q} = \prod_{p^\alpha \| q} \mathbb{1}_{v_p(d) \le \alpha} .$$

By similar manipulations, we reach

$$\sum_{\substack{d_1 \mid q, \\ d_2 \mid q, \\ (d_1, d_2)=\delta}} \mu(q/d_1)\mu(q/d_2) =$$

$$\sum_{d_1, d_2 \ge 1} \prod_{p^\alpha \| q} \left(\mathbb{1}_{v_p(d_1) \le \alpha} \mathbb{1}_{v_p(d_2) \le \alpha} \mathbb{1}_{v_p(d_1)+v_p(d_2)=v_p(\delta)} \mu(p^{\alpha-v_p(d_1)})\mu(p^{\alpha-v_p(d_2)}) \right)$$

from which the final formula follows readily.

Hint for 28-4-iii. It may be worth noticing that, for some fixed d_1 and d_2 that divide q and some fixed m and n such that $m \equiv n \pmod{(d_1, d_2)}$, we have

$$\sum_{\substack{b \bmod q, \\ m \equiv b[d_1], \\ n \equiv b[d_2]}} 1 = \frac{q}{[d_1, d_2]} = \frac{q(d_1, d_2)}{d_1 d_2} .$$

Hint for 28-5. We may restrict in (28.7) all the inner summations to $b = 0$. Infer from the resulting inequality that

$$\sum_{p \le z} \frac{1}{p} \left(\sum_{n \le N} G(n) - p \sum_{\substack{m \le N/p, \\ p \le z}} G(mp) \right) \ll \sqrt{\sum_{p \le z} \frac{1}{p} N \sum_{n \le N} |G(n)|^2}$$

and conclude.

──────────────── §*Chapter* 29 ────────────────

Hint for 29-1-i. Use the Ramanujan sum formula (1.9), then reduce to the large sieve inequality (28.6) by the Cauchy–Schwarz inequality.

Hint for 29-1-ii.
   First assume that $M \ge \sqrt{N}$ and then recall Exer. 10-4.

Hint for 29-2-vi. First notice that

$$\int_{-T}^{T} \left| \sum_{T^2 < p \le x} \frac{a(p) \log p}{p^{it}} \right|^2 dt \le \int_{-\infty}^{\infty} \left| \sum_{T^2 < p \le x} \frac{a(p) \log p}{p^{it}} \right|^2 \hat{K}(t) dt$$

and develop the inner square to introduce $K(\log(p_1/p_2)/(2\pi))$.

Hint for 29-3-ii. We only mention Examples $9, 11, 13$, then $27, 29, 31$ and $3^{41} - 4, 3^{41} - 2, 3^{41}$. We found twelve such examples below 10^{100} and none other with $n \le 3^{150}$.

Hint for 29-7. This is Exer. 73 from [2] and a proof can be found there! As a hint, we mention that a sum of two primes is likely to be... even except in a very special case.

Hint for 29-8-ii. Select x_0 congruent to 1 modulo 2, 1 modulo 7, 2 modulo 5, 2^3 modulo 17 and 2^{23} modulo 241 as we may by the Chinese Remainder Theorem. The progression x_0 modulo $2 \cdot 7 \cdot 5 \cdot 17 \cdot 241$ answers the question.

Hint for 29-9. Define $r(n)$ as in Exer. 29-6. The readers can start by considering $\Re = \sum_{n \le X} r(n)$ and use a Chebyshev lower bound to show that $\Re \gg (x/\log x)^2$. On the other hand, the Cauchy–Schwarz inequality gives us $\Re^2 \le A(X) \sum_{n \le X} r^2(n)$. One can proceed by majorizing the second sum via Exer. 29-6. The proof ends by finding an upper bound for $\sum_{n \le x} \mathfrak{S}(n)^2$; the readers should know by now how to do that.

Chart of Common Arithmetical Functions

Here is a quick reference card on how to compute some ofently used arithmetical functions, both in Pari/GP and in Sage.

Divisor Function $d(n)$: The number of (positive integer) divisors of n.
Pari/GP : numdiv(n) *Sage* : number_of_divisors(n)

Divisors of n: The list of the divisors of n.
Pari/GP : divisors(n) *Sage* : divisors(n)

Sum of the kth power of the divisors of n.
Pari/GP : sigma(n, k) *Sage* : sigma(n, k)

List of prime factors.
Pari/GP : factor(n)[,1]~
Sage : prime_divisors(n), prime_factors(n)

Number of prime factors $\omega(n)$: Number of prime factors of n.
Pari/GP : omega(n) *Sage* : len(prime_divisors(n))

Total Number of prime factors $\Omega(n)$: Number of prime factors of n counted with multiplicity.
Pari/GP : bigomega(n) *Sage* : pari(n).bigomega()

Moebius Function $\mu(n)$: Returns 0 if n is divisible by the square of some prime and otherwise $(-1)^{\omega(n)}$ where $\omega(n)$ is the number of prime divisors of n.
Pari/GP : moebius(n) *Sage* : moebius(n)

Liouville Function $\lambda(n)$: defined as $(-1)^{\Omega(n)}$.
Pari/GP : if(bigomega(n)%2,1,−1) *Sage* : (−1)^(pari(n).bigomega())

Euler Phi Function $\varphi(n)$: Number of integers in $[1, n]$ coprime with n.
Pari/GP : eulerphi(n) *Sage* : euler_phi(n)

Dedekind Psi Function $\psi(n)$: defined by $\psi(n)/n = \prod_{p|n}(1 + 1/p)$.
Pari/GP : vecprod([1 + 1/p | p <- factor(n)[,1]])*n
Sage : dedekind_psi(n)

© The Editor(s) (if applicable) and The Author(s), under exclusive license to Springer Nature Switzerland AG 2022
O. Ramaré, *Excursions in Multiplicative Number Theory*, Birkhäuser Advanced Texts Basler Lehrbücher, https://doi.org/10.1007/978-3-030-73169-4

Radical of n**,** $k(n)$: Product of the primes dividing n.
Pari/GP : vecprod(factor(n)[,1])   *Sage* : radical(n)

Legendre symbol $(\frac{n}{p})$: Returns 1 when n is a square modulo prime p and 0 otherwise.
Pari/GP : kronecker(n, p)
Sage : legendre_symbol(n, p), kronecker(n, p)

Ramanujan τ**-function** $\tau(n)$: nth coefficient of the power series expansion of
$q \prod_{k \geq 1} (1 - q^k)$.
Pari/GP : ramanujantau(n)   *Sage* : pari(n).ramanujantau()

Ramanujan sum $c_q(n)$: One definition is $c_q(n) = \displaystyle\sum_{d \mid (q,n)} d\mu(q/d)$.
Pari/GP : vecsum([d*moebius(q/d) | d <- divisors(gcd(q, n))])
Sage : sum([d*moebius(q/d) for d in divisors(gcd(q, n))]

p**-adic valuation of** n**,** $v_p(n)$: the largest power of the prime p that divides n.
Pari/GP : valuation(n, p)   *Sage* : valuation(n, p)

Is n **a power?**: Characteristic function of the set of integers of the shape m^k for some
integer m and some integer $k \geq 2$.
Pari/GP : ispower(n, 1)   *Sage* : bool(pari(n).ispower()[0]>1)

Is n **a square?** $\mathbb{1}_{X^2}$: Characteristic function of the squares.
Pari/GP : issquare(n)   *Sage* : is_square(n)

Is n **powerful?**: Characteristic function of the set of integers n that such that, if a
prime p divides n, then so does p^2.
Pari/GP : ispowerful(n)   *Sage* : bool(pari(n).ispowerful())

Is n **square-free?** $\mu^2(n)$: Characteristic function of the set of integer that are not
divisible by the square of any prime.
Pari/GP : issquarefree(n)   *Sage* : is_squarefree(n)

Is n **a prime?**
Pari/GP : isprime(n)   *Sage* : is_prime(n).

Is n **a prime power?**
Pari/GP : isprimepower(n)   *Sage* : is_prime_power(n)

Is n **a totient?**: Answers whether n is a value taken by the Euler phi function.
Pari/GP : istotient(n)

Greatest Common Divisor of m **and** n**.**
Pari/GP : gcd(m, n)   *Sage* : gcd(m, n)

Least Common Multiple of m **and** n**.**
Pari/GP : lcm(m, n)   *Sage* : lcm(m, n)

Two remarks are called for:

- Several other Sage functions may be found on the web page

doc.sagemath.org/html/en/reference/rings_standard/sage/arith/misc.html

  Similarly, several other Pari/GP functions may be found by asking "?5" to the GP interface.
- The underlying mechanism to compute most of the above functions depends on factoring integers; the very large factors are only checked to be "strong pseudoprimes". Hence they are *not provably prime*, though we believe they *are* prime. To avoid this (tiny!) uncertainty, the readers should use default(factor_proven, 1) in Pari/GP and proof.arithmetic(True) in Sage.

Loops and iterators

We simply enumerate some, as the names are rather explicit, and let the readers find the documentation in the classical repositories.

Pari/GP:	*Sage*:
primes(),	prime.range(),
nextprime(),	primes(),
forprime,	primes_first_n(),
forprimestep,	prime_powers(),
forfactored,	moebius.range(),
forcomposite,	squarefree_divisors(),
forsquarefree,	next_prime(),
fordiv,	next_prime_power(),
fordivfactored,	next_probable_prime(),
forsubset.	previous_prime(),
	previous_prime_power().

Zeta and L-functions

We again simply enumerate some functions' names, as they are rather explicit, and let the readers find the documentation in the classical repositories. Some examples may also be found during the exposition.

Pari/GP:	*Sage*:
zeta(),	zeta(),
lfun(),	zetaderiv(),
lfuncreate(),	Dokchitser(),
zetahurwitz(),	hurwitz_zeta(),
zetahurwitz(),	
dirzetak().	

Pari/GP has also the three functions dirmul(), dirdiv() and direuler() to handle Dirichlet series.

Getting Hold of the Bibliographical Items

Many references to external research papers occur throughout the text. They have all been published in journals that need some income to work, but it is still possible to have access to many of these papers for free. The style they adopt is surely more rigid than the one we have adopted here, but the reader may still gain from their content. The language chosen is often English, or what passes out to be English in our community, though some may be in French, German, Russian or Chinese. When in English, usually a rather rudimentary mastery of the language is enough to read them. The main issue remains: how to get to these papers?

Several of them may be found by simply asking a search engine with the title of the paper followed by the name of the author. When this does not work, we mention that:

1. Many countries have *Digital Mathematical Libraries*. The search engine EuDML at https://eudml.org, developed and maintained by European instances, is a powerful tool to browse through many of them at once. The project of a *Global Digital Mathematical Library* is being carried by the *International Mathematical Union*, but is still at an early stage.
2. The digital base NumDam of *ancient* (meaning not brand new!) papers and monographs at www.numdam.org is very rich. It is developed and maintained by Mathdoc, an entity that depends on the CNRS and on the University Grenoble Alpes.
3. It is possible to have access directly to some journals via their web page, like "L'enseignement Mathématique" at https://www.unige.ch/math/EnsMath/. The (many) journals for the Polish academy of sciences can be found at http://pldml.icm.edu.pl/pldml/ and the publications of the Banach Center are gathered at www.impan.pl/en/publishing-house/banach-center-publications
   A great many papers are downloadable for free there provided you use them for studying purpose.
4. The Arxiv preprint server at https://arxiv.org is also an essential tool, but a search via a usual search engine incorporates already these results.

© The Editor(s) (if applicable) and The Author(s), under exclusive license to Springer Nature Switzerland AG 2022
O. Ramaré, *Excursions in Multiplicative Number Theory*, Birkhäuser Advanced Texts Basler Lehrbücher, https://doi.org/10.1007/978-3-030-73169-4

5. More preprints can be found on the *Hal* preprint server at https://hal.archives-ouvertes.fr/, while many Ph.D. memoirs are on the TEL and these.fr servers, respectively, found at https://tel.archives-ouvertes.fr/ and at http://www.theses.fr/en/accueil.jsp.

6. The database https://mathnet.ru developed and maintained by the Steklov Mathematical Institute of the Russian Academy of Science is good for papers published in Russian journals.

7. Concerning Sage [3], the forum ask.sagemath.org is a good place to ask questions.

8. All the papers of Pál Erd may be found at https://users.renyi.hu/~p_erdos/Erdos.html. It is maintained by the Alfréd Rényi Institute of the Hungarian Academy of Sciences.

9. All the paper of may be found at https://www.imsc.res.in/~rao/ramanujan/contentindex.html. This is maintained by the Institute of Mathematical Sciences in Chennai, India.

10. References for papers can be found from the first of January 2021 in the very extensive database Zentralblatt MATH located at https://zbmath.org. It is edited by the *European Mathematical Society*, the *Heidelberg Academy of Sciences and Humanities* and *FIZ Karlsruhe*.

11. Last but not least, a great many mathematicians have their own web pages and you can have a look whether or not they have put a version there of the paper you want.

12. The On-Line Encyclopedia of Integer Sequences is a major reference for sequences. It can be found at http://oeis.org. Sequence referenced, for instance, A064533 is simply found at http://oeis.org/A064533.

13. Concerning the life of mathematicians, we refer the readers to the rich *and accurate* MacTutor website http://www-history.mcs.st-and.ac.uk/. The Hungarian base História - Tudósnaptár (*History - Scholars Calendar*) http://tudosnaptar.kfki.hu/historia/olvassel.php has also interesting informations... though in Hungarian!

The above indications specify websites, and some addresses may get stale when time passes. We have, however, restricted our comments to sites that are around for more than a decade, and that we expect to remain for a long time. The description added should help the readers find the proper address if it were to change.

References

[1] O. Ramaré. "Explicit average orders: news and problems". In: *Number theory week 2017*. Vol. 118. Banach Center Publ. Polish Acad. Sci. Inst. Math., Warsaw, 2019, pp. 153–176 (cit. on pp. 106, 109).

[2] W. Sierpinski. *250 problems in elementary number theory*. Modern Analytic and Computational Methods in Science and Mathematics, No. 26. American Elsevier Publishing Co., Inc., New York; PWN Polish Scientific Publishers, Warsaw, 1970, pp. vii+125 (cit. on p. x).

[3] The Sage Developers. *SageMath, the Sage Mathematics Software System (Version 9.0)*. 2019. www.sagemath.org (cit. on p. viii).

Name Index

O. Ramaré, *Excursions in Multiplicative Number Theory*, Birkhäuser Advanced Texts
Basler Lehrbücher, https://doi.org/10.1007/978-3-030-73169-4

Index

© The Editor(s) (if applicable) and The Author(s), under exclusive license to Springer
Nature Switzerland AG 2022
O. Ramaré, *Excursions in Multiplicative Number Theory*, Birkhäuser Advanced Texts
Basler Lehrbücher, https://doi.org/10.1007/978-3-030-73169-4

Printed in the United States
by Baker & Taylor Publisher Services